MODERN
PRACTICAL BOTANY

MODERN PRACTICAL BOTANY

Volume - 1

(ALGAE, FUNGI, BACTERIA, LICHENS, VIRUSES AND BRYOPHYTES)

Dr. S. Sundara Rajan
Professor and Head
Dept. of Post Graduate Studies and Research in Botany
St. Joseph's College
Bangalore - 560 025

ANMOL PUBLICATIONS PVT. LTD.
NEW DELHI - 110 002 (INDIA)

ANMOL PUBLICATIONS PVT. LTD.
4374/4B, Ansari Road, Daryaganj
New Delhi - 110 002

Modern Practical Botany

First Edition, 2000

ISBN 81-261-0690-5 (Set)

PRINTED IN INDIA

Published by J.L. Kumar for Anmol Publications Pvt. Ltd., New Delhi -110 002 and Printed at Mehra Offset Press, Delhi.

CONTENTS

PART TWO - FUNGI

PART THREE - BRYOPHYTES

Preface

Laboratory study is an integral part of the knowledge that a student acquires in the college. This is true not only for botany but for all science subjects. Theoretical study of a subject has to be supplemented with laboratory experiments to provide a comprehensive understanding.

One of the cheif advantages of laboratory or practical study is thefirst hand knowledge that a student acquires about the material of study.

A systematic, structured and disciplined method of work is necessary in the laboratory to obtain maximum benefit. In order to help the students realize their goal practical manuals indeed play an important role.

Modern practical botany aims to provide the students, a useful hand book to guide them in the proper study of the subject material. Theoretical details are kept at the minimum and more emphasis is laid on understanding the structural details to enable the student a proper identification. A large number of properly labelled original diagrams are incorporated in the book as they are integral to the understanding of the subject.

Modern Practical Botany is a series of three volumes.

Volume One deals with Algae, Fungi, Bacteria, Lichens, Viruses and Bryophytes.

Volume Two deals with Pteridophytes, Gymnosperms and Anatomy and Embryology of Angiosperms.

Volume Three deals with Taxonomy, Economic Botany, Ecology, Physiology and Cytogenetics.

I have taken utmost care in the presentation of matter, style of writing and choice of illustrations so that the volumes are user friendly. The examples chosen for study are such that the books should cater to the needs of all Indian Universities.

Constructive criticisms and suggestions are most welcome.

Dr. S. SUNDARA RAJAN
Professor and Head

Dept of Botany
St. Joseph's College
Bangalore - 560 025.

CHAPTER 1

INTRODUCTION

The study of science subjects is carried out at two levels,viz., *Theoretical* and *Practical*. Theoretical study is carried out in lecture classes when the students are imparted all the details regarding a particular topic. This has to be supplemented by practical or Laboratory studies, when the students come into active contact with the objects about which they would have heard in lecture classes. Therefore it may be safely said that laboratory studies form an integral part of a comprehensive study of any topic in science subjects.

In the study of plants, laboratory assumes a special significance as the students acquire a meaningful and firsthand knowledge about the plants which they study in lecture classes. In fact as far as botany is considered any study of a plant without being supplemented by laboratory observation is futile. It is only in the laboratory study that the student can dissect, observe and analyze the various features of the plant and it is only a study of this type that makes the student acquire a meaningful knowledge about the pant. Besides observation and study in the laboratory plants are also to be studied in field to obtain an account of their habitat, habit, association and such other details.

In this chapter an attempt is made to explain some of the essentials that a student should understand before he begins a practical work.

Clednliness: This is the most important point that a student should keep in mind while in the laboratory. The working table and the surroundings should be kept dry and spotlessly clean.. All the requirements needed by the students such as petri dishes, slides, reagent bottles, stains etc. must be properly labeled and arranged neatly. The microscope and particularly the optical parts must be cleaned with a white lint free cloth. After the completion of the assigned work, the student should take care to clean the place, remove any debris of the plant material and put it into the dust bin.

The students should work in silence and independently unless the assigned experiment is to be carried out by a group. When a particular work is assigned to a student, he must carefully plan and divide the available time for experimentation, observation and recording of the results. Whenever a doubt arises, it is always better to consult the teacher in charge, than a classmate. It is only by a proper and systematic work, that a student can derive maximum benefit from the practical classes.

Laboratory requirements

The department provides certain amenities, instruments etc to each student for the purpose of practical work. But a student also should carry some of the requirements in order to be effective in the practical class. One cardinal principle that every student should remember in a practical class is to have one's own requirements and not to depend on anyone. Never lend or borrow any requirements. Borrowing or lending a pencil, needle or blade will disturb the work of both.

In general every student should individually possess the following requirements

1. *Two sharp pointed needles* mounted on stainless steel or plastic handle. This is essential for the purpose of separating and teasing the microscopic material.

2. *A pair of forceps:* This is essential to remove the material either from the bottle or from petri dish.

3. A sharp properly ? razor to cut sections of the plant material.

4. A new safety razor blade. This will help either to cut section or to trim the material.

5. Two fine hair paint brushes. This will be of use in lifting the sections of plant (with out damage) from the petri dish on to the slide.

6. Two sharpened pencils (H, Hb or 2b).

7. A pencil eraser.

8. A pencil sharpener.

9. Glass droppers.

10. A piece of thin, white, lint free cloth to clean the optical parts of the microscope.

11. An observation book to note down immediately the points of interest.

12. A practical record.

13. Glass slides and cover slips(if they are not provided by the department).

Scheme of Practical Work

Practical work in the laboratory will be most useful and effective if the student keeps in mind the following:

1. Maintain silence during working. All the requirements for the practical work must be absolutely clean.

2. Every student should work independently, unless the experiment involves the participation of others.

3. Before beginning the practical work, listen to the instructions of the teacher attentively and work accordingly.

4. Follow a systematic procedure in a study.

5. If the work involved is the morphological study of an organism, follow the procedure given below:

(a) Study of external characters - for example whether a plant is microscopic or macroscopic, unicellular or multicellular, filamentous or colonial etc.

(b) Study of internal structure like details of cell tissues etc. If the material requires sectioning, sections are to be cut and internal details should be observed.

(c) Study the reproductive details like asexual and sexual reproductive organs.

6. Mount the material after properly teasing and separating the filaments, mycelia etc. Stain the material wherever necessary and cover the preparation with a cover glass.

7. Observe the slide carefully under the microscope, first under low power and then under high power.

8. Make sure that you have observed all the details as instructed by the teacher.

9. After all the observations are over, record your findings, structural details etc. in the practical records in the form of diagrams. Draw proportionate diagrams in the proper place on the drawing sheet. Diagrams should never be drawn to the very edge of the drawing sheet. Write suitable explanatory notes pertaining to the diagrams on the interleaf.

Laboratory Equipments

In addition to the individual requirements carried by the student, the department generally provides the following equipments/instruments etc.to the students. Sometimes these equipments may be given to each student .

These are :

(1) Students compound microscope
(2) Dissection microscope.
(3) Binocular Research microscope.
(4) Watch glasses.
(5) Cavity blocks.
(6) Slides and cover glasses.
(7) Petri dishes.
(8) Enamel tray.
(9) Polythene/glass wash bottle.
(10) Spirit lamps.
(11) Glass marking pencil.
(12) Microtome.
(13) Staining /bottles.
(14) Stone/ for honing the razor.
(15) Chemicals and stains.
(16) Culture media(to be prepared by students).

In the list of requirements mentioned above, we will discuss some in detail.

Microscope

Of all the instruments used by students in the botany laboratory, the one that is most indispensable to the students of biology is the 'Microscope'. It is not exaggeration to say that no biological study in the laboratory is possible without the microscope.

The function of the microscope is to enlarge the small objects, so that they become visible to the human eye. In reality the magnification power of the microscope lies in its resolving capacity between one particle and another. A microscope distinguishes one particle from another which are situated very closely. A human eye cannot resolve particles that are lying closer than 0.025 mm. In its function a microscope can be compared to a human eye, but much more effective than that.

There are different types of microscopes, but all of them are based on the same principle with variations in their construction. Some of the types of microscopes used in the laboratory are:

(1) Compound microscope.
(2) Binocular Research microscope.
(3) Dissection microscope.
(4) Stereoscopic dissection microscope.
(5) Phase contrast microscope.
(6) Electron microscope etc.

Of the six types of microscopes mentioned above the compound microscope and the dissection microscope are the ones that are most frequently used by students. The Electron microscope is very costly and a very highly sophisticated instrument.

In this book the compound microscope and the dissection microscope are explained in great detail as the students should be familiar with them.

Compound Microscope

The microscope has a strong 'U' shaped or 'V' shaped base. It is also called the horse shoe shaped base. Attached to the base is a bent limb. This limb can be tilted in any direction as it is attached to the base with an inclination joint.

Attached to the limb is a stage at its lower end and a body tubeat its upper end. The body tube is attached to the limb with the help of a rack and so that it can move up and down. To help in the movement of the tube during focusing, there are 2 knobs. Viz.,coarse adjustment knob and fine adjustment knob.

The body tube at its top end consists of the ocular or eye piece lens. The eye pieces usually have three magnifications 5x,10x,and 15x, and can be used to get a low or high magnification. Attached to the lower end of the tube is a revolving nose piece with three objective lenses low (10x) medium (40x) and high (60x). Some times there will be a fourth objective of the magnification of 100x. This is also called the oil immersion lens (while viewing through an oil immersion lens a drop of oil-liquid paraffin has to be put on the viewing slide).

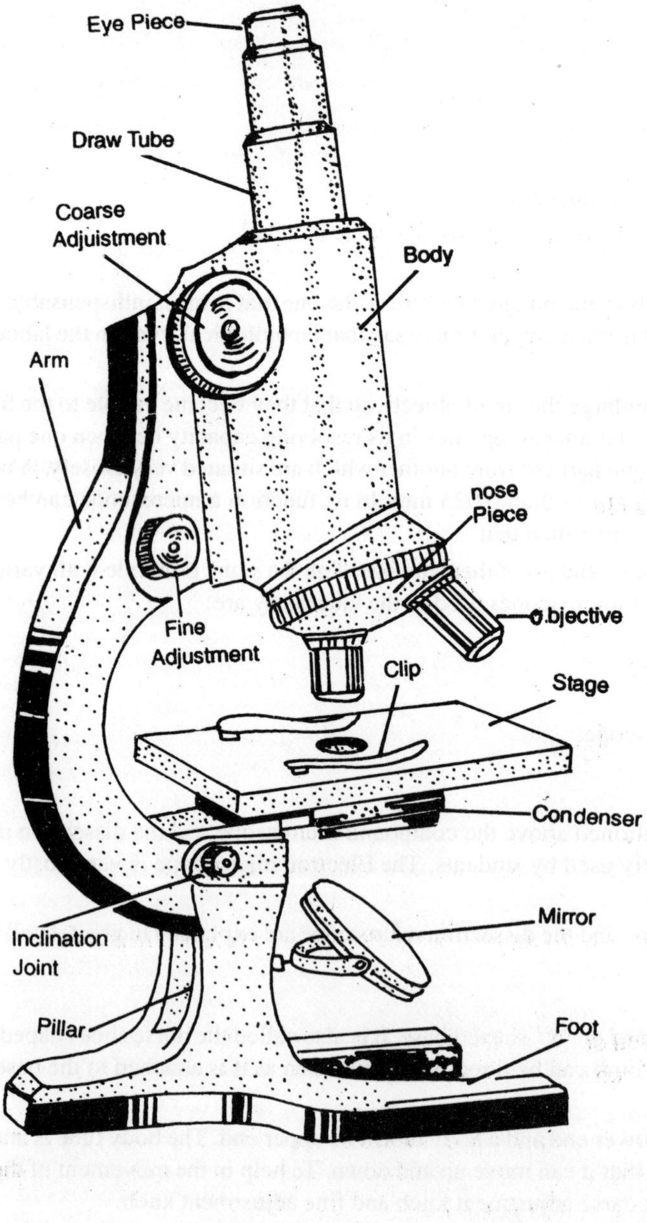

Fig. 1.1 A compound microscope (Diagrams 1-9 from pra.bot vol 1. SRIVASTAVA)

Below the objectives and attached to the limb is the *stage* of the slides are kept for viewing. The stage may be circular, square or rectangular in shape. In order to keep the slides properly placed on the stage, there are two stage clips. In some of the costly microscopes, the stage has movable sliding system operated by knobs to move the slides. This is called the mechanical stage. The mechanical stage permits the movement of the slide backtowards or forwards, left or right with the turning of the knobs provided for the purpose. A hole is provided in the stage to receive the light from below and illuminate the object. Attached to the limb, below the stage is a condenser (fixed or movable up and down) lens. Below the condenser lens is an iris diaphragm to regulate the light passing through. Iris diaphragm is a metallic disc with an aperture. The operator can be narrowed or widened by turning the lever. The condenser which is a system of lenses condenses the light rays passing through and converges them on the object.

The source of illumination in a student's microscope is a mirror which reflects the natural light or electric light. The mirror has two faces plain and concave and is attached to the limb below the condenser. The mirror is adjustable and can be turned to the direction of light. When day light is the source of illumination, the concave side of the mirror is usually used to converge the rays of light. In some of the costly microscopes, instead of the mirror there is a built in illumination ,This is a tube with a lamp (9v or 12v) attached to a transformer, which can be plugged to any AC source.

Operation of the Microscope

Microscope should be placed in a well illuminated area but avoiding the direct light as it is harmful to the eyes. Generally northern light is preferable. When ' built in illumination' is employed, it is always better to use a blue or green fitter to subdue the light.

The light has to be first focused on the condenser by adjusting the mirror towards the source of light. The amount of light entering the condenser may be regulated by closing or opening the iris diaphragm. The slide to be viewed is kept on the stage and fixed securely with the help of the stage clips. The object is first observed with a low power objective and then turned on to medium or high power, the specimen or section has to be covered with a cover slip. The slide should be observed with both the eyes open even while observing with monocular microscope.

Care of the Microscope

1. Clean all the parts of the microscope before and after use. The lenses and the mirror are to be cleaned with a lint free clothe or tissue paper.
2. Do not pull the microscope to the very edge of the table.
3. While lifting the microscope
4. The slide to be observed must be clean and dry (specially the under surface of the slide).
5. Do not remove the eyes piece or objectives from the microscope.
6. The microscope should be covered when not in use.

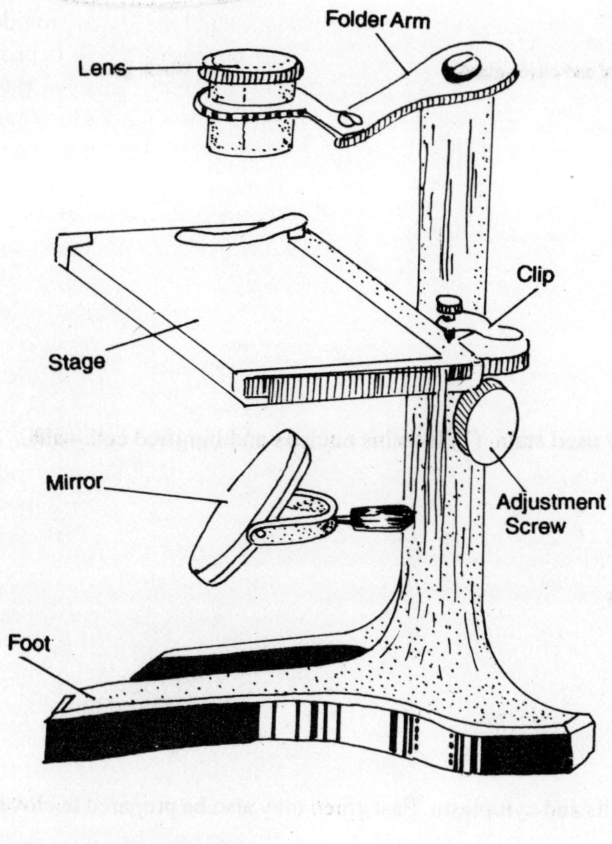

Fig. 1.2 A dissecting microscope

Binocular Research Microscope

This is like the ordinary compound microscope except that, it has binocular head for providing a natural stereoscopic view.

Dissection Microscope

This is a very simple microscope with only one set of lenses. Microscopes of this type will not give a very high magnification, but are very useful in dissecting flowers, plant parts etc.

The microscope has a 'V' shaped base and a limb. The stage is made up of a transparent glass plate. Below the stage is a mirror to focus the light. Attached to the limb is a mechanical arm which can be moved up and down with the help of a rack and pinion movement. The arm can also be moved horizontally. Attached to the free end of the arm is a holder to accommodate the lens (10x,15x etc.).

Operation: The specimen to be observed / dissected is placed on the stage and located with the help. Floral parts or others may be then dissected with the help of two needles, while still observing the specimen.

Stereoscopic Binocular Dissection Microscope

This is a combination of compound microscope and dissection microscope. While it has the higher magnification power of a compound microscope with two sets of lenses, it has retained the versatility and the manoeuvourablity of a dissection microscope.

Phase contrast and Electron microscope are not normally used by students.

Other Laboratory Equipments

Stains: These are colouring fluids. The specimens to be observed (algal or fungal material or sections of plant parts would be generally decolourised due to preservation. These are to be coloured in order to observe them clearly. The following are the stains that are generally used in the laboratory.

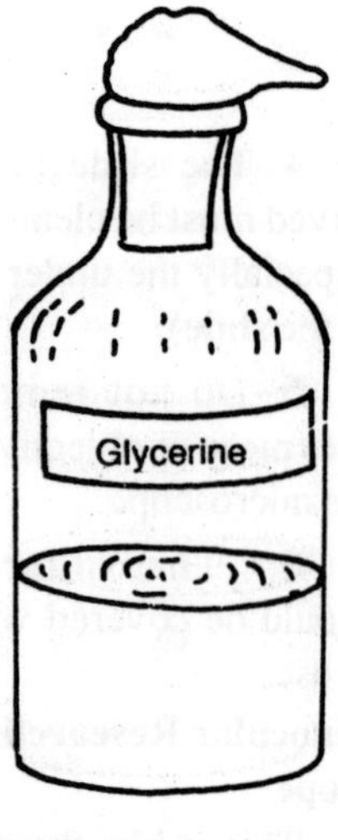

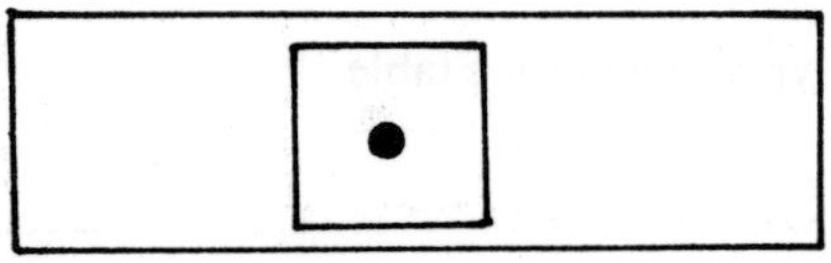

Slide and cover-glass

Watch glass

Fig. 1.3 Dropping bottle

(a) Saffranin: This is the most commonly used stain. (1) It stains nucleus and lignified cell walls.

Preparation —

Saffranin —1gm.

95% (1) Ethyl alcohol –50ml.

Distilled water – 50ml.

An aqueous stain may also be prepared.

Saffranin – 18ml

Distilled water – 100ml.

(b) Fast Green

Fast green – 1gm.

90% Ethyl Alcohol – 100ml.

This stain is used to colour parenchyma cells and cytoplasm. Fast green may also be prepared in clove oil (75ml clove oil+25ml 100% alcohol).

(c) Cotton blue.

Cotton blue –1gm.

Lactophenoil – 100ml.

(d) Haematoxylin

The dye is obtained from the heartwood of the tree *Hametoxylon campechianum,* There are two methods of preparation of the stain. These are:

1. *Heidenhains Hametoxylin*

Dissolve 0.5 or 1.0 gm of the powder in 100 cc of distilled water. Filter and store in a dark bottle away from light. The stain is ready for use after 4-5 days of ripening. Ripening is not necessary if the stain is prepared in 70% ethyl alcohol.

2. *Delafields Haematoxylin*

Sol.1 Aqueous solution of ferric ammonium sulphate 100ml).

Sol.2 One gm dye dissolved in 6ml of absolute alcohol.

mix solutions 1 and 2. To this mixture add 25 ml of glycerine and 25 ml absolute alcohol. The solution should be allowed to stand until it becomes dark red.

(e) Anilin blue

Also called cotton blue or China blue. It is prepared as follows:

Aniline blue –1gm.

Ethyl alcohol (90%)

or — 100ml.

Distilled water

Hydrochloric acid – 2drops.

(f) Gentian Violet (Crystal Violet)

Gentian Violet – 1gm.

90% Ethyl Alcohol 100ml.

(g) Erythrosine

Erythrosine –1gm.

90% Ethyl alcohol – 100ml.

Erythrosine may also be prepared in clove oil (Absolute alcohol 0.5 ml +95ml Clove oil).

(h) Light green

Light green –1gm.

Clove oil –95ml.

Absolute alcohol –5ml.

Light green may also be prepared without clove oil (in 100ml Absolute alcohol).

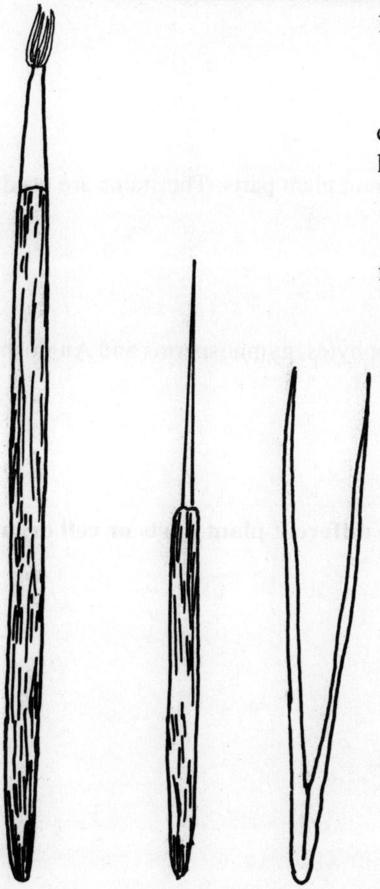

Fig. 1.4 Brush Needle Forceps

(i) Acetocarmine: The stain is prepared as follows:

Add 1 gm of stain to 100 ml of boiling acetic acid (45%)or propionic acid. Cool and decant, A few drops of saturated ferric acetate solution is added and the mixture is cooled over ice. The solution is allowed to stand for about 12 hours. Filter and store in a dark coloured bottle under refrigeration.

Fig. 1.5 Staining rack

Specificity of Stains

Usually specific stains are employed in colouring different plant parts. The stains are used singly or in combination. The most popular stain combination are:

Saffranin fast green.
Haematoxylin Erythrosin.
Haematoxylin light green.

While stain combinations are employed to study pteridophytes, gymnosperms and Angiosperms, single stains are used for Avascular cryptogams.

Green or Saffranin for Algae and Bryophytes.

Cotton blue for fungi.

The following table gives the specificity of stains to different plant parts or cell organelles.

Structure	Stain
Cellulose	Anilin blue Haematoxylin Fast green Light green
Cutinized cell wall	Gentian violet Erythrosine Saffranin
Lignified cell wall	Gentian violet Saffranin
Suberised cell wall	Saffranin
Cytopalsm	Fast green Light green Aniline blue

Chitin	Saffranin
Callose	Aniline blue
Achromatic figure	Aniline blue Erythrosine
Proteins	Saffranin
Nucleus	Heamatoxylin Saffranin Gentian violet
Chromonemata	Heamatoxylin
Chromosomes	Saffranin
Plastids	Gentian violet Haematoxylin
Mitochondria	Gentian violet
Flagella	Cotton blue

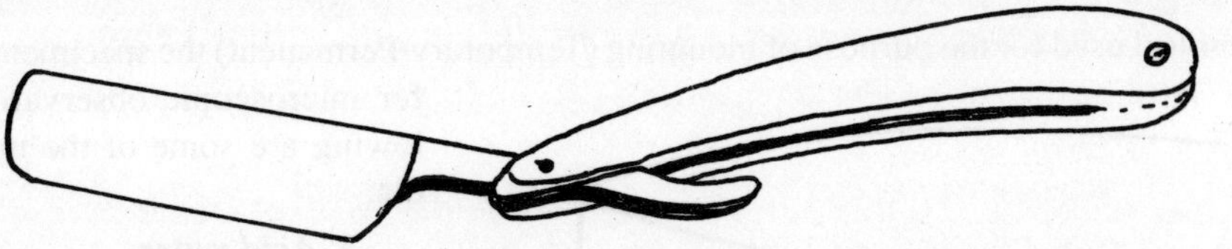

Fig. 1.6 Razor

Fixing / Preserving Fluids

The plants or plant parts collected in nature are to be killed and preserved in the laboratory for the purpose of storage and study.

In killing the specimen, the various life processes are stopped at that instant, so that after preparation for microscopic study various stages may be observed.

When ever a material is collected, it is better to kill it in the field, so that there will be minimum or no distortion. Alternatively the material collected in the plastic bags or bottles should be brought to the laboratory and killed / preserved. In preservation, the material must be completely immersed in the fluid. The following are some of the killing/preservative fluids generally used:

1. *Formaline acetic alcohol (F.A.A)*

This is the most commonly used killing /preserving agent. It is popularly called standard preservative.

Ethyl alcohol (70%)	- 90ml.
Glacial acetic acid	- 5ml.
Formalin	- 5ml

2. *Formalin - Propiono - Alcohol*

Formalin	- 5ml
Propionic acid	- 5ml.
Ethyl alcohol 70%	- 90ml.

3. Carnoy's Fluid

Ethyl alcohol	- 30ml
Glacial acetic acid	- 10ml

4. *Carony's Fluid*

Ethyl alcohol	- 30ml.
Glacial acetic acid	- 30ml.
Chloroform	- 15ml.

Carony's fluid is widely used in killing root tips, flowerbuds etc. for the purpose of cytological study.

5. *Randolph's Modified Navashin Fluid*

This consists of two solutions A and B.

Solution A

Chromic acid	- 5.0 gms.
Glacial acetic acid	- 50ml
Distilled water	- 320ml.

Solution B

Commercial formalin	- 200ml.
Distilled water	- 175 ml.
Saponin	- 3gms.

Mix A and B in equal volumes before use.

Mounting media

These are chemical used for the purpose of mounting (Temporary/Permanent) the specimens on the slide for microscopic observation. The following are some of the mounting media.

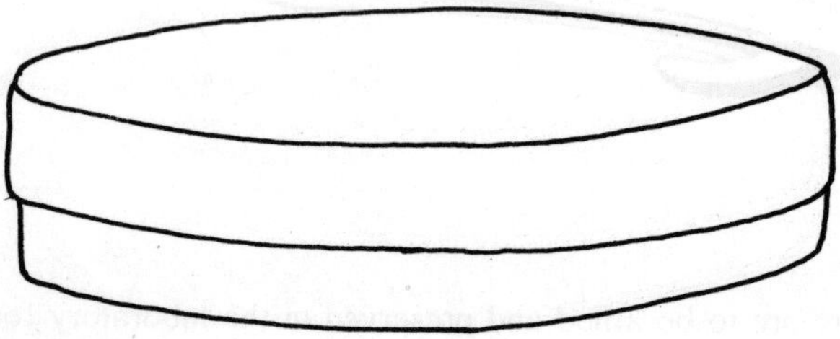

Fig. 1.7 Petri dish

1. ***Acid water***

HCl – 10-15 drops

water – 100 ml.

2. ***Lactophenol***

Phenol	-100gms
Lactive acid	-100 ml.
Glycerine	-100 ml.
Distilled water	-100ml.

3. ***Glycerine***

Glycerine	- 15-20 drops
Formaline	- 1-2 drops
Distilled water	- 100ml.

4. ***Canada Balsam:*** This is a gum obtained from Gymnosperms. This fluid is readily available in the market.

5. ***Euparol:*** This is available in the ready to use form in the market.

Canada balsam and Euparol are used for permanent mounting, while the rest are used for temporary preparations.

Staining rack: This is a wooden rack used for holding the staining bottles.

Dropping bottles: This is a grass bottle having a narrow mouth fitted with a stopper. The stopper is grooved and has a beak. When the bottle is upturned the stain/fluid comes out in drops.

Slides: These are readily available in the market. The usual size(25mmx75mm). The thickness is about 1mm. Special types of slides called cavity slides are available for specific study.

Cover slips (Cover glasses)

These are meant to cover the materials when the preparation is ready for observation. The coverslips may be round, square or rectangular.

Methods of Study of Plant Material

The following is the procedure for a scientific study in the laboratory.

Morphological study

The specimen given to external detail study should be studied either with or without a magnifying glass. Whatever details can be seen are to be recorded. If the specimen requires microscopic study, it should be suitably prepared for the purpose of microscopic observation. There are two types of microscopic preparation.

1. Whole Mount Technique

This is suitable for the purpose of external study of algal and fungal material. Take a little of the material, stain it with a suitable stain (staining procedure is explained later.) Wash the excess stain in water / alcohol. Mount in a suitable medium (temporary or permanent) and observe under the microscope.

2.Section cutting

For the purpose of anatomical study, sections are to be cut of plant parts. In algae, and globule of chara, stolons of caulerpa., receptacles of Sargassum etc., can be cut. Similarly fruiting bodies of fungi, thalli of bryophytes also many be cut in a suitable plane. In higher plants cylindrical plant parts like root, stem, petiole etc. may be cut for the study of internal structure.

Planes of Cutting: In order to study the internal structure from various angles, sections are cut in various planes.

Transverse :The section is cut at right angles to the longitudinal axis.

Longitudinal : The section is cut parallel to the long axis (right angles to the transverse axis. The longitudinal section may be cut along a radius (Radial longitudinal section-RLS) or along a tangent (Tangential longitudinal section-TLS).

Cylindrical organs like roots and stems are cut transversely or longitudinally, while dorsiventral organs like thalli of bryophytes, leaves etc. are usually cut in vertical transverse plane.

Types of Section cutting

There are usually 2 types of section cutting methods employed for plant parts. These are (a)Free hand sectioning and (b) Microtome sectioning .

(a) Free hand Sectioning

For the purpose of laboratory study where slides are not made permanent, usually free hand sectioning methods is employed. Free hand sections are cut mostly with the help of a good sharp razor (sometimes safety razor blades also may be used).

Honing the razor : The razor used for sectioning must have a sharp, even edge. There should be no nicks. Before cutting sections, the razor edge has to be honed to maximum sharpness either on a stone or on a leather sharp. Oblique strokes, with the edge forward are to be given on the sone or strap. Usually oil drops are used on the honing medium to give maximum sharpness.

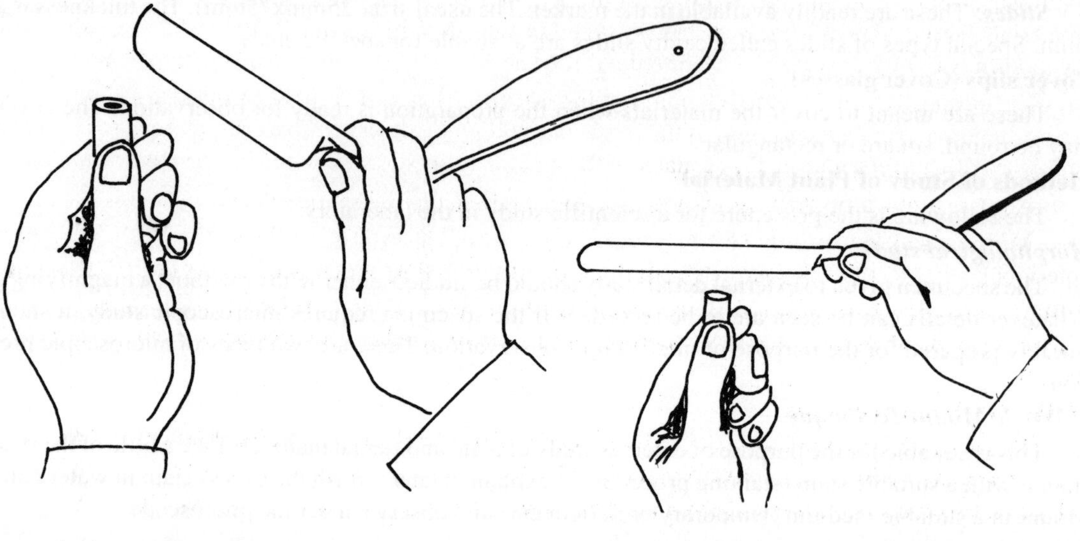

Fig. 1.8 Method of holding pith Method of holding razor Method of section cutting

Method of cutting : Open out the razor and bend the handle backwards. Hold the razor between the index finger and thumb. The handle should be free. The remaining fingers of the hand will be touching the backedge of the razor blade.

In order to cut a transverse section of a cylindrical part like stem or root, follow the procedure given below.

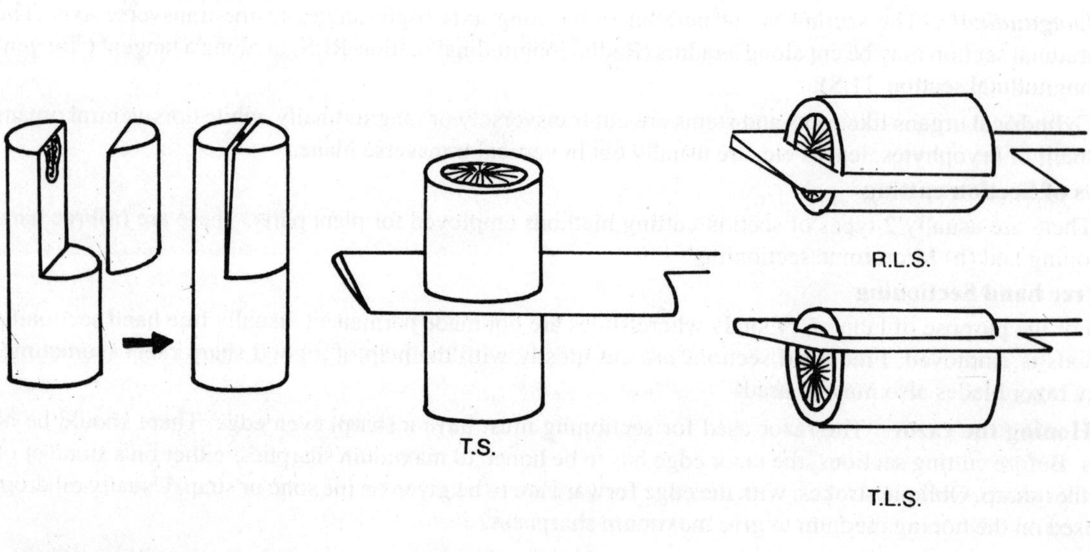

Fig. 1.9 Method of using pith of insertion of material T.S. T.L.S

Planes of section cutting

The stem or root should be cut into a 2-3" long piece. It should be held in the hand in an erect position between the thumb and index finger. The material held in the hand should be exactly at right angles to the razor's edge. The thumb holding the specimen should be slightly at a lower level. Flood the edge of the razor with water. Now using the index finger as the platform move the razor quickly several strokes across the material. During this process sections are cut and they will be collected on the blade of the razor. The sections now may be carefully transferred with the help of a fine brush in to a petridish containing water.

For cutting sections of delicate and fragile materials like thalli or leaves a suitable supporting medium has to be used. For this purpose a pith, carrot or potato may be used. These are to be cut into rectangular cubes and then split a tone end. The material for sectioning should be inserted in the split and sections are cut

After section cutting the razor should be wiped dry. It should be greased and encased.

Microtome Sectioning: This is generally employed very frequently in the study of bryophytes, pteridophytes and other higher plants. The microtome sectioning is a long and complicated process. It takes anywhere between 15-20 days to process a material and to obtain the sections.

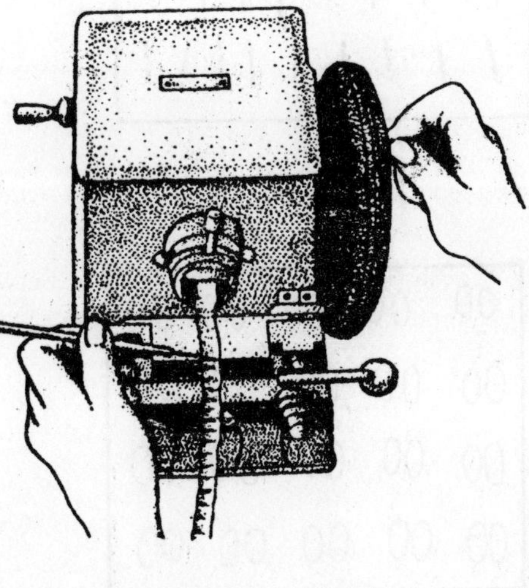

Fig. 1.10 Rotary microtome (SANTRA)

Microtome sectioning involves the following steps:

1. Killing and preserving the material.
2. Dehydration and infiltration.
3. Paraffin embedding .
4. Section Cutting.

The material like stem, leaf, flower etc. is to be killed and preserved in a suitable fluid.

(a) Cut the material into bits of suitable size.

(b) Place the material in small vials containing a fixative or preservative.

(c) Label the vial giving the details of place of collection , name of the material, preservative used and date of collection.

The material should be in the killing fluid at least for six hours (preferably a day).

Dehydration: In order to dehydrate the material, remove it from the preservative and wash thoroughly in water and transfer it to a suitable tube or vial.

Dehydrate the material in alcohol series (Ethyl alcohol) as follows

Stage 1	50% Alcohol	First day
Stage 2	70% Alcohol	Second day
Stage 3	90% Alcohol	Third day
Stage 4	100% Alcohol	Fourth day
Stage 5	Absolute alcohol and xylol (50/50)	Fifth day
Stage 6	Absolute alcohol and xylol (25/75)	Sixth day
Stage 7	Xylol	Seventh day

When the material is transferred to pure xylol, there should be no fogging (i.e the solution should not turn milky). If there is fogging repeat the procedure starting from 4th stage.

The material in xylol will be completely dehydrated and looks transparent. It is now ready for infilteration by paraffin wax. The purpose of infilteration is to see that the wax should enter into the material completely so that it can easily be cut.

Infilteration: Add a few pieces of paraffin wax everyday for about 4-5 days transfer the vial into a hot air oven. Whole temperature is adjusted too the melting point of wax. Everyday add a few pieces of wax. Adding of wax should be continued until all the xylol in the vial is replaced by paraffin wax. When the tube is smelt, there should be no smell of xylol. The material is now ready for embedding.

Paraffin embedding: In order to cut the material, it has to be embedded into a paraffin block. Prepare a paper boat or take a porcelain rectangular dish and smear glycerine to the inner surfaces. Pour molten paraffin into the porcelain dish and empty the contents of the vial into the dish. Before the paraffin soldifies, arrange the material in proper rows and in such a manner as desired. Leave it for cooling for 6-8 hours. For quick cooling the dish may be transferred to a bowl containing ice water. Carefully remove the solidified paraffin block from the dish and cut it into suitable pieces. The material is now ready for sectioning.

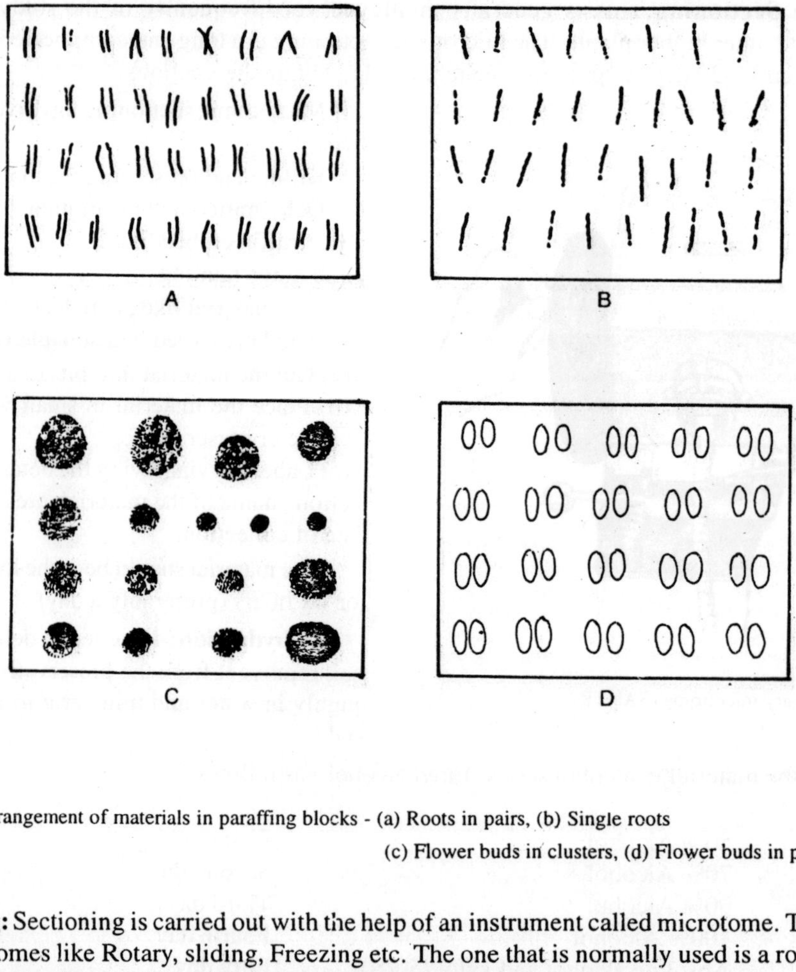

Fig. 1.11 Arrangement of materials in paraffing blocks - (a) Roots in pairs, (b) Single roots (c) Flower buds in clusters, (d) Flower buds in pairs.

Sectioning: Sectioning is carried out with the help of an instrument called microtome. There are various types of microtomes like Rotary, sliding, Freezing etc. The one that is normally used is a rotary microtome. The following procedure can be adopted in the case of the rotary microtome.

1. The paraffin block is to be cut into small pieces each having the embedded material.

2. The pieces of paraffin containing the material is to be fixed to wooden 'riders'. For this purpose, heat the top of the rider and the base of the paraffin piece with a hot knife and fix. Allow it to cool. The paraffin piece is now firmly fixed to the rider. In order to make the edges of the paraffin even, cut the edges and make them parallel.

3. Fix the wooden rider into the microtome so that the paraffin piece with the material is exactly parallel to the edge of the microtome knife.

4. Adjust the knob of the mivrotome to give the required thickness to the sections.

5. Rotate the wheel of the microtome in a clckwise direction. The cylinder containing the rider moves up and down on the knife edge. Sections are cut in a sequence and get attached to each other forming a paraffin ribbon.

6. Keep a needle in your left hand to hold up the ribbon as it lengthens. When the ribbon is sufficiently long, cut it and transfer it on to a black cardboard. Cut the ribbons into suitable length and transfer to the slide.

7. Smear the slides with an adhesive like egg albumen or gelatin and rub it well.

8. Pour some water onto the slide with the help of dropping bottle.

9. Transfer the paraffin ribbons on to the slide and drain excess water.Retain a little water on the slide.

10. Place the slide on a slide warmer. The ribbon expands, and so also the sections of the material. Now draw out the water completely and allow the slides to cool.

11. The slides should be allowed to stand for atleast 24 hours whereupon the sections will stick to the slide permanently.

The slides are now ready for staining.

Staining procedure

The plant material (whole mount 41 sections) are to be stained and mounted properly before they are ready for microscopic observations.

Algal filaments, fungal myucelia, pollen grains and spores, prothalli of pteridophytes are usually stained with a single stain saffranin or green. In some case Aniline blue is used to stain the filaments of Algae and mycelia of Fungi. The staining procedure is as follows.

Place the material either in a watch glass or cavity slide and add a few drops of stain. The period of staining varies with different materials. But usually a 5-10 minute period is sufficient. Drain out the stain and observe the material under the microscope. If the material has taken les stain, it may be kept in stain for some more period, or if it has taken excess stain it can be removed by washing either with water or 50% ethyl alcohol.

After the material is suitably stained pass it through a series of alcohol grades to dehydrate it. Usually different grades of alcohol (70%,90%,100%) taken in a filler are put on the material. After passing through absolute alcohol the material is kept in xylol and it is now ready for mounting.

Fungal mycelia can be stained with aniline blue. Usually the material taken on a clean slide is stained with a drop or two of the stain prepared in lactophenol. The material can bemounted straightway as lactophenol is also a mounting medium. If the stain is not prepared in alctophenol the material may be transferred to fresh lactophenol and mounted.

Double Staining Procedure

In the case of most of the algae and fungi where there is little or no tissue differentiation and hence single staining is sufficient. In some case of plants with relatively more tissue differentiation, a single stain is not sufficient to mark out the various tissues clearly. In such cases 2-3 stain combinations are used to stain the tissues. For example lignified and non-lignified tissues are stained with a combination of acidic and basic dyes.This type of staining is also called differential staining where different tissues take up different stains.

The following are some of the popular stain combinations.

Main stain	Counter stain
Haematoxylin	Saffranin
Haemootoxylin	Erythrosine

Haematoxylin	Eosin
Haematoxylin	Orange G.
Saffranin	Fast green
Saffranin	Light green
Saffranin	Aniline blue
Saffranin	Crystal violet
Crystal violet	Erythrosine

Staining Procedure for free hand sections

(a) Procedure for temporary preparations

Free hand sections of root, stem etc., should be washed thoroughly in distilled water in a watch glass. After repeated washing the sections should be transferred to a watch glass containg the main stain. The sections should be kept in the stain for a period ranging from 30-40 minutes. After this period the section should be transfereed to a watch glass containing distilled water. The section should be washed for 2-3 minutes whereupon excess of stain is removed. The sections should be observed under the microscope. If the main stain is saffranin only lignifed tissues should retain the stain, others should be destined. If the main stain isssued to specific to non lignified tissues, only they should retain the stain. Then the sections should be washed thoroughly in water and repeatedly observed under the microscope. The sections can be mildly stroked with a fine hair brush, where as excess of stain comes out with distilled water. The sections are now ready for counter-staining. They should be transferred to a watch glass containing the counter stain (fast green etc). Counterstain acts quickly on the tissues. Hence one to two minutes of staining should be sufficient. The sections are now observedunder sections with 15-20%glycerine. After the final stianing the tissues whould be demarcated clearly with no mixing of the stains. The sections are now ready for mounting.

(b) Procedure for permanent preparaions

If the sections of plant materials are to be mounted and made into permanent slides, the procedure should be slightly altered so as to include dehydration and clearing with suitable agents. The procedure is as fallows:

1.Stain with the principle stain (saffranin, Haemotoxylin etc).

2.Allow the sections to remain in stain for about 15-20 minutes.

3.Transfer the sections to water and wash thoroughly. The sections should be stroked mildly until no mopre stain comes out, and the water should remain clear. If the main stain like saffranin is produced in 50% alcohol then desatining (washing) should be done in 50% alcohol.

4.The sections should be now dehydrated in alcohol grades of 50%, 70% and 100%.

5.In each grade of alcohol keep the stin for about 5 minutes.

6.Transfer the sections from one grade of alcohol to the other with the help of a fine brush. Never us needles as they injure the sections.

7.Transfer the sections to the counter stain (fast green etc.)The counter stains are usually nprepared in 90% alcohol.

8.Keep the section in counter stain for about a monute. Wash the sections in 90% alcohol and transfer them to 100% alcohol.

8a. In some case counter stains are prepared in clove oil. In such cases counter staining should be done after the sections are completely dehydrated with absolute alcohol.

1. Sections are transferred compldgely dehydrated to a solution of 50:50 Absolute alcohol and xylol followed by 25:75 Absolute alcohol and xylol and ultimately transfer the sections to pure xylol. Now xylol should not turn milky or turbid. If it turns milky dehydrate the sections again through combination of absolute alcohol:xylol series.

The sections are now ready for mounting.

Staining procedure for microtome sections

The slides with the paraffin ribbons adhering to them have sections of plant material.The processing and staining is as follows:

1. Immerse the slides in a coupling Jar containing xylol. Leave them for 10 minutes . The paraffin dissolves in xylol and only the sections remain on the slides.

2. The slides are now kept in 50:50 absolute alcohol and xylol.

3. Pass the slides through a down grade series of alcohol. 100%® 90%® 70%

4. Transfer the slides to pure water and wash them thorughly to remove traces opf alcohol.

5. Keep the slides in amordant like 4% iron alum. This is necessary obly when Haematoxylin is used. Afterabout 10 minutes wash the slides in water/

6. Transfer the slides to the main stain Haematoxylin. Saffranin etc. and keep them for 15-30 minutes.

7. Remove the slides from stain and wash in water.

8. Observe the slides under microscope. Usually the sections are overstained.

9. If the main stain used in Hametoxylin, destaining can be done with 2% Iron alum or with water containing few drops of concentrated HCl/ Repetaedly observe the slides under the microscope. Stop destaining when only the nuclei have retained the stain.

Caution: When destaining in done with HCl not more than 5-6 drops should be used for 75ml of water. Excess macid will eat away the adhesive and the sections come out of the slide.

10.Wash the slides in water overnight if the destining agent is Iron alum, otherwise washing for 30 minutes is enough.

11.Now pass the slides through an upgrade series of alcohol (5 minutes in each grade) 70% 90% 100%

12. When the sections are completely dehydrated stain them with the counter stain (Fast green etc) for à minute or two. Wash the excess stain in clove oil and transfer the slides to pure xylol. The slides are now ready for mounting.

Mounting

(a) Mounting Media

Mounting media are of two categories: Temporary mounting media and permanent mounting media.

A temporary mounting medium may be just water or water mixed with glycerine. If the temporary sections are to be retained for 2-3 hours dilute glycerine (15-20%) is recommended.

Permanent mounting media are used when the slides are to be retained indefinitely. The following are some of the permanent media.

(1) **Canada Balsam:** This is the most widely used mounting medium in botanical preparations. It is abvailable in ready to use bottles. It is a gummy substance. If the solution is very thick add little xylol.

Canada balsam is a resinous product of a gymnosperm, Abies balsamea/ The gun has very good refractive index and as such it is highly recommended.

(2) **Lactophenol:** Lactophenol can be prepared in the laboratorty. Mix equal quantites of phenol cystals, lactic acid glycerine and distilled water.

(3) **Glycerine gelly:B This is prepared as follows:**

Gelatine 1 part.
Water 6 parts.
Glycerine 7 parts.

Glycerine jelly is suitable even for mounting fresh specimens (even without dehyderation).

Add gelatine to water stir thoroughly and warm. Add glycerine and cool the solution. After sometime the solution settles into a jelly.

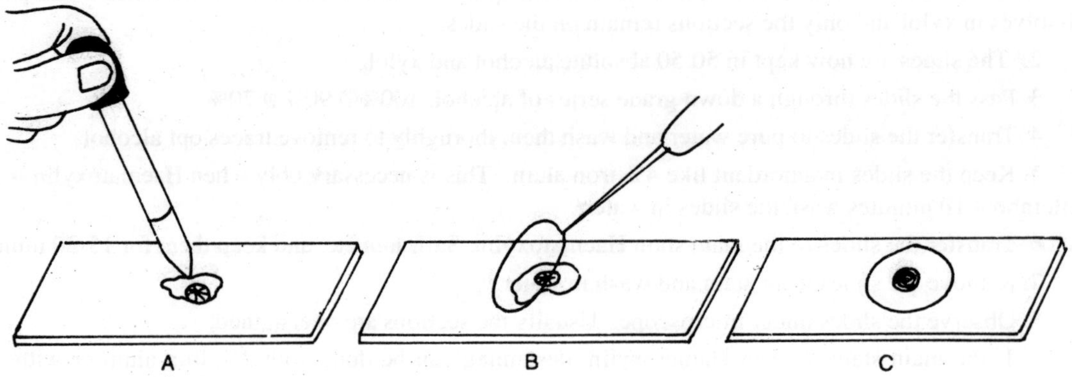

Fig. 1.12 Method of mounting the material

(b) Mounting Method

The sections (free hand) are to be mounted on tyo clean slide in a specific mounting medium. If the sections are microtome preparations they would be already adhering to the slide. In any case the sections are to be covered with a thin coverslip or cover glass are readily available in the market. They may be round, squasre or rectangular. Select the one that is suitable for your preparation.

The sections or objects should be mounted on the centre of the slide. In oreder to ensure placing of tne object on the centre of the slide, the following procedure may be adopted. On a white paper draw a rectangle of the same sizxe as the slide. Join the opposite corners by drawing diagonal lines. The centre of the slide lies where the two lines cross. Place the slide on the rectangel and material or objects may be kept on the central point.

Place a drop or two of the mounting medium. Hold the coverslip at the edges (see the figure) and allow it to touch the edge of the mounting medium at an angle of approximately 45o slowly lower the cover glass on the mounting medium with the help of a needle Take care to see that no air bubbles enter the mounting medium. If there are air bubbles, lift the cover glass and mount again.

Excess mounting medium may be carefully wiped wioth the help of a clean tissue. Allow the slides to dry. If necessary the slides may be transferred to a hot air oven for drying. The slides should be labeled at one end as shown in the figure. They are now ready for microscopic observation.

Ringing of Slides

Temporary preparations may be kept for few days if the borders are sealed. This procedure is known as ringing . Ringing is done by applying wax, canadabalsam or even nail polish to the edges. Ringing may be dfone manuyally or by using a ringing table having a rotary metallic disc

Record of Practical Work

Practical work carried out in the laboratory should be recorded. The botany practicval record book has two types of sheets viz., Interleaves and drawing sheets. Interleaves are meant to write the notes pertaining to the diagrams. The drawing sheets which are thicker than the interleaves are used for diagrams. The drawing sheets which are thicker than the interleaves are used for diagrams. The following points may be useful in maintaining a good record.

1. Use good pencils (Hsnf HB) with sharp points and a good quality eraser.
2. Leave a minimum of one inch margin on all the Sides.
3. Alway write diagrams of only one plant in a page. Do not mix up diagrams of other plants.
4. On the left hand corner of the drawing sheet write the date of the experiment.
5. Give the complete systematic position of the member on the top of the drawing sheet.
6. For microscopic organisms draw first the habit sketch followed by low power and high power vies\ws/ High power view should show the cellular details.
7. In the case of anatomical drawings draw the ground plant first, followed by a sector enlargement to show the cellular details.
8. All parts of the diagrams must be properly labelled.
9. Stages of reproduction, if any, must be drawn after the details of vegetative structure.
10. The description corresponding to the diagrams must be written on the opposite interleaf.

Field Collection of plants

Specimens of l\algae, fungi and bryophytes can be collected in the field for the purpose of laboratory study. Field trip is an integral part of botanical study and all students must participate in it. It is in the field study that the students will come to understand the natural growth, habitat and environmental association of plants. While a field trip to collect specimens, the following items are to be carried:

1. Pocket lens.
2. Fixative F,A.A or Carnoy's fluid.
3. Collection bottles.4. Blotting paper.
4. Polythene bags
5. Field note book
6. Labels.

When the specimens are collected they can be fixed and preserved for the purpose of laboratory study. Use the required preservative and immerse the specimens in it. Label the bottle showing details of

Date if collection

Location

Preservative fluid (name) used.

Name of the plant

As it is necessary for the purpose of identification details of habitat, associations with other plants if any, colour of the fresh specimen etc. should be recorded in the field note book.

CHAPTER 2

CLASSIFICATION OF THE PLANT KINGDOM

The living world consists of plants and animals of various kinds. Evern within the plant world many different types of plants have been recognized and perhaps many more are yet to be discovered. If such huge number of individuals are not properly arranged, their study will not only be difficult but also meaningless because proper comprehension is impossible. So, in order to bring an arrangement and discipline plants have been grouped into larger and smaller groups based on resemblances and differences (mostly morphological). This system of dividing plants into groups is known as "Plant classification or Systematics."

In general there are mainly two systems of classification, viz. (1) Artificial system and (2) Natural system.

In the artificial system of classification, as proposed by earlier botanists, importance is given to minor, superficial and frequently variable characters. For example divisions of plants based on their habit - herbs, shrubs and trees and using this as a criterion in the classification. The artificial system does not indicate the relationship existing among different plants.

The natural system of classification is based on the natrual phylogenetic affinities existing between various types of plants. Natural system lays stress on the similarity in reproductive structures including anatomical character, cell structure and physiology. Such structures are more stable than the ordinary vegetative characters and are not pone to frequent alterations by the environmental influence.

According to the natural system, all the plants have been included in the plant kingdom. The plant kingdom is divided into divisions, each division into several sub-divisions each subdivision into classes, each class into orders, each order into families, each family into genera and each genus into species, and in some cases each species into several varieties.

Various natural systems of classification have been proposed time and again. It is not the idea of present attempt to give a detailed survey of all those syetems. Only a few systems have been briefly discussed below so as to give an understanding to the students the relative positions of different groups of plants in the evolutionary sequences of plant kingdom so that they will be in a better position to understand and appreciate the different groups of non-vascular and vascular plants.

In general, two main categories can be identified in the natural systems. They are:

(1) Old or Traditional systems.
(2) Revised or Modern systems.

Traditional systems

Of all the old systems perhaps the one that is followed even today is that of Eichler (1866). He proposed that the plant kingdom can be divided into:

(A) Cryptogamae and (B) Phanerogamae

(A)**Cryptogamae:** Cryptogamae have their reproductive structures embedded in the plant body and do not bear flowers or seeds.

They include three divisions viz.,

(1) Thallophyta
(2) Bryophyta and
(3) Pteridophyta

Thallophyta: The division thallophyta is characterised by plants

a) With an undifferentiated flat plant body called thallus.
b) having no vascular tissue.
c) with unicellular reproductive structures.
d) without embryonal stage in the life cycle.
e) members generally aquatic (not a rule)

Thallophyta is further divided into 2 classes (1) Algae and (2) Fungi

Algae: The class Algae includes the members majority of which are:

(a) Aquatic in habitat.

(b) Autotrophic because of the presence of chlorophyll (green photosynthetic pigment) and other pigments.

(c) Having unicelluar or multicellular plant body.

Fungi: This includes all achlorophyllous thallophyta. As the members are pigmentless they are not autotrophic. Their mode of nutrition is heterotropic. Fungi follow two methods of hetertrotropic nutrition viz.

(1) Parasitic and (2) Sapropytic.

In the parasitic method fungi obtain their nutrition from living organisms (plants or animals) while in the saprophytic method fungi derive nutrition from dead and decaying organisms (for details see introductory chapter on fungi).

Bryophyta: Plants are amphibious in their habitat. They differ from thallophytes in the following respects:

(1) Possession of multicellular sex organs.
(2) Presence of an embryonal stage in the life cycle.
(3) Occurrence of heteromorphic alternation of generations in the life cycle.
(4) Sporophyte being dependent on the gametophyte.

Pteridophyta: Pteridophytes are called vascular cryptogams because of all the cryptogams, only they have vascular tissue or conducting tissue, which helps in the transportation of food materials and also offers mechanical support to the plant body. They are seedless. Unlike in Byrophytes sporophyte is the predominat plant body but nevertheless the gametophyte is independent (with very few exceptions).

(A)**Phanerogams:** These are the plants in which the reproductive structures are not embedded. They produce seeds. Phanerogams are divided into two divisions viz.

(1) *Gymnospermae*: Seeds are naked, not covered by any envelope, no fruit formation.

(2) *Angiospermae*: Seeds enclosed by fruit wall, fruit being formed by the mature ovary which is a special envelop for the ovule (seed). Angiosperms are divided into two classes viz.,

(1) *Dicotyledonae :* seeds with two cotyledons.

(2) *Monocotyledonae* : seeds with one cotyledon.

Some botanists divided the entire plant kingdom into four major divisions viz.,

(1) Thallophyta.
(2) Bryophyta.
(3) Pteridophyta.
(4) Spermatophyta (includes Gymnosperms and Angiosperms).

Modern System of Classification

According to one of the modern systems there are only three major divisions in the plant kingdom. They are:

(1) Thallophyta
(2) Bryophyta
(3) Tracheophyta

The last division includes all the vascular plants, which are classified into the following sub-divisions.

(1) Psilopsida
(2) Lycopsida
(3) Sphenopsida
(4) Pteropsida

Pteropsida includes 3 classes viz.,

(a) Filicineae
(b) Gymnospermae
(c) Angiospermae

Angiospermae has two sub classes

(1) Dicotyledonae and
(2) Monocotyledonae.

In another system only Thallophyta and Embryophyta have been recognised as two major divisions. In Embryophyta there is the formation of an embryonal stage in the life cycle while it is absent in thallophytes. Embryophyta includes bryophytes, pteridophytes, gymnopserms and Angiosperms (Details not discussed).

Many modern botanists consider that the plant kingdom consists of 18 major divisions.

They are:

(1) Chlorophyta
(2)Charophyta
(3)Euglenophyta
(4)Pyrrophyta
(5)Chrysophyta
(6)Phaeophyta
(7)Cyanophyta
(8) Rhodophyta
(9)Schizomycophyta (Schizomycota)
(10)Myxomycophyta (Myxomycota)
(11)Eumycophyta (Eumycota)
(12)Bryophyta
(13)Psilophyta
(14)Lycophyta (Lepidophyta)
(15)Arthrophyta (calamophyta)
(16)Filicophyta (pterophyta)
(17)Gymnophyta
(18)Angeophyta (Anthophyta)

PART ONE
ALGAE

CHAPTER 3

GENERAL ACCOUNT AND CLASSIFICATION

The study of Algae is called phycology. Algae have been known to human beings since time immemorial, though their scientific study is recent. Early Indian, Chinese, Roman and Greek literatures contain numerous references to this group of plants. In Amarakosa, a Sanskrit lexicon of the 4th or 5th century A.D., Algae are mentioned and are called *'jalanili'* or *'shivala'* meaning waterscum. Ancient Hawaiians used Algae for food. At present the study of Algae is one of the important fields of Botanical science having far reaching implications from the point of view of human welfare.

Introduction

According to the present trend of plant classification however Algae do not constitute any compact group. The term Algae refers to about eight divisions of the plant kingdom. From the point of view of convenience however the term has till been retained. All Algae are chlorophyllous, thallophytes and hence autotrophic. They are non-vascular plants with predominant aquatic distribution. There are about 30,000 species of Algae.

Occurrence of Habit

The members are aquatic being distributed both in fresh and salt waters. Fresh water Algae abound in stagnant waters, tanks, ponds, ditches, slow running steams, rivers, etc. Some of the terrestrial forms grow abundantly wherever nature offers a damp surface. They grow on wet trunks and branches of trees or even on rocks. They may also live as epiphytes or ephizoophytes (Algae growing on animals, specially molluscan shells). Fresh water Algae may be found free floating or attached to rocks, stones or other higher water plants like *Elodea, Ottelia* etc.

Some of the Algae are endophytic growing in company of higher plants. Some of the Cyanophycean Algae like *Nostoc* and *Anabena* which are known to possess the capacity of nitrogen fixation are found forming symbiotic association with plants like *Anthoceros, Iseotes, Cycas* etc.

Marine Algae are commonly called seaweeds occupying a large area (sometimes several thousand miles). They may be free floating or attached. The free floating marine Algae are known as "phytoplanktons". Those which are attached found growing in shallow waters form "Benthons".

Range of Vegetative Structures

The thallus organisation in Algae may be classified into the following types:

(1) *Motile types:* This is the simplest type. Plant body is motile and unicellular throughout. In some cases motile forms may be colonial with little or no division of labour. Example *Chalamydomonas, Volvox* etc.

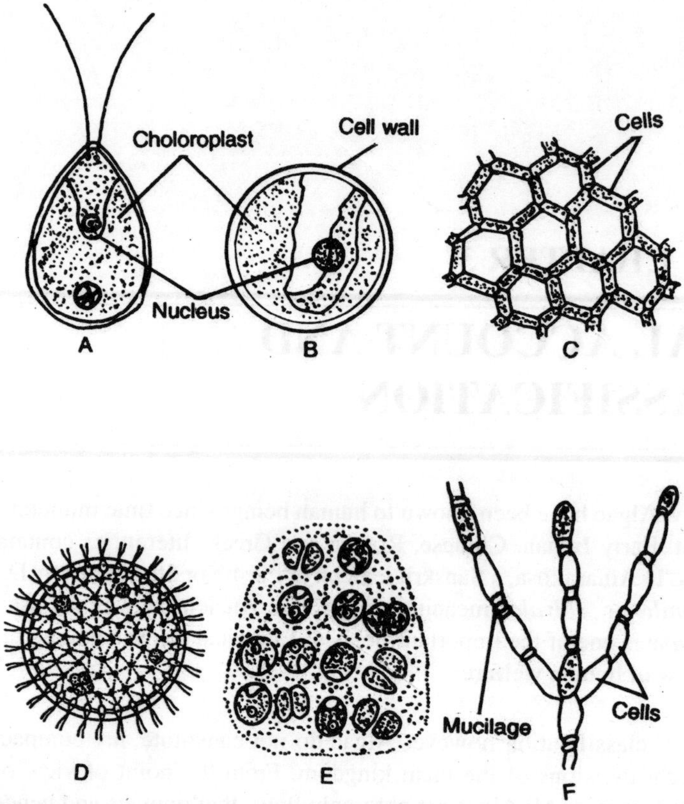

Fig. 3.1 A few algae
a. Chlamydomonas; b. Chlorella; c. Hydrodictyon; d. Volvox; e. Tetraspora; f. Prasinocladus.

(2) *Palmelloid type and Dendroid types:* Palmelloid or non-motile stage is found as a temporary state even in motile forms. But in some forms like *Tetraspora*, this is a permanent feature. Here the cells are surrounded by mucilage covering. Dendroid type is a slight variation of the palmelloid type. For *Tetraspora, Prasinocladus* etc.

(3) *Coccoid forms*:In these forms the locomotory organs (flagella) are lost and the cells become rounded. For example *Chlorcoccum.*

(4) *Filamentous types* : The filamentous habit is believed to have originated from motile forms. Loss of motility and restriction of cell division to one plane has resulted in the filamentous types. The filaments may be free floating or attached. All the cells in the filament may be alike. One of the lower cells may be modified to help in attachement (hold fast, hapteron etc). The filaments may be branched or unbranched e.g. *Oedogonium, Cladophora* etc.

(5) *Siphonous habit*: The plant body consists of tube like filaments (branched also) but no cross walls are found. Hence the entire plant body is comparable to a single cell. Cross walls appear only to demarcate the reproductive structures or to separate of the dead portions of the thallus e.g., Vaucheria, Caulerpa etc.

(6) *Advanced types:* They may be monosiphonous or polysiphonous filamentous types or even true parenchymatous e.g., *Batrachospermum, Polysiphonia* etc.

Reproduction

Algae reproduce by two methods viz:

(1) Asexual (includes vegetative propagation also).

(2) Sexual

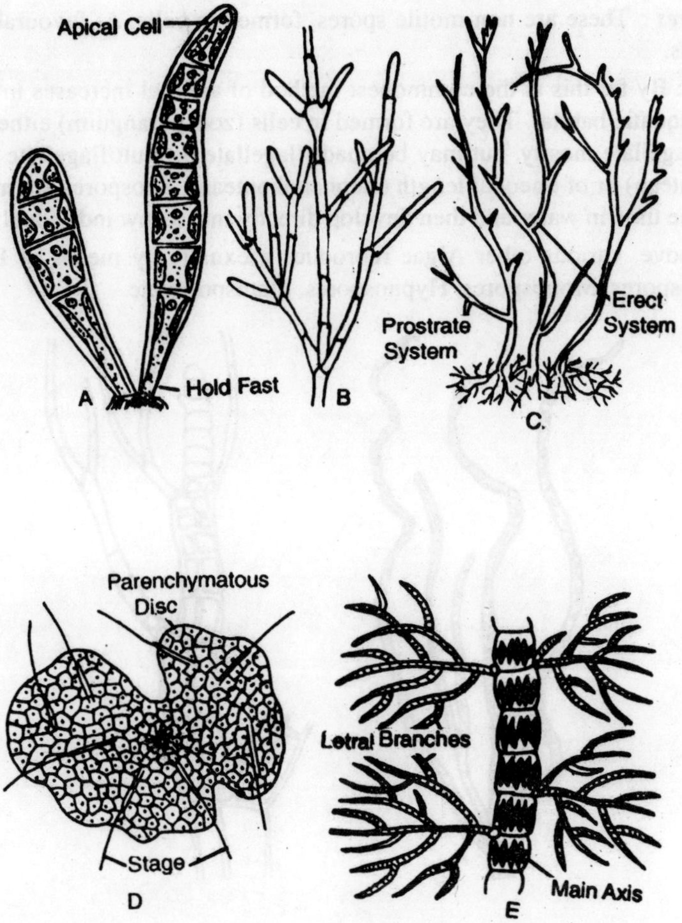

Fig. 3.2 A few Algae
a. Ulothrix; b. Cladophora; c. Stigeoclonium; d. Coleochaete; e. Draparnaldia.

Asexual Reproduction

This is brought about by a variety of methods

(1) *Cell division* : Cell division as a means of multiplication of the individuals is seen in several unicellular members e.g., *Pleurococcus* etc.

(2) *Fragmentation :* Accidental break up of the plant body results in fragments. Each fragment develops into a new individual e.g., *Spirogyra, Ulothrix* etc.

(3) *Hormogones and Hormocytes:* This type of vegetative multiplication is found in some cyanophyceae. Filament breaks up into 3-4 celled bits called hormogones (thick walled ones are called hormocytes) which develop into a new individual e.g., *Oscillatoria.*

(4) *Protonema:* These are thread like bodies helping in vegetative multiplication e.g., *Chara.*

(5) *Amylum stars:* These are starch filled stellate structures helping in asexual reproduction e g., *Chara.*

(6) *Akinetes:* These are non-motile oval shaped structures formed in cells usually to withstand the adverse environmental conditions. They are formed in many members of chlorophyceae.

(7) *Aplanospores :* These are non-motile spores, formed in cells. At favourable season these develop into new individuals.

(8) *Zoospores* : By far this is the commonest method of asexual increases in Algae. These are motile spores adopted to aquatic habitat. They are formed in cells (zoosporanguim) either singly or in large numbers. They are biflagellate mostly, but may be quadriflagellate or multiflagellate also. Flagella may be of equal length (Iskontean) or of unequal length (stephenokontean). Zoospores are mostly pyriform in shape. They swim for some time in water and then develop directly into a new individual.

Besides the above, various other Algae reproduce asexually by means of Endospores, Autospores, Auxospores, Carpospores, Monospores, Hypnospores, Statospores etc.

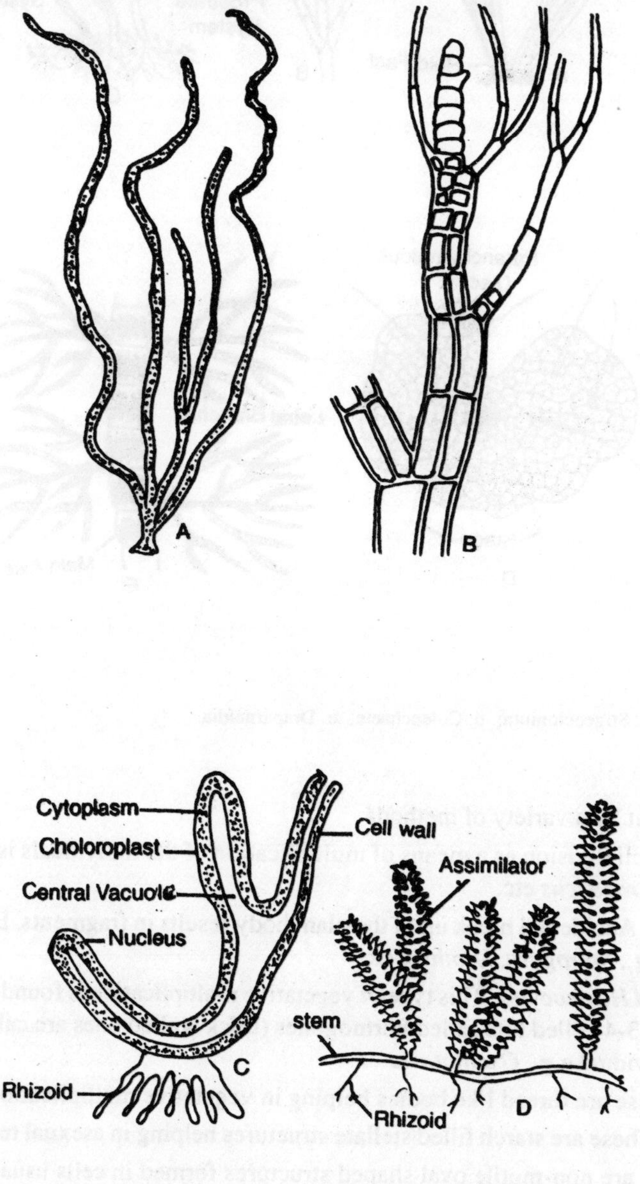

Fig. 3.3 A few Algae
a. Nemalion multifidum ; b. Polysiphonia flexicaulis ; c. Vaucheria sp. ; d. Caulerpa taxifolia.

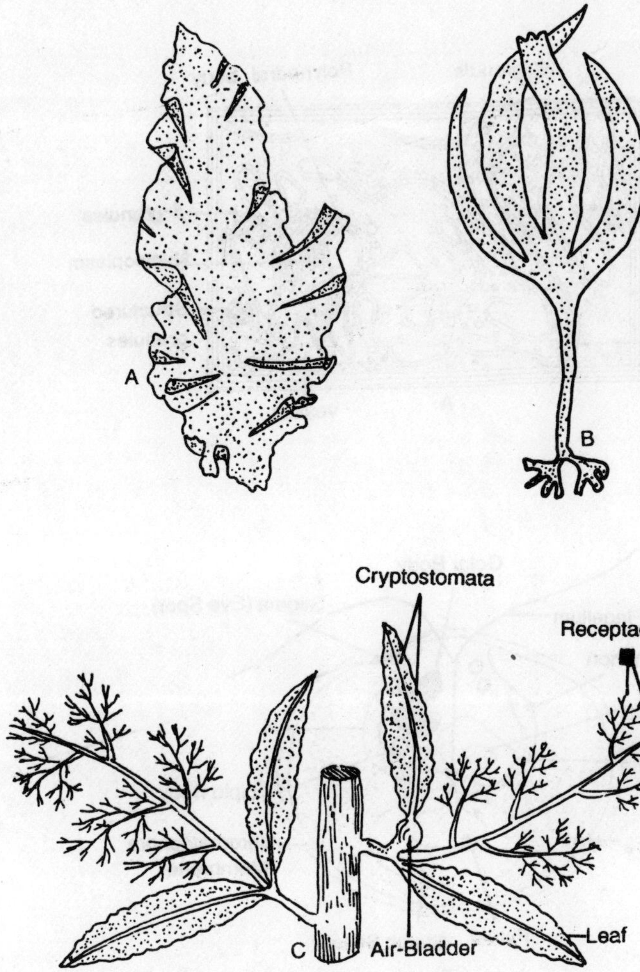

Fig. 3.4 A few Algae
a. Ulva ; b. Laminaria ; c. Sargassum.

Sexual Reproduction

This is brought about by the fusion of sex cells known as gametes which are produced in gametangia.

There are three types. viz. (1) Isogamy (2) Anisogamy and (3) Oogamy. The oogamous type of reproduction is the most advanced type.

Isogamy

In this type there is no morphological difference between the fusing gametes. The gametes are motile and come out of the gametangia. Fusion takes place outside.

Anisogamy

In this, two types of gametes are formed in gametangia. Some gametangia produce large number of smaller gametes (male) and some others produce less number of large sized gametes (female). Both however are liberated out. Fusion is external. Large gametes are called Macrogametes and small gametes are called microgametes.

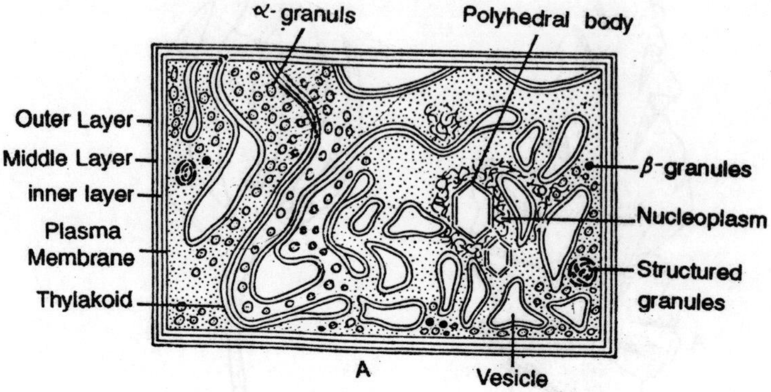

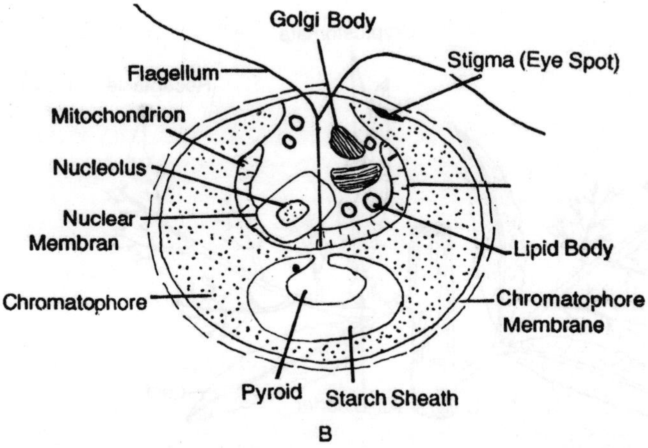

Fig. 3.5 A: Diagrammatic plan of a procaryotic cell of a blue-green Alga, B : Eucaryotic cell organisation (*Heteromasticx rotunda*).

Oogamy

In this, the male and the female gametangia are different. The former is called Antheridium (spermatangium) and the latter, oogonium. Antheridium produces flagellate gametes. Oogonium produces only one non-motile female gamete (egg) which never comes out. Fertilization is internal.

Alternation of Generations

In some Algae, the life history comprises of two district generations viz., sporophyte and gametophyte. The former is asexual and diploid, the latter sexual and haploid . Haploid gametophyte reproduces sexually and the fused product (zygote) directly develops into the diploid sporophyte (No alternation of generations if the zygote undergoes reduction division). This diploid sporophyte reproduces asexually by spores. During spore formation reduction division takes place, hence the spores are haploid. They develop into gametophyte. If both gametophyte and sporophyte are alike morphologically, alternation of generations is said to be isomorphic and if they are different, heteromorphic.

Only few Algae however exhibit alternation of generations.

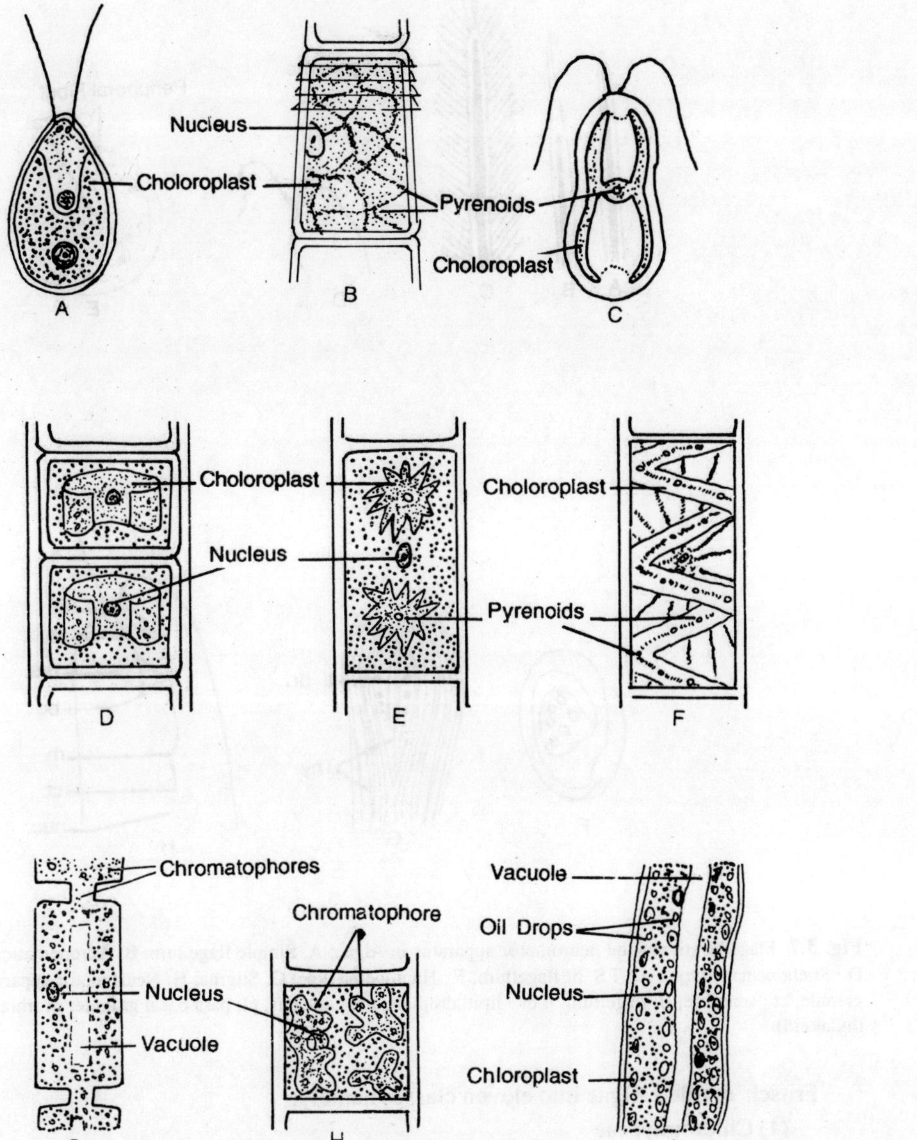

Fig. 3.6 Different types of chloroplasts in Aigae. A: cup shaped (*Chlamydomonas eugametos*); B. reticulate (*Oedogonium sp) ;* C : H-shaped (*Chlamydomonas alpina*) ; D: girdle shaped (*Ulothrix* sp); E. star shaped (*Zygnema* sp.) ; F. ribbon shaped (*Spirogyra* sp.) ; G. discoid (*Polysiphonia* sp.); H. irregular (*Ectocarpus* sp.) ; I. discoid (*Vaucheria* sp.).

Classification (All systems have not been discussed).

Before 20th century only four classes were recognised in Algae. They are:

(1) Chlorophyceae
(2) Phaeophyceae
(3) Rhodophyceae and
(4) Cyanophyceae.

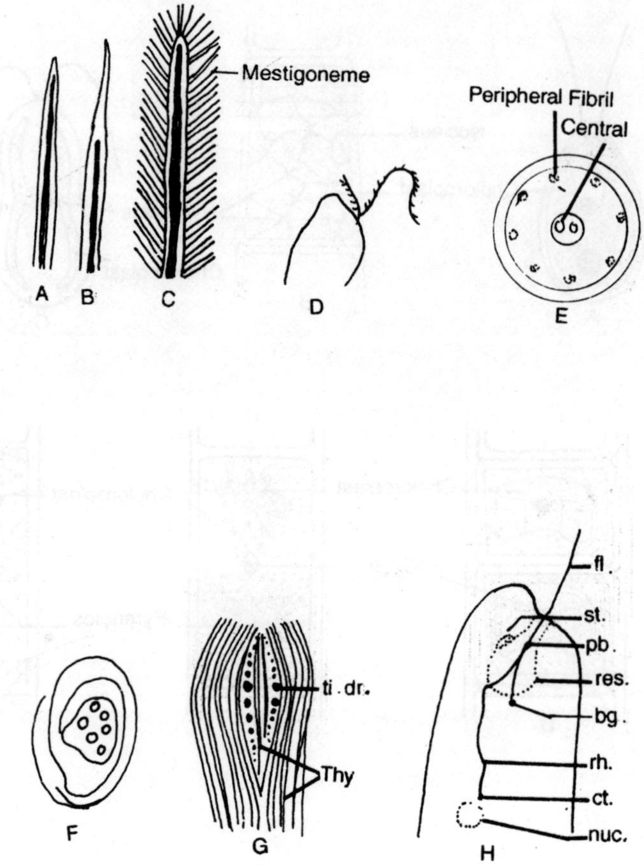

Fig. 3.7 Flagella, stigma and neuromotor apparatus in Algae; A. Simple flagellum; B: Acronematic type ; C: Pantonematic type ; D : Stichonematic type ; E : T.S. of flagellum; F : Haptonema type; G. Stigma; H: Neuromotor apparatus of *Euglena* (bg: basal granule; ct : centriole; fl: flagellum; li dr : lipid droplets; nuc: nucleus; pl: para basal granule; rh: rhizoplast; st: stigma; thy : thylakoid).

Fritsch divided Algae into eleven classes namely:

(1) Chlorophyceae
(2) Xanthophyceae
(3) Chrysophyceae
(4) Bacillariophyceae
(5) Cryptophyceae
(6) Dinophyceae
(7) Chloromonodineae
(8) Euglenophyceae
(9) Phaeophyceae
(10) Myxophyceae
(11) Rhodophyceae

In the classification proposed by Fritsch importance is given to pigmentation, nature of plant body, reproductive features, metabolic products, locomotory organs etc.

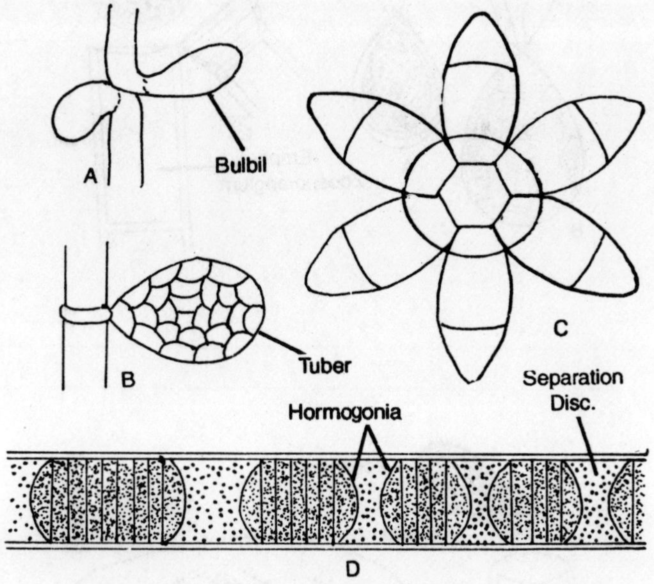

Fig. 3.8 Vegetative reproduction,
A: Bulbil of *Chara*; B: Tuber of *Chara*; C: Amylum star of *Chara*; D : Hormogonia of *Lyngbya* sp.

Glibert M.Smith classifies Algae as follows:

Division : (1) *Chlorophyta*
Class : (a) Chlorophyceae
(b)Charophyceae
Division : (2) *Euglenophyta*
(a) Euglenophyceae
Division : (3) *Pyrrophyta*
(a) Desmophyceae
(b) Dinophyceae
Division : (4) *Chryosophyta*
(a) Chryosophyceae
(b) Xanthophyceae
(c) Bacillariophyceae
Division : (5) *Phaeophyta*
(a) Isogeneratae
(b) Heterogeneratae
(c) Cyclosporeae
Division : (6) *Cyanophyta*
(a) Myxophyceae
Division : (7) *Rhodophyta*
(a) Rhodophyceae.

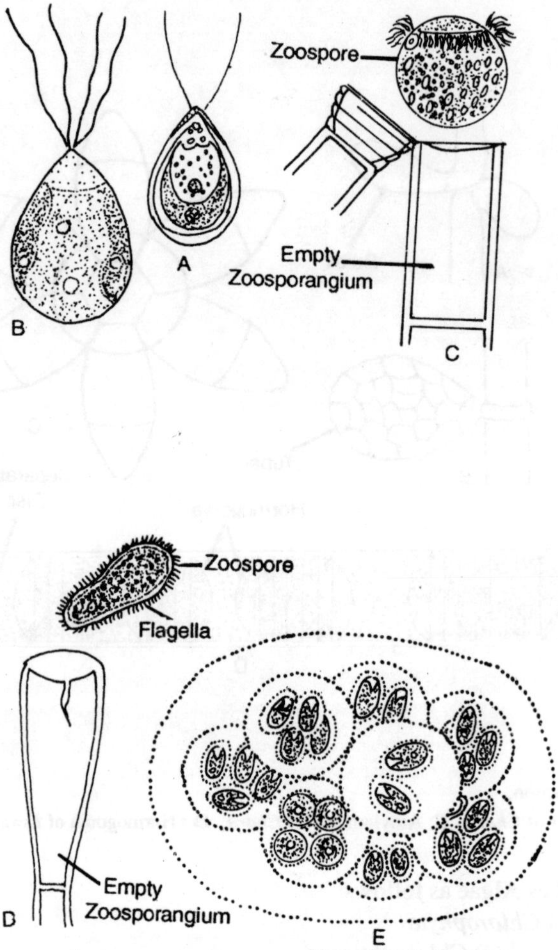

Fig. 3.9 A sexual reproduction, A: Zoospore of *Chlamydomonas* sp, B: Zoospore of *Ulothrix* sp, C: Zoospore of *Oedogonium* sp, D : Zoospore of *Vaucheria* sp, E : Palmella stage in *Chlamydomonas* sp.

Algae of uncertain systematic position

(1) Chloromonadales (2) Cryptophyceae

Papenfuss divides Algae into eight phyta. They are:

(1) Chlorophycophyta
(2) Charophycophyta
(3) Euglenophycophyta
(4) Chrysophycophyta
(5) Pyrrophycophyta
(6) Phaeophycophyta
(7) Schizophyta
(8) Rhodophycophyta

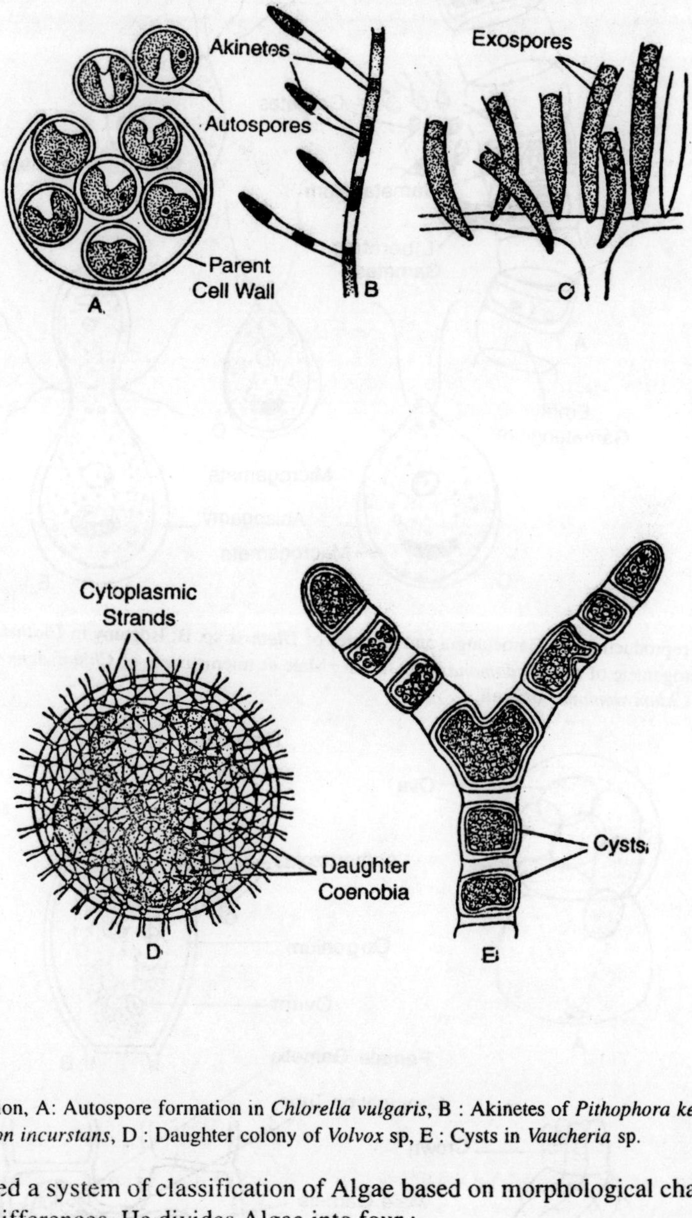

Fig. 3.10 Asexual reproduction, A: Autospore formation in *Chlorella vulgaris*, B : Akinetes of *Pithophora kewensis*, C : Exospores in *Chaemosiphon incurstans*, D : Daughter colony of *Volvox* sp, E : Cysts in *Vaucheria* sp.

Chapman (1926) proposed a system of classification of Algae based on morphological characters, pigments and biochemical differences. He divides Algae into four :-

(1) Euphycophyta
(2) Myxophycophyta
(3) Chrysophycophyta
(4) Pyrrophycophyta

He includes charophyceaee, chlorophyceae, phaeophyceae and Rhodophyceae under Euphycophyta.

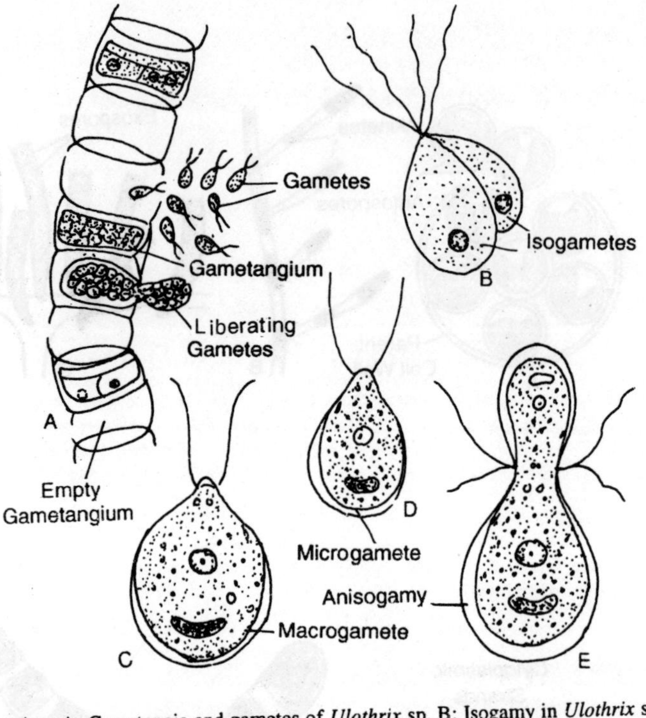

Fig. 3.11 Sexual reproduction, A: Gametangia and gametes of *Ulothrix* sp, B: Isogamy in *Ulothrix* sp, C : Female or macrogamete of *Chlamydomonas braunii*, D : Male or microgamete of *Chlamydomonas braunii*, E : Anisogamy in *Chlamydomonas braunii*.

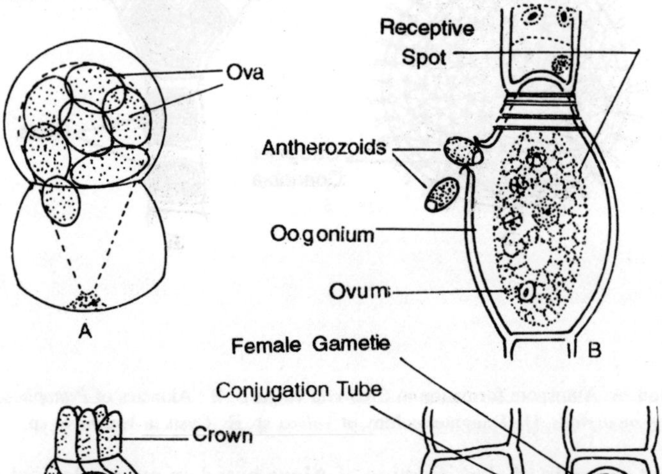

Fig. 3.12 Sexual reproduction, A : Eggs or ova of *Fucus* sp, B : Oogamy in *Oedogonium* sp, C: Nucule of *Chara* sp, D : Aplanogamy in *Spirogyra* sp.

CHAPTER 4

MYXOPHYCEAE (Cyanophyceae)

The members of the class Myxophycae can be identified by the following general characters.

1. Members are blue green in colour due to specific pigmentation.

2. Plastids are absent.

3. There is no motile phase in the life history.

4. Sexual recombination as understood in other higher members is absent. But genetic recombination has been reported in several members.

5. The cell interior is divided into an outer coloured chromoplasm and an inner colourless centroplasm representing the incipient nucleus. (There is no nuclear membrane).

The class consists of about 165 genera and more than 1500 species. The members are commonly called blue green algae, because the cell has bluish green pigment phycocyanin and the red pigment phycoerythrin in addition to chlorophyll, myxoxanthin, myxoxanthophyll and carotenoids.

Occurrence and Distribution

Blue green algae posses a wide range of tolerance to climatic fluctuations and are found inhabiting almost all types of ecological niche of the environment.

A good majority occur in a variety of fresh water habitats like hard water lakes, streams and wet soil, rocks, damp barks etc.

Some of the members are planktonic (free floating) and marine in their habitat. Some members are endophytic higher plants, where they help in nitrogen fixation. Some myxophycean members also enter into lichen association.

Thallus construction: Among the algae, the blue greens are the most primitive and possess a very simple plant body. The plant body is of three types, namely:

1. Unicellular forms.
2. Non-filamentous colonial forms.
3. Filamentous forms.

1. Unicellular forms: The plant body is made up of only one cell, elb., *Chroococcus.*

2. Colonial forms: The plant body is made up of aggregates of cells held together by a gelatinous matrix, e.g., *Merismopaedia.*

3. Filamentous forms: The plant body is multicellular. The filament is built up by the end to end joining of cells. The filament may be branched or unbranched.

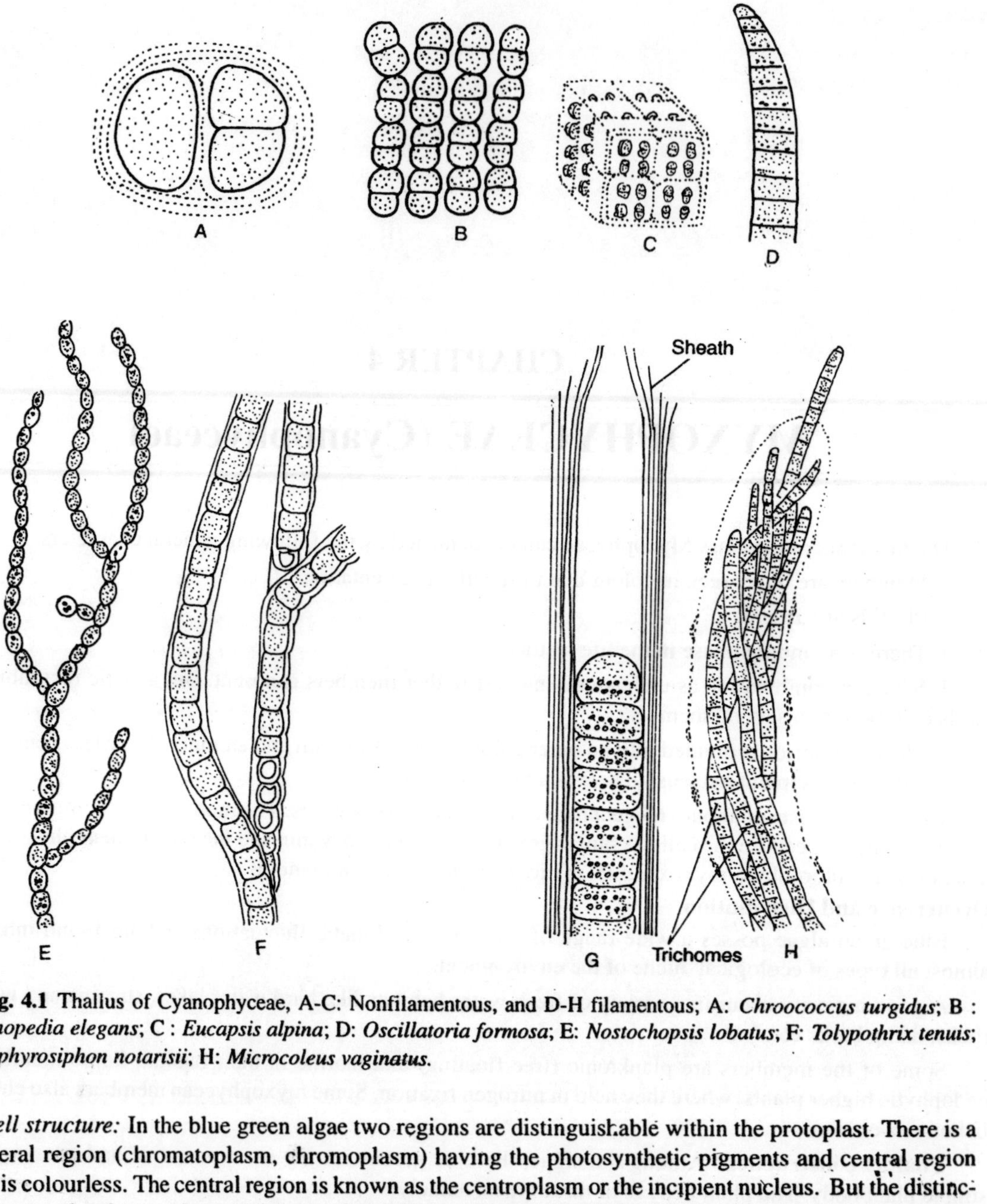

Fig. 4.1 Thallus of Cyanophyceae, A-C: Nonfilamentous, and D-H filamentous; A: *Chroococcus turgidus*; B : *Merismopedia elegans*; C : *Eucapsis alpina*; D: *Oscillatoria formosa*; E: *Nostochopsis lobatus*; F: *Tolypothrix tenuis*; G: *Porphyrosiphon notarisii*; H: *Microcoleus vaginatus.*

Cell structure: In the blue green algae two regions are distinguishable within the protoplast. There is a peripheral region (chromatoplasm, chromoplasm) having the photosynthetic pigments and central region which is colourless. The central region is known as the centroplasm or the incipient nucleus. But the distinction between the chromoplasm and centroplasm may not be always marked as in for eg. *Chroococcus turgidus*, where the pigments are scattered throughout the protoplast.

The cell is externally bounded by a cell wall made up of cellulose and pectic compounds. The wall is surrounded by gelatinous sheath which may be stratified.

The chromoplasm is either alveolar in nature or homogenous. It has various pigments like chlorophyll, carotene, myxcoxanthin, myxoxanthophyll, flavisin phycocyanin, xanthophyll and phycoerythrin.

The centroplasm is composed on DNA, RNA and proteins. There is neither any nucleolus nor a nuclear membrane.

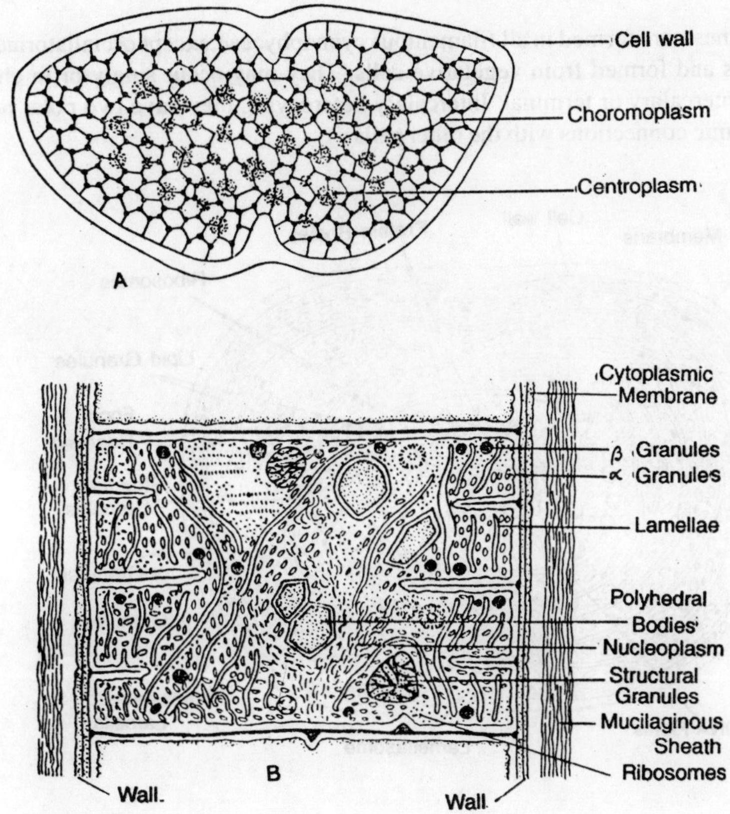

Fig. 4.2 Cell structure in Cyanophyceae; A: *Chroococcus turgidus*; B: Diagrammatic sketch of *Symphloca muscorum* based on electron micrograph.

Reproduction: It is mostly asexual only in a few members like *Anacystis nidulans* (Kumar H.D, 1962), *Cylindrospermum majus* (Singh R.N. and Singh R.N., 19965)etc, genetic recombination has been reported.

Asexual reproduction takes place by means of vegetative propagation and spore formation.

Vegetative Propagation:

1. *Cell division:* In unicellular members, the two daughter cells, after division mature into individual members.

2. *Hormogones:* These are formed in filamentous species. Hormogones represent portions of filaments between two dead cells. They get separated from the filament and develop into new forms.

3. *Hormocysts:* Hormogones formed at the tip of the filament, having thick walled, nutrient rich cells are called *Hormocysts.*

Spore Formation

1. *Akinetes, Resting spores or Arthospores:* These are non-motile spores, generally formed to tide over adverse environmental conditons. Usually these spores have a thick wall and a large amount of reserve food.

2. *Endospores:* These are spherical or angular asexual cells produced in the order Chaemosiphonales.

3. *Exospores:* These are formed externally by obstiction e.g., *Chaemosiphon.*

4. *Nannospores or Nannocysts:* The young daughter cells, immediately after cell division in *Gloeocapsa* are called Nannospores.

5. *Heterocysts:* These are formed in all filamentous cyanophyceae,except osciallatoriaceae. Heterocysts are thick walled cells and formed from vegetative cells. They may occur singly or in chains. In position heterocysts may be intercalary or terminal. Intercalary heterocysts, they have two polar nodules to help in maintaining cytoplasmic connections with the other cells.

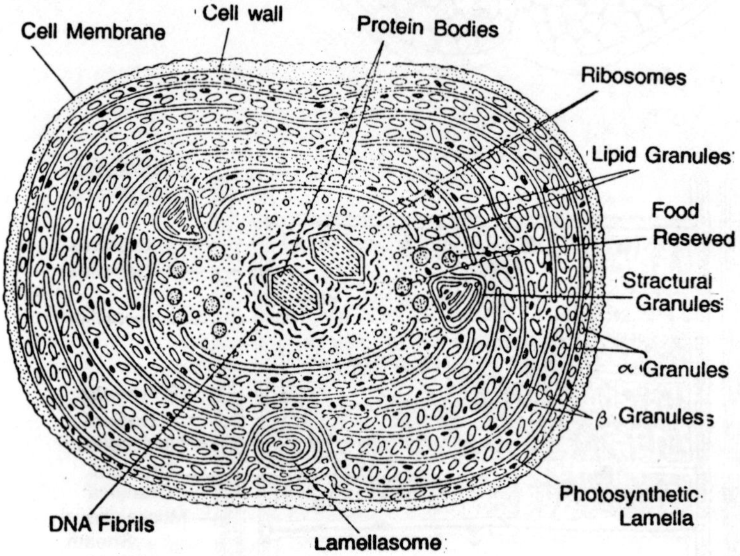

Fig. 4.3 General structure of a blue-green algal cell based on electron micrograph

Classification: Fritsch (1935) has classified the Class into five orders, namely, Chroococcales, Chaemosiphonales, Pleurocapsales, Nostacales and Stigonematales.

Smith divides the Class into three orders, namely, Chroococcales, Chaemosiphonales and Hormogoanles.

In this book the following members are discussed:

Gloeocapsa,Nostoc, Anabaena, Lyngbya, Oscillatoria, Rivularia, Gloeotrichia, Scytonema and *Stigonema.*

GLOEOCAPSA
(Kuetzing 1843)

Common species: *G. decorticans, G. calcarea, G. rupestris* etc.

Distribution : Wet rocks or soil, marine or fresh water.

Study: 1. External features.

2. Reproductive structures.

3. Systematic position and identification.

External features

1. Cells are arranged in colonies of 2-8. Sometimes unicellular also.
2. Colonies are surrounded by many envelopes of mucilage. These may or may not be coloured

3. Each mucilage sheath is lamellated or stratified.
4. Cells may be spherical or oval in shape.
5. Each cell in the colony has its own mucilage sheath.
6. Cell structure typical to a cyanophycean cell and has outer chromoplasm and inner centroplasm.
7. Cell wall is two layered -outer pectin and inner cellulose.

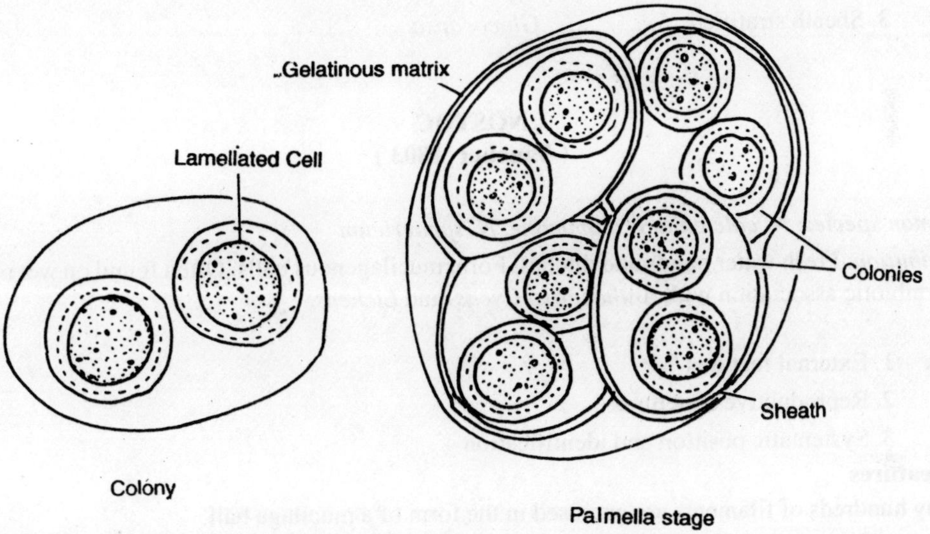

Fig. 4.4 Gloeocapsa

Reproduction

1. Sexual reproduction has not been reported.
2. Asexual reproduction takes place by the following methods:
   - (a) Cells divide and the daughter cells develop into new individuals.
   - (b) Nannospores or Nannocytes also help in asexual reproduction.
   - (c) In some cases spore formation also has been observed.

Systematic Position and Identification

A. 1. Chlorophyll present
   2. Plant body a thallus
   3. Cell wall is cellulosic .... *Algae*

B. 1. Blue green pigments present
2. Plastids absent
3. Protoplasm divided into chromoplasm and centroplasm
4. Sexual reproduction generally not seen ....*Myxophyceae*

C. 1. Plants unicellular or colonial
2. Nannocytes present ....*Chroococcales.*

D. 1. Plants unicellular or colonial
2. Filamentous plant body absent ....*Chrooceocaceae*

E. 1. Cell spherical or oval in shape
2. Each cell has its own mucilage sheath
3. Sheath stratified ....*Gloeocapsa.*

NOSTOC
(Vaucher 1803)

Common species: *N. calcicala, N. commune, N. sphaericum*

Distribution: Fresh water ponds and ditches. Form mucilagenous balls. Often found on wet rocks, also found in symbiotic association with *Anthoceros, Cycas* and *Lichens.*

Study: 1. External features
2. Reproductive structures
3. Systematic position and identification

External features

1. Many hundreds of filaments are enclosed in the form of a mucilage ball.
2. The ball like colony has a thick mucilage envelop and may be coloured.
3. The plant body is a filament and is enveloped by a mucilage sheath.
4. The filaments are unbranched and do not have a base or apex.
5. Each filament is made up of a number of bead like cells joined end to end.
6. Here and there in the filament certain thick walled cells are seen. These are called heterocysts and may be intercalary or terminal in position.
7. Heterocysts are thick walled and pale yellow coloured. Terminal heterocyst has a single polar module, while intercalary heterocysts have two polar nodules.
8. Vegetative cells have a typical cyanophycean structure. Protoplast having centroplasm and chromoplasm. Cell wall is cellulosic.

Reproductive features

1. Sexual reproduction has not been reported.
2. Asexual reproduction is by the formation of hormogones and akinetes.
3. Hormogones are fragments of the filament, that generally separate from the main filament near a heterocyst.
4. Hormogones develop into new individuals after separation from the main filament.
5. Akinetes are thick walled cells rich in reserve food materials. They separate from the plant and develop into a new individual.

6. Occasionally heterocysts (*N. commune*) may also germinate to form new individuals.

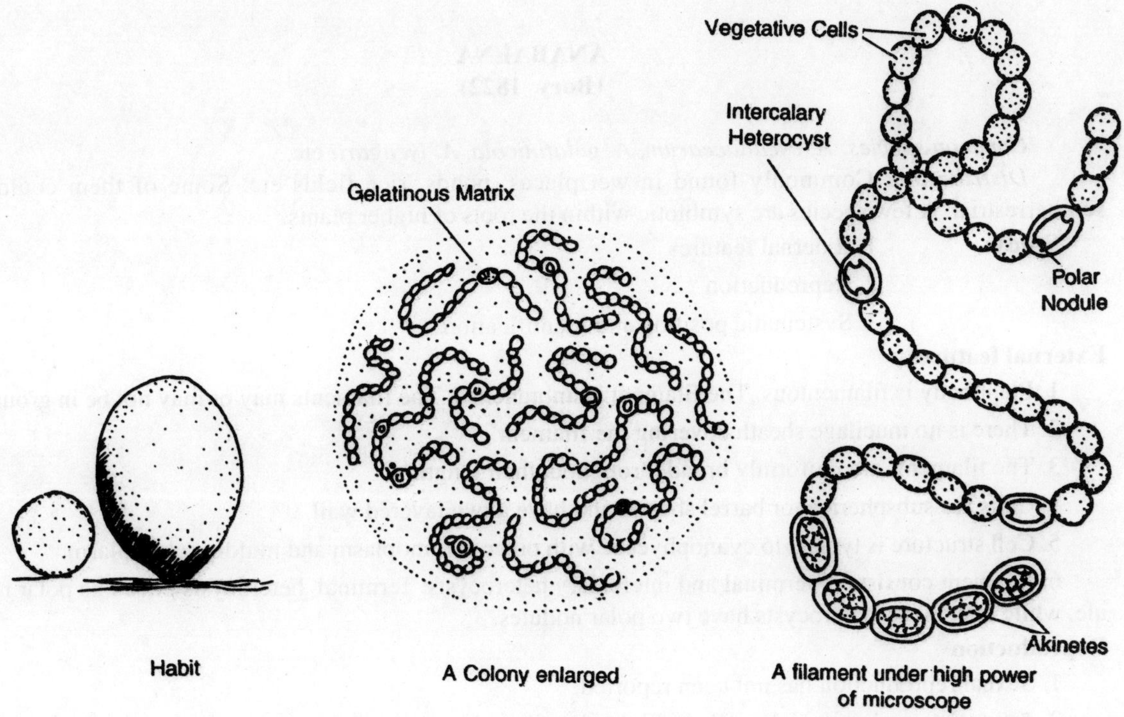

Fig. 4.5 Nostoc

Systematic position and identification

A. 1. Chlorophyll present
2. Plant body a thallus
3. Cell wall cellulosic... *Algae*

B. 1. Blue green pigments present
2. Plastids absent
3. Protoplast divided into chromoplasm and centroplasm.
4. Sexual reproduction generally not seen...*Myxophyceae*

C. 1. Plant body is filamentous
2. Filaments unbranched or show false branching
3. Heterocysts present...*Nostocales.*

D. 1. Filaments do not show differentiation into base and apex.
2. Filaments unbranched and uniseriate.
3. Heterocysts present ...*Nostocaceae*

E. 1. Heterocysts generally intercalary
2. Mucilage balls constitute a colony.
3. Reproduction by Hormogones...*Nostoc.*

ANABAENA
(Bory 1822)

Common species: A. cycadacearum, A. gelatinicola, A. iyengarii etc.

Distribution: Commonly found in wet places, ponds, rice fields etc. Some of them could be semiterrestrial. A few species are symbiotic within the roots of higher plants.

Study: 1. External features
2. Reproduction
3. Systematic position and Identification.

External features

1. Plant body is filamentous. The filament is unbranched. The filaments may or may not be in groups.
2. There is no mucilage sheath covering the filament.
3. The filaments are uniformly broad throughout their length.
4. Cells are subspherical or barrel shaped and have a two layered wall.
5. Cell structure is typical to cyanophyceae with outer chromoplasm and middle centroplasm.
6. Filament consists of terminal and intercalary heterocysts. Terminal heterocysts have one polar nodule, while intercalary heterocysts have two polar nodules.

Reproduction

1. Sexual reproduction has not been reported.
2. Asexual reproduction is usually brought about by Akinetes.
3. Akinetes are thick walled cells, much larger in size than the vegetative cells. They occur either singly or in a row of 4-6. They may or may not be adjacent to a heterocyst. Akinetes germinate into a new filament.

Systematic Position and Identification

A. 1. Chlorophyll present
2. Plant body is a thallus
3. Cell wall cellulosic...*Algae*

B. 1. Blue green pigments present
2. Plastids absent
3. Protoplast divided into chromoplasm and centroplasm
4. Sexual reproduction generally not seen....*Myxophyceae.*

C. 1. Plant body filamentous
2. Filaments unbranched or show false branching
3. Heterocysts present...*Nostoçales*

D. 1. Filaments do not show differentiation into base and apex
2. Filaments unbranched and uniseriate
3. Heterocyst present...*Nostocaceae*

E. 1. Mucilage sheath absent

2. Akinetes many times larger than the vegetative cells and are usually associated with heterocysts…*Anabaena*.

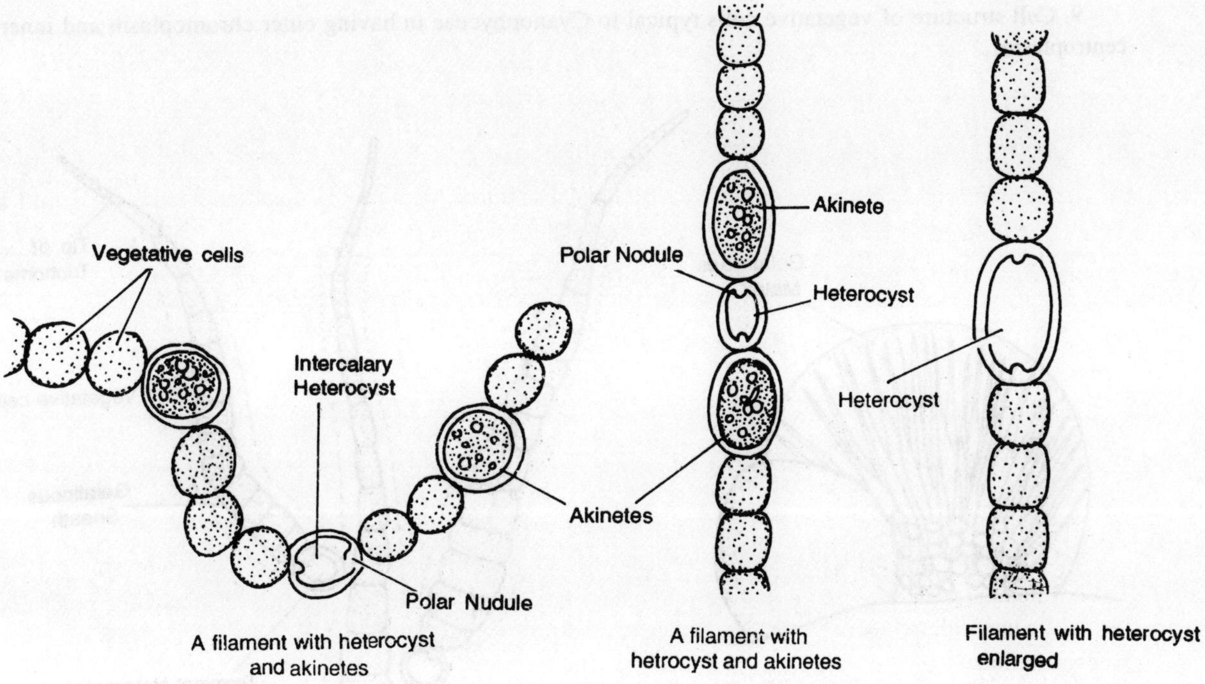

Fig. 4.6 Anabena

RIVULARIA
(Agardh 1842)

Common species: R.atra

Distribution: Found in the form of shiny mucilage balls floating in water or sticking to aquatic plants such as *Elodea, Ottelia* etc. Occasionally found in salt marshes also.

Study 1. External features.
2. Reproduction
3. Systematic position and Identification

External features

1. Colonies are irregular masses, yellowish brown in colour with a greenish tinge.

2. Each colony has hundreds of filaments which radiate from the centre. The filaments are held together by the mucilage.

3. The filaments have a broad base which gradually tapers towards the apex.

4. The filaments are usually completely enveloped by the mucilage.

5. Each filament consists of a basal heterocyst. Occasionally, the heterocyst may be intercalary also.

6. Frequently, the filaments show false branching occurring beneath an intercalary heterocyst.

7. The branching is not evident as the branches separate immediately.

8. Radiating filaments of the colony result due to false branching.

9. Cell structure of vegetative cells typical to Cyanophyceae in having outer chromoplasm and inner centroplasm.

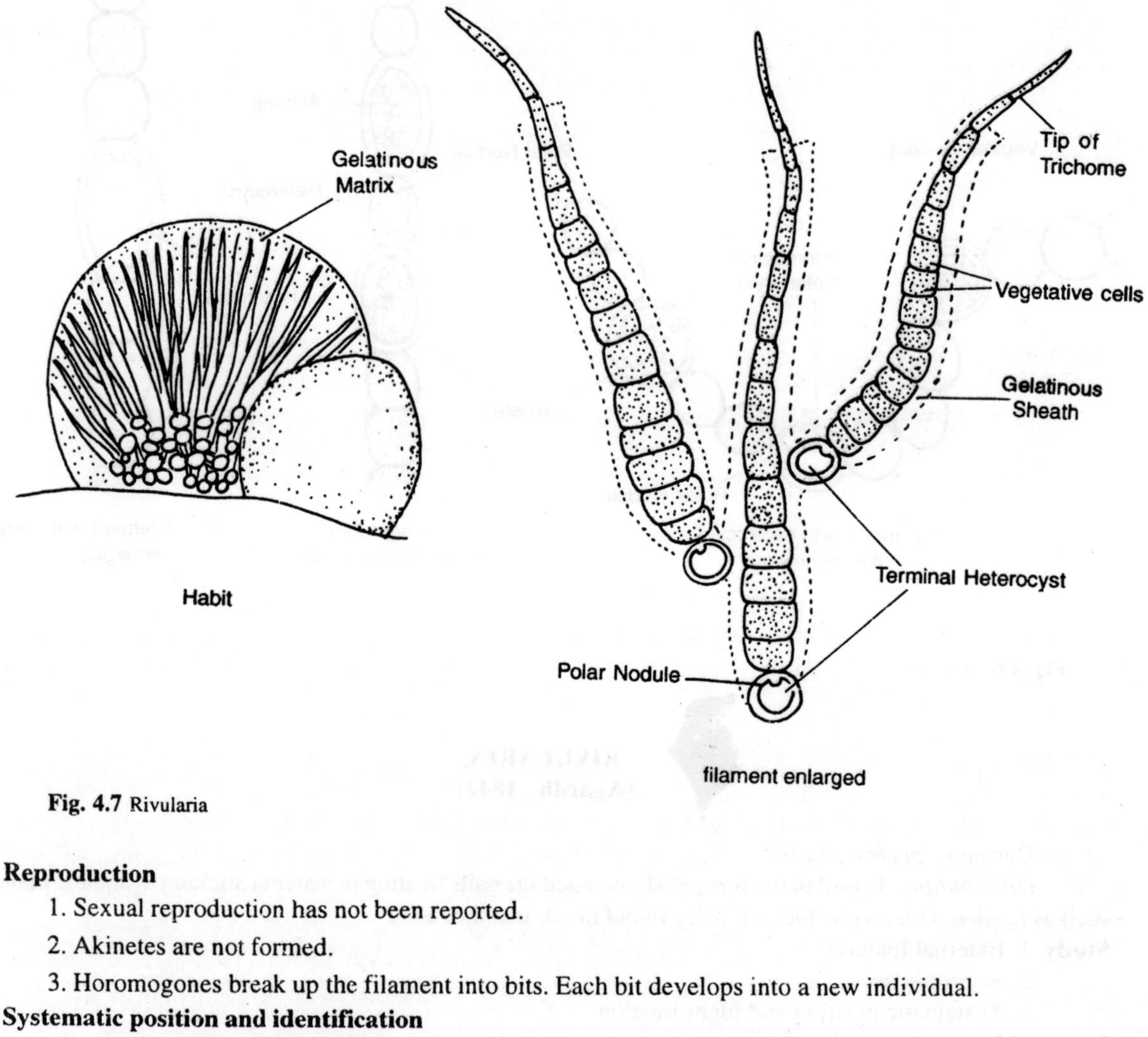

Fig. 4.7 Rivularia

Reproduction

1. Sexual reproduction has not been reported.

2. Akinetes are not formed.

3. Horomogones break up the filament into bits. Each bit develops into a new individual.

Systematic position and identification

A. 1. Chlorophyll present
   2. Plant body a thallus
   3. Cell wall cellulosic...*Algae*

B. 1. Blue green pigments present
   2. Plastids absent

3. Protoplast divided into chromoplasm and centroplasm
4. Sexual reproduction not seen...*Myxophyceae*

C. 1. Plant body filamentous
2. Filaments unbranched or show false branching
3. Heterocysts present...*Nostocales*

D 1. Plant body filamentous and unbranched
2. Heterocyst basal
3. Filaments broad at the base tapering towards the apex...*Rivulariaceae.*

C. 1. Akinetes absent
2. Filament mpletely enclosed by the mucilage sheath
3. Heterocyst usually basal
4. Filaments radiate from the centre in a colony...*Rivularia*

GLOEOTRICHIA
(Agardh 1842)

Common species: *G.natans, G. pisum* etc.
Distribution : Freshwater ponds, ditches immediately after rains.
Found in the form of mucilagenous masses.

Study : 1. External features
2. Reproduction
3. Systematic position and Identification

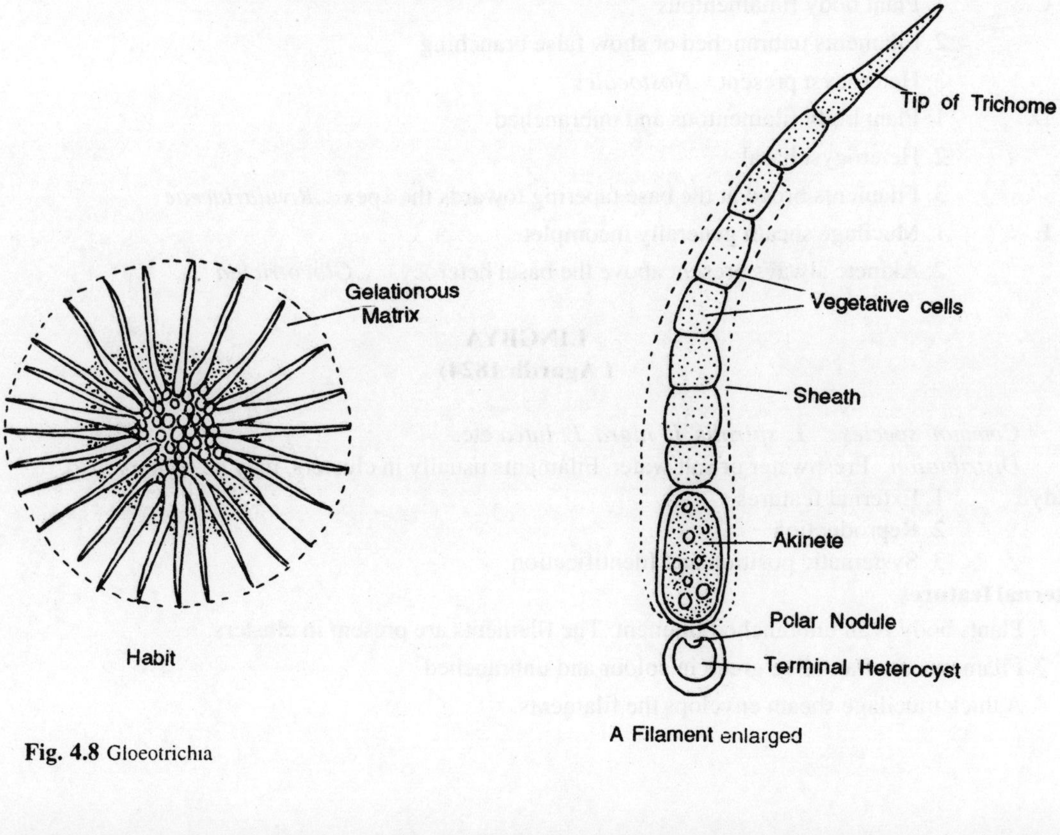

Fig. 4.8 Gloeotrichia

External features

1. Found in the form of irregular bluish green mucilagenous masses. Colonies are completely ensheathed by mucilage.

2. Many hundreds of filaments are found in each colony arranged radiately.

3. Fialments show false branching and have a mucilage sheath which incompletely envelops them.

4. Heterocyst is basal.

5. Above the heterocyst is an Akinete. Akinete is very large; 4-5 times bigger than the vegetative cells (Akinete not seen in Rivularia).

6. Cell structure typical to cyanophyceae in having an outer chromoplasm and a central centroplasm.

Reproduction

1. Sexual reproduction has not been reported.

2. Asexual reproduction brought about by hormogones and akinetes.

3. Hormogones are bits of filaments which on separation develop into a new individual.

4. Akinetes separate from the plant and develop into a new individual.

Systematic position and identification

A. 1. Chlorophyll present
2. Plant body a thallus
3. Cell wall cellulosic...*Algae*

B. 1. Blue green pigments present
2. Plastids absent
3. Protoplast divided into chromoplasm and centroplasm
4. Sexual reproduction generally not seen ...*Myxophyceae*

C. 1. Plant body fimamentous
2. Filaments unbranched or show false branching
3. Heterocyst present...*Nostocales*

D. 1. Plant body filamentous and unbranched
2. Heterocyst basal
3. Filaments broad at the base tapering towards the apex...*Rivulariaceae*

E. 1. Mucilage sheath generally incomplete
2. Akinete always present above the basal heterocys ....*Gloeotrichia*

LINGBYA
(Agardh 1824)

Common species : *L. spiralis, L. nigra, L. lutea* etc.

Distribution: Freshwater or salt water. Filaments usually in clusters, may be solitary also.

Study : 1. External features
2. Reproduction
3. Systematic position and Identification

External features

1. Plants body is an unbranched filament. The filaments are present in clusters.

2. Filaments are flat, olive green in colour and unbranched.

3. A thick mucilage sheath envelops the filaments.

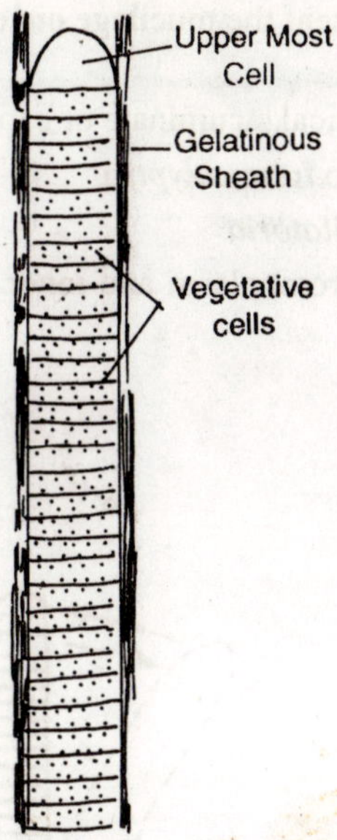

Fig. 4.9 Lyngbya. External features

4. The filament is made of cells joined end to end.

5. The cells are broader than long and the terminal cell is dome shaped.

6. Cell structure typically cyanophycean with peripheral chromoplasm and central centroplasm.

7. Cyanophycean granules and starch present.

Reproduction

1. Sexual reproduction has not been reported.

2. Asexual reproduction by means of hormogones and separation discs. They (hormogones) develop into new individuals.

Systematic position and Identification

A. 1. Chlorophyll present
2. Plant body a thallus
3. Cell wall cellulosic...*Algae*

B. 1. Blue green pigments present
2. Plastids absent
3. Protoplast divided into chromoplasm and centroplasm
4. Sexual reproduvtion generally not seen...*Myxophyceae*

C. 1. Plant body fimamentous
2. Filaments unbranched or show false branching ...*Nostocales*

D. 1. Filaments uniseriate
2. Cells broader than long
3. Asexual reproduction by hormogones...*Oscillatoriaceae*

E. 1. Filaments enclosed by a mucilage sheath
2. Calyptra at the tip absent...*Lyngbya*

OSCILLATORIA
(Vaucher 1803)

Common species: *O.martini, O. nigra, O. princepis* etc.

Distribution : A very common blue green alga, found in ponds, stagnant gutter water, ditches etc. Some times the alga is found on damp soil forming dark blue green patches.

Study : 1. External features
2. Reproduction
3. Systematic position and Identification

External features

1. Plant body is an unbranched filament. The filaments may be solitary or many of them interweave into knots or patches.

2. Filaments are unbranched with cells joined end to end.

3. Cells are broader than long and have a cellulosic cell wall.

4. Filaments may or may not have a mucilage envelope. When present the mucilage envelope is very thin.

5. The terminal cell of the filament is tapering. It may be hemispherical, acuminate or capitate.

6. In some species the apical wall of the terminal cell is thickened to form *calyptra.*

7. The filaments show oscillating movements hence the name *Oscillatoria*

8. Cell structure typical to cyanophyceae; has outer pigmented chromoplasm and inner centroplasm. Cyanophycean granules and starch present.

9. Here and there in the dead cells are present biconcave discs

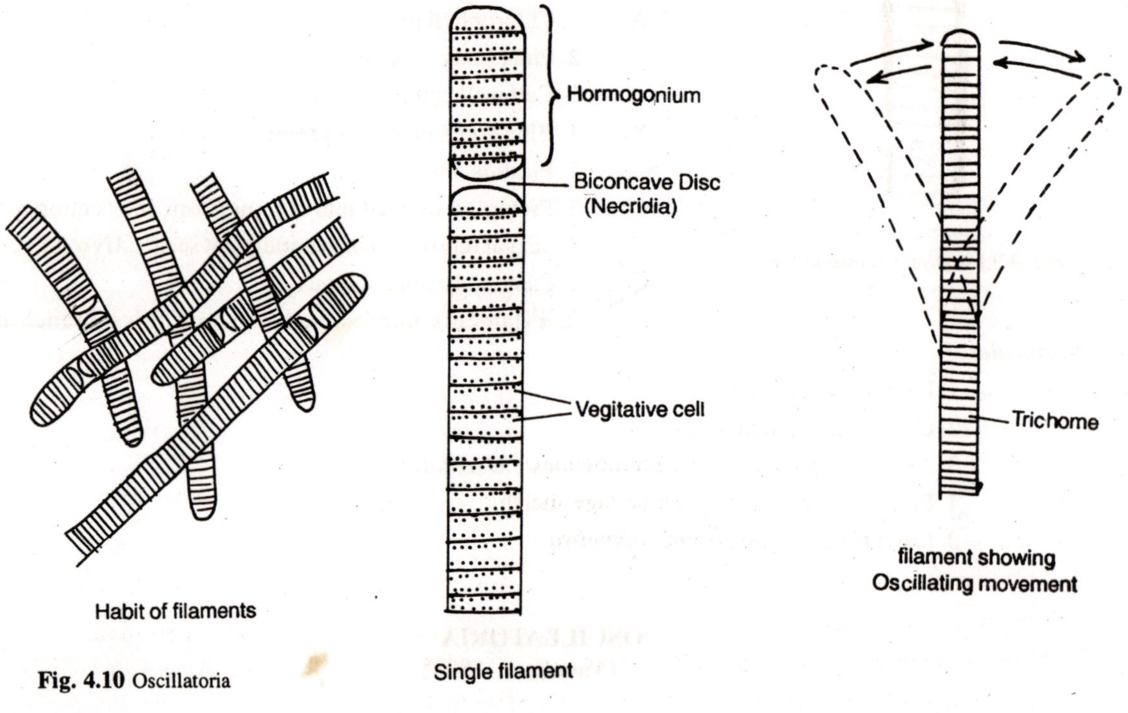

Fig. 4.10 Oscillatoria

Reproduction

1. Sexual reproduction has not been reported.

2. Asexual reproduction by hormogones.

3. Hormogones are bits of filaments (5-6 cells long) and present between the two biconcave discs. The hormogones separate at these discs and grow into a new filament.

Diagrams to be drawn

1. Habit sketch

2. A few filaments enlarged.

3. Hormogone and biconcave discs.

4. Detailed cell structure

Systematic position and Identification

A. 1. Chlorophyll present

2. Plant body a thallus

3. Cell wall cellulosie...*Algae*

B. 1. Blue green pigments present

2. Plastids absent

3. Protoplast divided into chromoplasm and centroplasm

4. Sexual reproduction generally not seen...*Myxophyceae*

C. 1. Plant body filamentous

2. Filaments unbranched or show false branching ...*Nostocales*

D. 1. Filaments uniseriate with a terminal tapering cell

2. Heterocysts absent

3. Mucilage sheath thin...*Oscillatoriaceae*

E. 1. Terminal cell of the filament forms a calyptra

2. Filaments show oscillating tips

3. Biconcave discs present in the filament

4. Mucilage sheath either absent or very thin when present...*Oscillatoria*

SCYTONEMA
(Agardh 1824)

Common species: *S. rivulare, S. simplex,, S. iyengarii, S.varium* etc.

Distribution : Aquatic or semiterrestial. Found on the bottom of ponds, pools etc. or on damp soil. Occcurs also in symbiotic association in lichens.

Study : 1. External features (including branching)

2. Reproduction.

3. Systematic position and identification

External features

1. The filamentous plant body is surrounded by a mucilage sheath. The sheath is lamellated.

2. Hetrocysts present. They are terminal as well as intercalary.

3. Filament shows false branching. One or two branches arise at a point.

4. Branching results due to the death of an intercalary cell, on which the two cells on either side resume apical growth and develop into branches.

5. Heterocysts are pale yellow in colour and have one or two polar nodules depending on their location.

6. Cell structure is typically cyanophycean. It has a central centroplasm, surrounded by chromoplasm.

Reproduction

1. Sexual reproduction has not been reported.

2. Asexual reproduction brought about by means of hormogones, which are formed between two separation discs.

3. Akinetes have been reported in some species.

Diagrams to be drawn

1. Habit sketch

2. One or two filaments enlarged

3. A portion of the filament at the region of branching
4. Cell structure

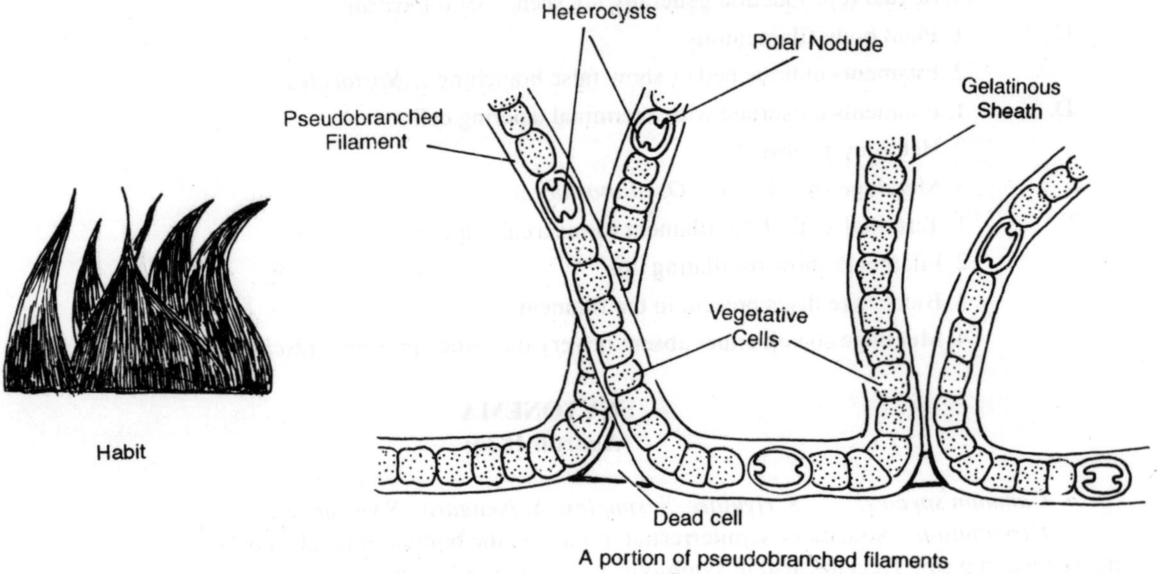

Fig. 4.11 Scytonema

Systematic position and Identification

A. 1. Chlorophyll present
2. Plant body a thallus
3. Cell wall cellulosic...*Algae*

B. 1. Blue green pigments present
2. Plastids absent
3. Protoplast divided into chromoplasm and centroplasm
4. Sexual reproduction generally not seen ... *Myxophyceae*

C. 1. Plant body filmentous
2. Filaments unbranched or show false branching...*Nostocales*

D. 1. Filaments exhibit false branching
2. Heterocysts mostly intercalary
3. Reproduction by hormogones...*Scytonemataceae*

E 1. False branching present
2. Mucilage sheath enveloping the filament...*Scytonema*

STIGONEMA

Common species: *S. arugineum, S. panniforme, S. ocellatum, S. dendroideum* etc.

Distribution: Mostly aquatic, could be semiterrestrial also. Sometimes found attached to aquatic weeds.

Study : 1. External features
2. Reproduction
3. Systematic position and identification

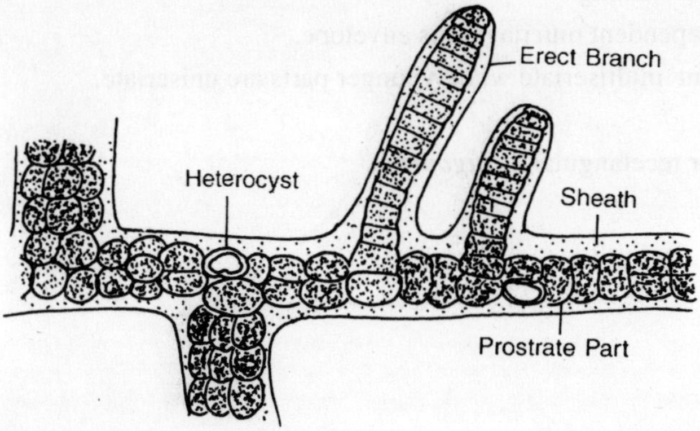

Fig. 4.12 Stigonema,
Thallus of Stigonema showing true branching

External features

1. The plant body is a filament. Many filaments accumulate to form brown mucilaginous lumps.
2. Each filament has an independent mucilaginous envelop.
3. Filaments are multiseriate at older regions but uniseriate in younger regions
4. Filaments are branched. Branching is true.
5. Cells are sub-spherical or rectangular., Sometimes broader than long.
6. Cell structure is typically cyanophycean i.e,. it has a periphereal chromoplasm surrounding a central centroplasm.
7. Growth is apical
8. Heterocysts present in some species.

Reproduction

1. Sexual reproduction has not been reproted.
2. Asexual reproduction is brought about by fragmentation, hormogones or heterocysts.
3. When the filament breaks up into bits accidentally, each bit develops into a new individual.
4. Few celled hormogones at maturity develop into new individuals.

Diagram to be drawn

1. Habit sketch.
2. A filament enlarged to show branching.
3. A portion of the filament enlarged at the region of branching.
4. Cell structure.

Systematic Position and Identification

A. 1. Chlorophyll present
2. Plant body a thallus
3. Cell wall cellulosic...*Algae*

B. 1. Blue green pigments present
2. Plastids absent.

3. Protoplast divided into chromoplasm and centroplasm.
4. Sexual reproduction generally not seen ...*Myxophyceae*

C. 1. Plant body is a filament
2. Filaments show true branching...*Stigonematales.*

D. 1. Plant body is a filament with true branches.
2. Filaments found in mucilaginous sheath.
3. Branching of the filament is lateral...*Stigonemataceae*

E. 1. Filament exhibiting true branching.
2. Each filament has an independent mucilaginous envelope.
3. Older parts of the filament multiseriate while younger parts are uniseriate.
4. Growth is apical
5. Cells are sub-spherical or recetangular...*Stigonema.*

CHAPTER 5

CHLOROPHYCEAE

(Chlorophyta)

The class Chlorophyceae includes about 360 genera and nearly 6500 species. The members are commonly called grass green algae because of their deep green colouration. All the members have their pigments arranged in plastids (chloroplasts). They possess a definite nucleus. The members may be motile or non motile. The reproductive cells are generally flagellate. Most of the members of chlorophyceae, possess the same phtotosynthetic pigments seen in higher plants.

Occurrence and Distribution: Approximately 90% of chlorophyceae are fresh water in distribution (Smith 1966). Some orders like Ulvales and Siphonales are predominantly marine, Ulotrichales predominantly fresh water, while Oedogoiales and Conjugales exclusively fresh water.

Some Chlorophycean members are terrestrial. Sime like *Trentepohlia* are epiphytic growing on wet barks of trees.

Range of Thallus Construction

The members of Chlorophyceae exhibit great diversity in form and structure. In the vegetative construction, they range from unicellular motile forms up to parenchymatous viz., motile forms, palmelloid forms, coccoid forms, filamentous forms, sipihonaceous habit, uniaxial type and parenchymatous type.

Cell structure: A few primitve chlorophycean members have naked protoplasts. But in a great majority the protoplast lies within a prominent cell wall. The cell-wall is usually 2-3 layered: the outer chitinous, middle pectic and inner cellulosic. Mucilage envelope is present in some members. Lime and silica may also be deposited over the cell wall.

The protoplast in unicellular forms consists of two small contractile vacuoles. In fialmentous forms, the cell usually has a large central vacuole traversed by a number of cytoplasmic strands. Such a cytoplasm is called Primordial utricle.

In motile forms locomotion is brought by flagella.

Chloroplasts: The members of Chlorophyceae exhibit a great range of variation in the structure and organisation of chloroplasts. Chloroplasts may be cup-sharped, discoid, stellate, spiral or reticulate, Pyrenoids (one to many) are present in chloroplasts. The pigments present in chloroplasts are chlorophyll a, chlorophyll b, carotene, and xanthophyll. Often a red coloured pigment called Haematochrome is also present.

The product of assimilation is starch and it is stored in pyrenoids.

Reproduction: Members of Chlorophyceae reproduce asexually as well as sexually.

Asexual Reproduction: It is brought about by the following methods:

Cell division: In unicellular forms cell-divisions is a means of reproduction. The two daughter cells after a cell division develop into new individuals.

Zoospore formation: The characteristic method of axexual reproduction in majority of the forms is the formation of either bi or guardriflagellate zoospores. They are usually produced in zoosporangium which may form one to many zoospores. Usually the protoplast divides to form a number of uninuleate bits which become motile. They are set free in water. Zoospores are generally formed when there is abundance of moisture. Usually zoospores are formed during night and liberated in the morning. After a sghort period of swarming , the zoospores develop into new individuals.

Fragmentation: This is the usual method of multiplication among filamentous forms. It is a filament, accidentally breaks, each bit will develop into anew individual.

Aplanospores and Akinetes: In some members, non-motile spores called aplanospores are produced. In some members resting spores called Aikinetes are formed.

Sexual Reproduction: It is fo three types Isogamy, Anisogamy and Oogamy (for details see General account of algae).

Type of Life Cycles

Three types of Life cycles are seen in Chlorophyceae. These are Haplontic type, Isomorphic alternation of generation and Hetermorphic alternation of generations.

The haplontic type, there is no differentiation into gametophyte and soprophyte. The same plant body produces gametes and spores at different times.

In a life cycle with isomorphic alternation of generations, sporophyte and gametophyte are identical morphologically, while in hetermorphic alternation of generations, they are dissimilar morphologically.

Classification: Various types of classification have been proposed for chlorophyceae. Of these the classification by Smith (1966) and Fritsch (1935) have been given below.

Smith's Classification: According to Smith, the division chlorophyta is divided into two classes namely-Chlorophyceae and Charophyceae. Chlorophyceae comprises of ten orders whereas Charophyceae hasonly one order.

Divisions **Chlorophyta**

Class **Chlorophyceae**

Orders:
1. Volvocales
2. Ietrasporales
3. Ulotrichales
4. Oedogoiales
5. Ulvales
6. Schizogoiales
7.Chlorococcales
8. Siphonales
9. Siphonocladales
10. Zygnematales

Class **Charophyceae**

Order Charales

Frithch's classification: The classification is based on the structure of vegetative and reproductive bodies. Fritsch divides the class into 9 orders viz.,

1. Volvocales
2. Chlorococcales
3. Ulotrichales
4. Cladophorales
5. Chaetophorales
6. Oedogoniales
7. Conjugales
8. Siphonales
9. Charales

In this book, the classification of Fritsch (1935) is followed.

CHLAMYDOMONAS
(Ehrenberg 1833)

Common species: *C. ehrenbergii, C.nivalis, C.reticulate, C.pluristigma* etc.

Distribution Found in stagnant water, pools, ponds and ditches.

Some species are found in saline water also.

Study 1. Cell structure
2. Palmella stage
3. Systematic position and Identification

Cell structure

1. The plant body in unicellular. Cells may be spherical or oval in shape.

2. Outermost layer is the cell wall, occasionally there might be mulcilage covering.

3. The cell can be distinguished into a broad posterior end and a narrow anterior end.

4. At the anterior end are found two long flagella, which help in locomotion. Each flagellum arises from a basal granule or blepharoplast.

5.The two basal granules are joined by a fibril known as Paradesmos. One of the basal ganule is connected to the nucleus thro' another thread called Rhizoplast. The whole structure is called Neuromotor Apparatus.

6. Near the basal granules are two contracrtile vacuoles, whose function is osmoregulation.

7. Towards the anterior end is also found the lens shaped stigma or Eyespot, which is photosensitive.

8. Internal to the cell-wall is the living plasma membrane.

9. Towards the broader end of the cell and occupying 2/3rd of the cell interior is a single cup shaped chloroplast, with a centrally embedded pyrenoid.

10. Filling the cup of the chloroplast is the cytoplasm.

11.The single nucleus lies suspended in cytoplasm.

12. Nutrition is autotrophic.

Reproduction: *Chlamydomonas* reproduces both asexually and sexually. However these stages cannot be easily observed in the laboratory. Only one method of Asexual reproduction viz., Palmella stage can be observed. The stage is given the name Palmella stage, because of its resemblance to the alga *Palmella*.

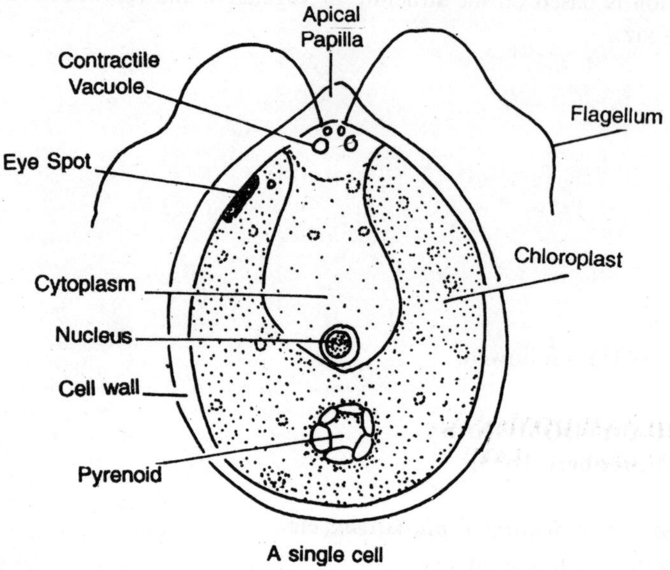

Fig. 5.1 Chlamydomonas

Palmella stage:

1. Palmella stage is usually seen when the dry season will be approaching fast.

2. The cells divide, but the protoplasts do not develop flagella and remain inactive within the parent cell.

3. Many groups of cells get surrounded by a common mucilage envelope.

4. Palemella stage consists of several generations of cells.

5. Plamella stage is a transition stage.

6. At the return of the favourable season, the mucilage envelope dissolves, and all the cells are set free. They develop flagella very soon.

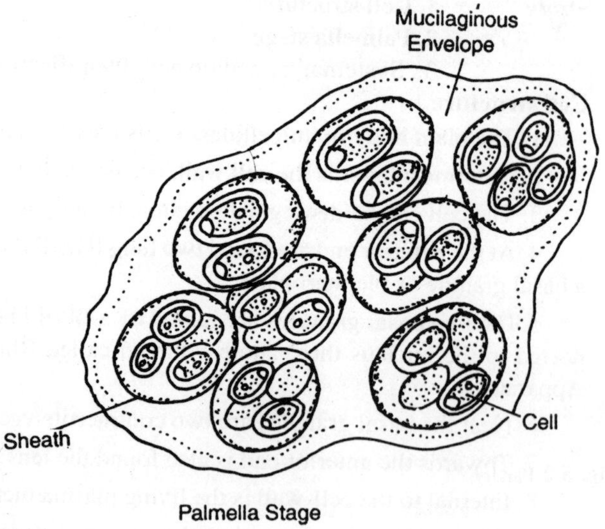

Diagrams to be drawn

1. Single cell
2. Anterior part of the cell enlarged.
3. Palmella stage

Systematic position and Identification

A. (1) Chlorophyll present
(2) Plant body is a thallus
(3) Cell-wall is celuosic ... *Algae*

B. (1) Plastids present
(2) Chlorophyll is the chief pigment
(3) True nucleus present
(4) Photosynthetic product is starch
(5) Reproductive cells motile... *Chlorophyceae*

C. (1) Plant body motile
(2) Chloroplast cup shaped... *Volvocales*

D. (1) Plant body unicellular
(2) Contractile vacuoles presents... *Chlamydomonadaceae*

E. (1) Cell pyriform in shape
(2) Eye spot present.
(3) Chloroplast cup-shaped with a centrally embedded stigma.
(4) Presence of Palemella stage during unfavourable conditions
(5) Cells uninucleate...*Chlamydomonas.*

PANDORINA
(Bory 1824)

Common species: *P. morum*
Distribution: Fresh water pools, ponds and ditches.
Study (1) External features of the colony
(2) Asexual Reproduction
(3) Sexual Reproduction
(4) Systematic position and Identification.

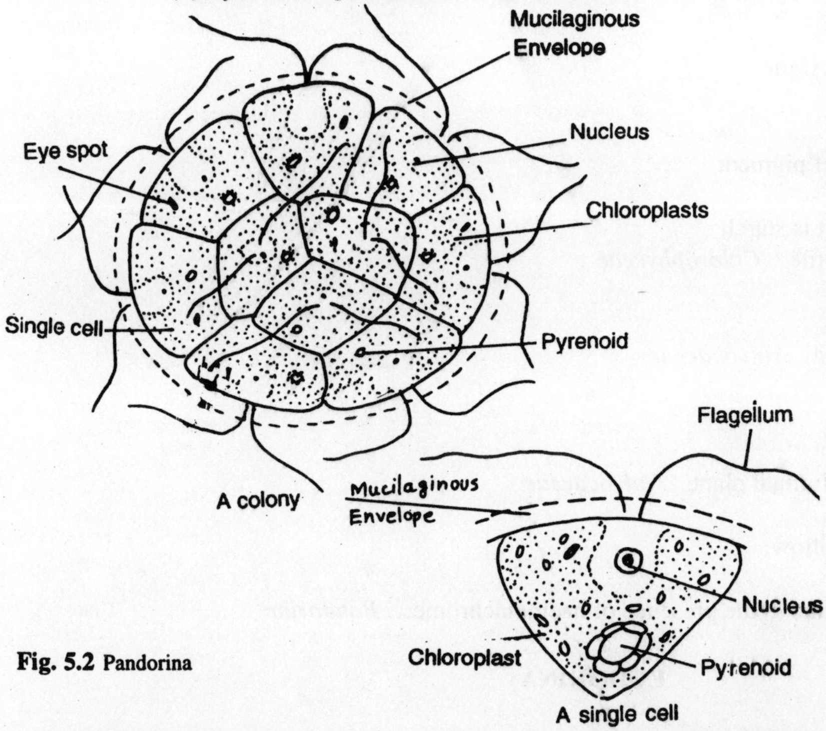

Fig. 5.2 Pandorina

External features

1. The plant body is a multicellular colony called the coenobium. It is a hollow sphere.

2. The coenobium is motile, because the cells of the colony are flagellate.

3. The number of the cells per colony range from 16-32. They are embedded in a mucilaginous matrix.

4. The cells are biflagellate and chlamydomanad (like *Chlamydomonas*) in structure.

5. The cells are arranged in tiers of four each.

6. Cells are pyrifrom having a broad posterior and narrow anterior end.

7. Cells are uninucleate with a cup shaped chloroplast.

8. There are two contractile vacuoles and an eye spot at the anterior end.

Asexual Reproduction

1. Each cell of the plant colony can divide and form daughter colonies.

2. Formation of daughter colonies takes place after the coenobia stop locomotion and settle down in a place.

3. After the formation of daughter colonies the mucilage envelop dissolves liberating them.

Sexual reproduction

1. Coenobia are dioecious

2. Reproduction is anisogamous.

3. All the cells of the colony take part in the formation of gametes.

4. Each cell of the colony divides to form a number of gametes.

5. All the cells of the colony either form macrogametes or microgamtes.

6, Fusion always takes place between a microgamete and macrogamete, resulting in the formation of a zygote. The zygote is red in colour due to haemotachrome.

7. The zygote develops into a new coenobium.

Diagrams to be drawn

(1) A colony enlarged

(2) A cell enlarged

(3) Colony showing daughter colonies

(4) Colony showing zygotes

Systematic position and Identification

A. 1. Chlorophyll present
2. Plant body is a thallus
3. Cell wall cellulosic... *Algae*

B. 1. Plastids present
2. Chlorophyll is the chief pigment
3. True nucleus present
4. Photosynthetic product is starch
5. Reproductive cells motile... *Chlorophyceae*

C. 1. Plant body motile
2. Chloroplast cup shaped... *Volvocaceae*

D. 1. Plant body is colonial.
2. Colonies motile
3. Cells divide in a longitudinal plane... *Volvocaceae*

E. 1. Colony spherical or hollow
2. Cells arranged in tiers
3. Zygote red in colour due to the presence of haemotachrome... *Pandorina*

EUDORINA

Common species: *E.indica,E.elegans, E.illinoiensis* etc.

Distribution: Commonly found in fresh water, ponds and ditches.
Abundant as Planktonic flora in lakes.

Study: (1) Structure of the colony
(2) Structure of the cell
(3) Asexual reproduction
(4) Sexual reproduction
(5) Systematic position and identification

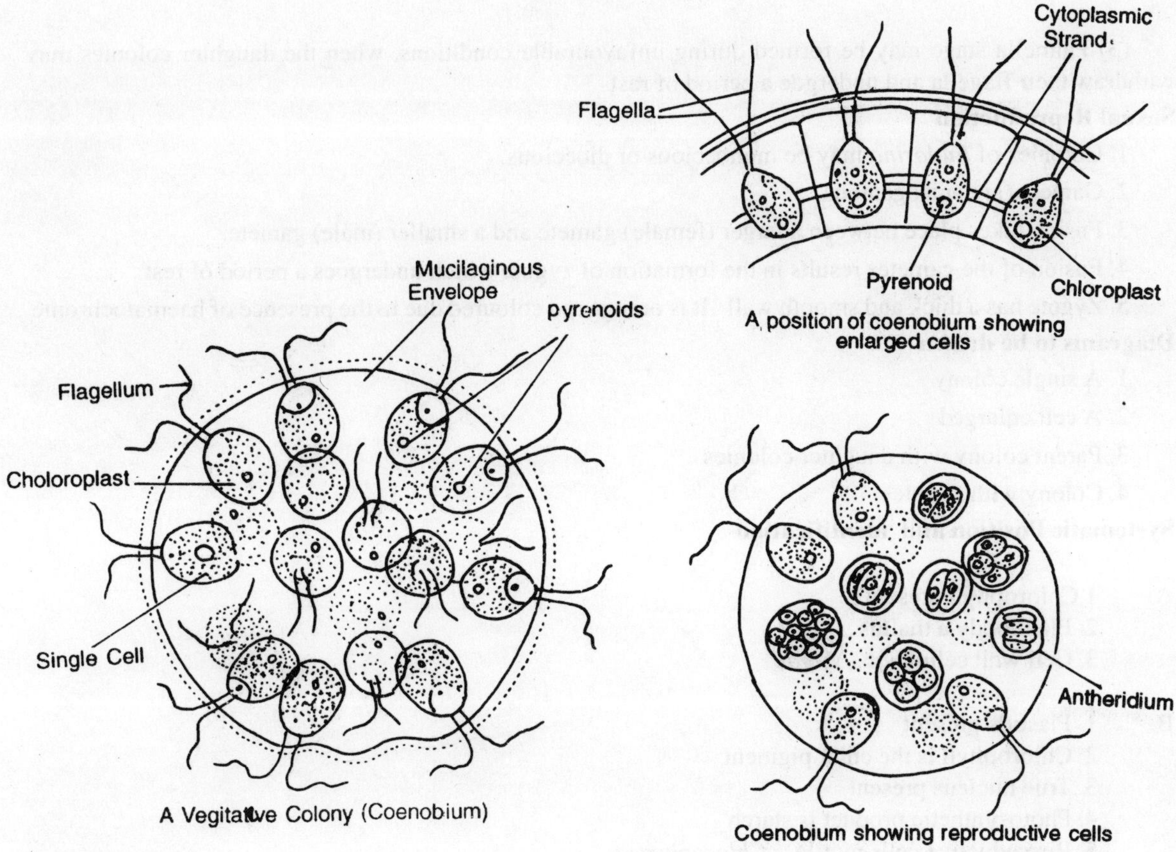

Fig. 5.3 Eudorina

Structure of colony

1. The plant body is a motile colony called Coenobium.
2. The colonies are subspherical or oval in shape.
3. The Coenobium is enveloped by a mucilage layer. It may be smooth or warty.
4. Cells are arranged in a single plane with the centre of the colony being hollow.
5. The number of the cells in the colony ranges from 16-64.
6. Cells are connected to one another by means of thin cytoplasmic connections.

Cell structure.

1. The cell structure is Chalmydomonad.
2. Cells are uninucleate and biflagellate
3. Each cell has a single cup shaped chloroplast.
4. At the anterior end there are two contractile vacuoles
5. A single eye spot is present near the anterior region.

Reproduction

Asexual

(1) Daughter colonies are formed in each cell of the colony or only a few cells may form daughter colonies.

(2) At maturity the mucilage envelops of the parent colony ruptures releasing all the daughter colonies.

(3) Palmella stage may be formed during unfavourable conditions, when the daughter colonies may withdraw their flagella and undergoe a period of rest.

Sexual Reproduction

1. Colonies of *Eudorina* may be monoecious or dioecious.
2. Gametes are anisogamous.
3. Fusion takes place between a larger (female) gamete and a smaller (male) gamete.
4. Fusion of the gametes results in the formation of zygote which undergoes a period of rest.
5. Zygote has a thick and smooth wall. It is orange red coloured due to the presence of haematochrome.

Diagrams to be drawn

1. A single colony
2. A cell enlarged
3. Parent colony with daughter colonies
4. Colony with zygotes

Systematic Position and identification

A. 1.Chlorophyll present
2. Plant body a thallus
3. Cell wall cellulosic...*Algae*

B. 1. Plastids present
2. Chlorophyll is the chief pigment
3. True nucleus present
4. Photosynthetic product is starch
5. Reproductive cells motile...*Chlorophyceae*

C. 1. Plant body motile
2. Chlorophyll cup shaped...*Volvocales*

D. 1. Plant body is colonial
2. Colonies are motile
3. Cells divide in a longitudinal plane...*Volvocaceae*

E 1. Colony subspherical or ovoid
2. Number of cells in a colony less than 256
3. All cells in the colony are of same size...*Eudorina.*

PLEDORINA

Common species: *P.california, P.indica, P.illinoiensis*

Distribution: Colonies are distributed extensively in fresh water and stagnant ponds.

Study:

1. Structure of the colony
2. Cell structure
3. Asexual reproduction
4. Sexual reproduction
5. Systematic position and identification.

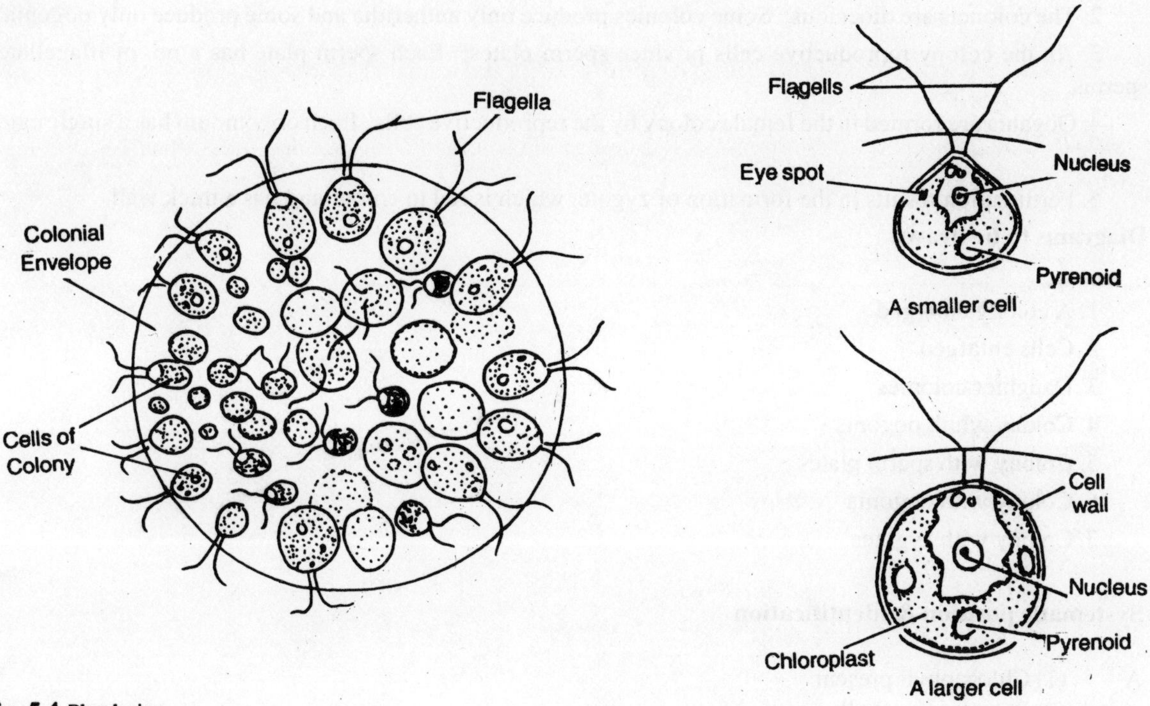

Fig. 5.4 Pleodorina

Structure of the colony

1. Plant body is a motile colony called coenobium.
2. The conenobia are spherical or occasionally ellipsoidal in shape.
3. Cells are arranged in a single layer at the periphery resulting in a hollow sphere. The cells in each colony range from 32,64, or 128.
4. The colony is enveloped by a mucilage layer.
5. Cells in the colony are of two sizes: Large non flagellate reproductive cells and small flagellate vegetative cells.
6. Vegetative cells occupy about 1/3rd of the colony, the rest being occupied by reproductive cells.

Cell structure

1. Cells are typically chlamydomonad in structure.
2. Each cell has two flagella, a cup shaped chloroplast a stigma, and a single nucleus.
3. Flagella are absent in the reproductive cells.

Asexual reproduction

1. Daughter colonies are formed by some of the large reproductive cells.
2. The daughter colonies are trapped inside the mother colony for quite a long time.
3. After a certain period of time the daughter colonies are released by the rupture of the mother colony.

Sexual reproduction

1. Colonies reproduct sexually by means of oogonia (female) & antheridia (male) oogamous method.
2. The coloneis are dioecious. Some colonies produce only antheridia and some produce only oogonia.
3. In the colony reproductive cells produce sperm plates. Each sperm plate has a no. of iflagellate sperms.
4. Oogania are formed in the female colony by the reproductive cells. Each oogonioum has a single egg.
5. Fertilization results in the formation of zygote, which is red in colour and has a thick wall

Diagrams to be drawn

1. A colony enlarged
2. Cells enlafged
3. Daughter colonies
4. Colony whith oogonis
5. Colony with sperm plates
6. Colony with oogonia
7. Colony with zygotes

Systematic position & identification

A (1) Chlorophyll present
(2) Plant body a thallus
(3) Cell-wall cellulosic...Algae

B. (1) Plastids present
(2) Chlorophyll is the chief pigment
(3) True nucleus presnt
(4) Photosynthetic product is starch
(5) Reproductive cell motile... Chlorophyceae

C. (1) Plant body motile
(2) Chloroplast cup-shaped...Volvocales
(3) Cells divide in a longitudibnal plane...Volvocaceae

D. (1) Plant body motile .
(2) Colonies are motile
(3) Cells divide in a longitudinal plane...volvocaceae

E. (1) Colonies are ellipsoidal in shape
(2) No. of cells in a colony do not exceed 256
(3) Division of labour is seen in the cells of the colony...Pleudorina

VOLVOX

(Linnaeus 1758)

Common species: *V.globator, V.africanus, V.prolificus*

Distribution: Fresh water, ponds and ditches, very common particularly after a spell of rain.

Study

(1) External features of the colony

(2) Cell structure

(3) Asexual reproduction

(4) Sexual reproduction

(5) Systematic position and Identification.

External features

1. The plant body is a coenobium found floating or revolving as small green balls of the size of a pin head in fresh water.

2. The colony is scarcely visible to the naked eyes and is a hollow sphere made up of cells ranging in number from 500-50,000.

3. The entire coenobium is enveloped by mucilage.

4. The colony being spherical or ovoid has no anterior or posterior end; but the end which has few or no reproductive cell is called anterior end.

5. The reproductive cells lack the flagella and the eye spot.

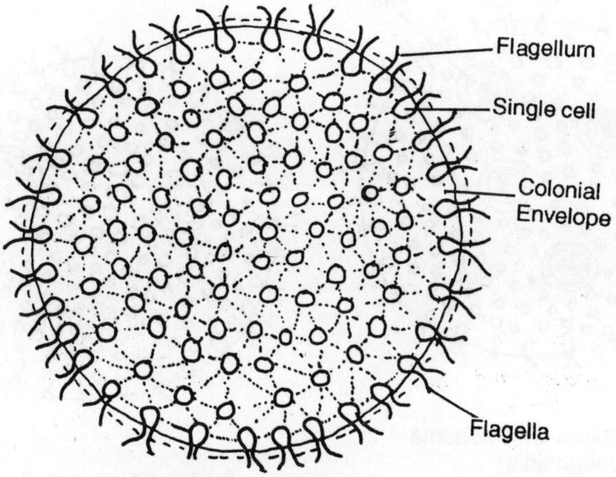

A vegetative coenobium

Fig. 5.5 Volvox

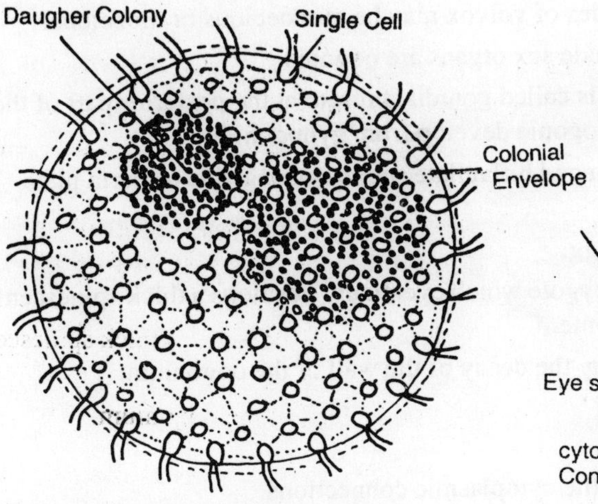

Coenobium showing daughter colonies

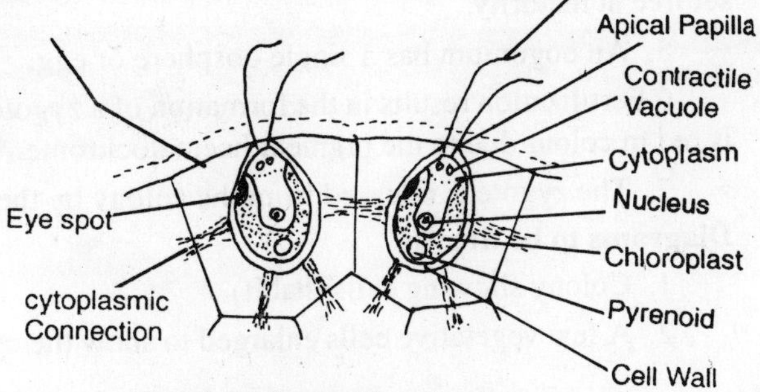

A portion of coenobium showing enlarged cells

Fig. 5.6 Volvox

Asexual reproduction

1. Asexual reproduction is brought about by means of daughter colonies.

2. The cells that give rise to daughter colonies are called 'parthenogonidia' They are bigger in size than the vegetative cells, have rich cytoplasm and a prominent nucleus.

3. The gonidium divides repeatedly and forms a daughter colony. The daughter colony is soon released into water.

4. In some cases the daughter colony is held trapped in the parent colony. It matures itself and some of its cells produce asexually another young colony called grand daughter colony.

5. In some cases an old colony of Volvox may have as many as 4-5 generations of colonies.

6. At a certain stage however the cell wall ruptures and all the colonies are set free. Whereupon they develop into independent colonies.

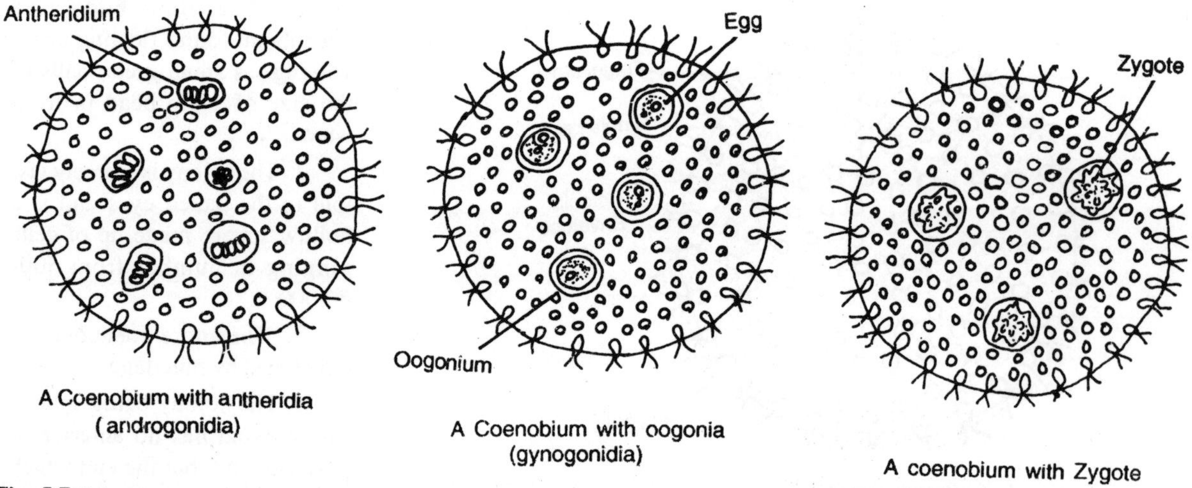

Fig. 5.7 Volvox

Sexual reproduction

1. This is of the oogamous type. The species of volvox may be monoecious or dioecious.

2. Male sex-organs are antheridia and female sex organs are oogonia.

3. The sex organs develop from special cells called gonidia, present at the posterior part of the colony. Antheridia develops from androgonidia while oogonia develop from gynogonidia

4. An antheridium has a number of cigar shaped biflagllate sperms arranged in a sperm plate. They are set free at maturity.

5. An oogonium has a single oosphere or egg.

6. Fertilization results in the formation of a zygote which eventually develops a thick spinescent wall. It is red in colour due to the pigment haematochrome.

7. The zygote is released from the colony by the decay of the wall of the oogonium.

Diagrams to be drawn

1. Colony showing cells (habit)

2. A few vegetative cells enlarged to show the cytoplasmic connections.

3. Colony with daughter colonies.
4. Colony with sperm plates.
5. Colony with oogonia.
6. Colony with zygotes.

Systematic position & Identification

A. (1) Chlorophyll present
(2) Plant body a thallus
(3) Cell wall cellulosic...*Algae*

B. (1) Plastids present
(2) Chlorophyll is the chief pigment.
(3) True nucleus present
(4) Photosynthetic product is starch
(5) Reproductive cells motile...*Chlorophyceae*

C. (1) Plant body motile
(2) Chloroplast cup shaped...*Volvocales*

D. (1) Plant body is a coenobium
(2) Colonies are motile
(3) Cells divide in a longitudinal plane...*Volvocaceae*

E. (1) Colony has more than 256 cells
(2) Division of labour seen in the cells of the colony
(3) Cells of the colony have structural and physiological differences...*Volvox*

CHLORELLA
(Beigerinok 1890)

Common spicies: *C.pyrenoidose, C.vuigaris, C.conglomerata* etc.
Distribution: Occurs in fresh water,ponds,damp soil, sides of garden pots etc.

Study

(1) Structure or the plant body
(2) Reproduction
(3) Systematic position and Identification

Structure of plant body

1. The plant body is unicellular and green in colour.
2. The cells are non motile, spherical or ellipsoidal in shape.
3. The cell wall does not have a mucilaginous envelope.
4. Each cell has a single cup shaped or band shaped chloroplast; there is no pyrenoid.
5. Cells are uninucleate.

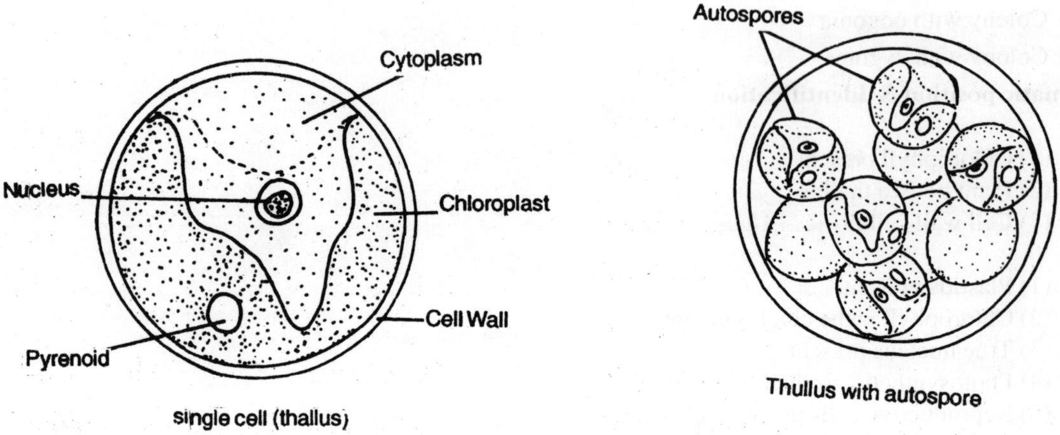

Fig. 5.8 Chlorella

Reproduction

1. Sexual reproduction has not been reported.
2. The cells divide and form spores called autospores.
3. The number of autospores per cell 2-8.
4. Autospores are uninucleate and develop into a new individual.

Diagram to be drawn

1. Structure of the cell
2. Cell with autopsores

Systematic position and Identification

A. 1. Chlorophyll present
2. Plant body a thallus
3. Cell wall cellulosic...*Algae*

B. 1. Plastids present
2. Chlorophyll is the chief pigment
3. True nucleus present
4. Photosynthetic product is starch...*Chlorophyceae*

C. 1. Cells are uninucleate and non-motile.
2. Reproduction by spores which are non-motile (motile pores very rare)...*Chlorococcales*

D 1. Plant body unicellular
2. Cells with a cup shaped chloroplast...*Chlorellaceae*

E. 1. Cells spherical and Uninucleate
2. Nucleus located at the centre
3. Chloroplast cup shaped ...*Chlorella*

PEDIASTRUM
(Meyer 1829)

Common species: *P.duplex, P.simplex, P.ovatum* etc.
Distribution: Found commonly in stagnant fresh water ponds and lakes.
Study

1. Structure of plant body
2. Reproduction
3. Systematic position and Identification

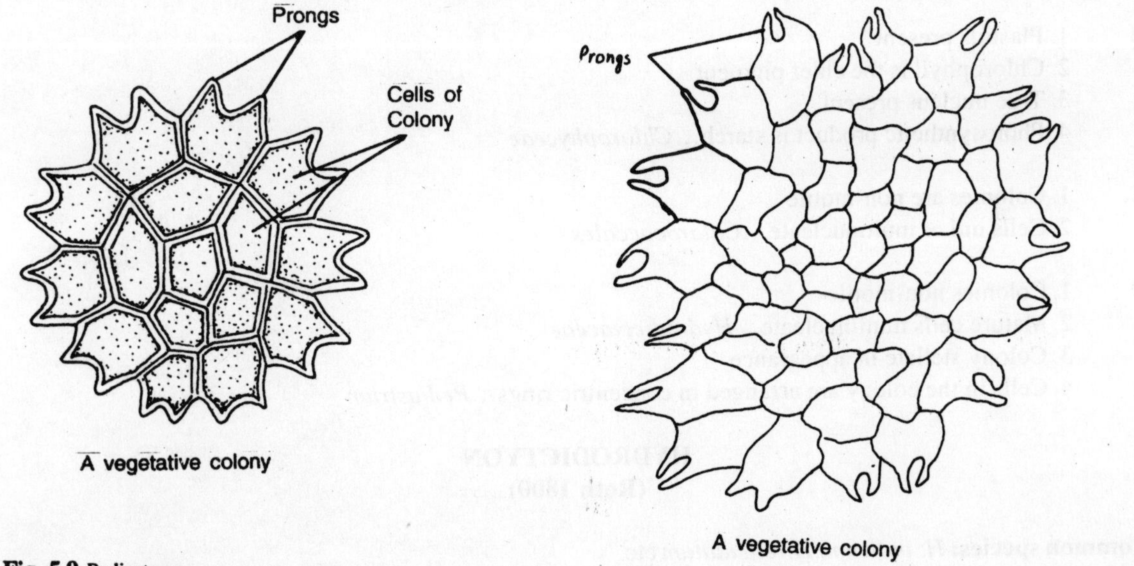

Fig. 5.9 Pediastrum

Structure of the Plant body

1. The plant body is a colony found freely floating on water.
2. The marginal cells of the colony have elongated projections giving the appearance of a star.
3. The cells are non motile and range in number per colony from 4-128.
4. The cells are arranged in cell plates. It is the cell plate (the margin of) which forms the projections giving a stellate appearance.

5. The cells have specific arrangements, which are as follows:
6. 8 celled colony-1 in the centre, 7 in outer ring.
7. 16 celled colony-1 in the centre, 5 in the first ring and 10 in second ring.
8. 32 celled colony 1 in the centre, 5 in the first ring, 10 in second ring and 16 in outer most ring.
9. Cells are polyhedral and arranged closely without leaving any intercellular space.
10. Cells have a cellulosic cell wall which may be smooth or granulate.
11. Young cells are uninucleate, but may become bi or tetra nucleate.
12. Chloroplast is parietal and has one or two pyrenoids.

Reproduction

1. Zoospores help in asexual reproduction

Diagrams to be drawn

1. A colony enlarged.
2. Few cells enlarged to show the arrangem,ent and other details.

Systematic position and Identification

A. 1. Chlorophyll present
2. Plant body a thallus
3. Cell-wall cellulosic...*Algae*

B 1. Plastids present
2. Chlorophyll is the chief pigment
3. True nucleus present.
4. Photosynthetic product is starch...*Chlorophyceae*

C. 1. Colonies are non-motile
2. Cells uni or multinucleate...*Chlorococcales*

D 1. Colonies non-motile
2. Mature cells mutlinucleate...*Hydrodictyaceae*
3. Colony stellate in appearance
4. Cells in the colony are arranged in concentric rings...*Pediastrum*

HYDRODICTYON
(Roth 1800)

Common species: *H. indicum,.H.reticulatum* etc.
Distribution: Found floating as a net in tanks, ponds & pools

Study

1. Structure of the colony
2. Cell sructure
3. Reproduction
4. Systematic position and Identification

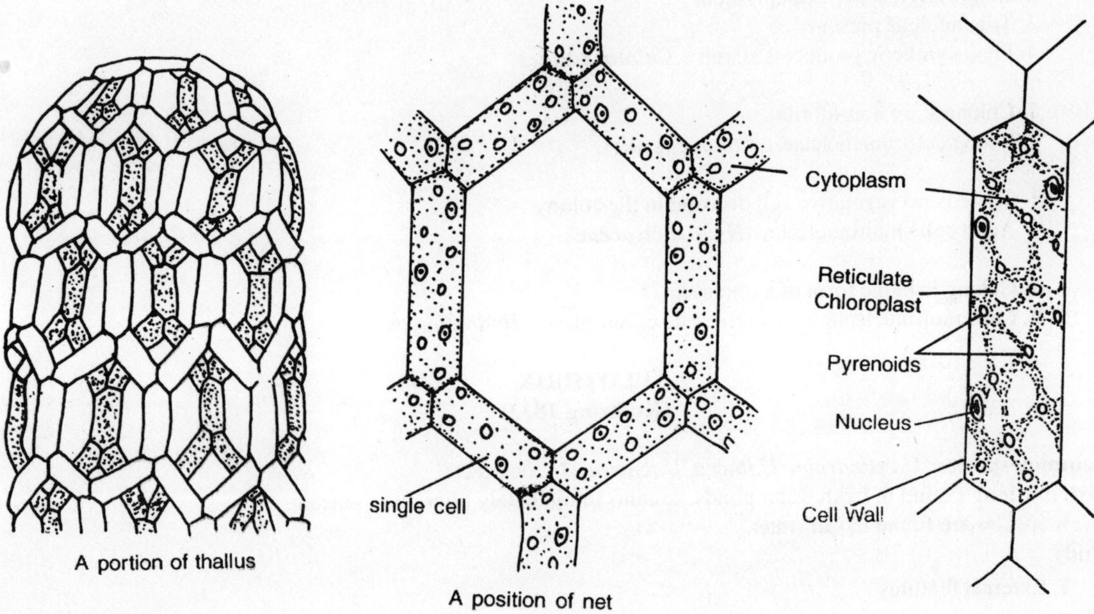

Fig. 5.10 Hydrodictyon

Structure of the colony

1. The plant body is a reticulate colony (commonly called waternet) found floating on water.
2. The çolony consists of many hundred cells attached to one another to form a reticulum.
3. At every joint usually three cells are joined, but occasionally four cells may also join at a point.
4. The meshes of the network appear like a hexagon.
5. The cells are many times longer than broad and do not hve a mucilage envelope.
6. Each cell is multinucleate and has a reticulate chloroplast with a number of pyrenoids

Reproduction

1. There is no vegetative cell division, the number of cells per colony is determined.
2. Daughter colonies are formed in the vegetative cells by the formation of zoospores which do not go out, but arrange themselves inside the cell (to form a daughter colony).
3. Sexual reproduction is Isogamous.

Diagrams to be drawn

1. Colony (Habit sketch)
2. Few cells enlarged to show the attachment of cells.
3. A few cells enlarged
4. A cell with daughter colony.

Systematic position Identification

A
1. Chlorophyll present
2. Plant body a thallus
3. Cell wall cellulosic...*Algae*

B
1. Plastids present
2. Chlorophyll is the chief pigment
3. True nucleus present
4. Photosynthetic product is starch...*Chlorococcales.*

C
1. Colonies are non-motile
2. Young cells unicucleate...*Chlorococcales.*

D
1. There is no vegetative cell division in the colony
2. Adult cells multinucleate...*Hydrodictyaceae*

E
1. Colony is in the form of a reticulum
2. Cells multinucleate with a reticulate chloroplast...*Hydrodictyon.*

ULOTHRIX
(Kuetzing 1833)

Common species: *U.cylindrium, U.flacca, U.tenissia, U.zonata* etc.
Distribution: Found in fresh water ponds, streams immediately after rainy season
A few species are found in salt water.

Study

1. External features
2. Reproduction
3. Systematic position and Identification

External features

1. The plant body is an unbranched filament
2. Filaments are thread like, fragile and formed by a uniseriate row of cells.
3. The basal cell of the filament is modified into a holdfast to help in attachment. Holdfasts are hyaline or brownish in colour.
4. The cells are broader than long, with a two layered cell wall - outer pectic and inner cellulosic.
5. Cells are uninucleate and have a single girdle shaped chloroplsst. There are 2-3 pyrenoids in each cell
6. There is a large vacuole in the centre of the cell.
7. The hold fast is colourless. It has no chloroplast

Reproduction-Asexual

1. All the cells of the filament except the hold fast can produce zoospores.
2. Zoospores are either bi or quadriflagellate.
1. Sexual reproduction is Isogamous
2. All cells except the hold fast and the terminal cell can form gametes.

Diagrams to be drawn

1. Habit sketch
2. A filament enlarged
3. A cell enlarged

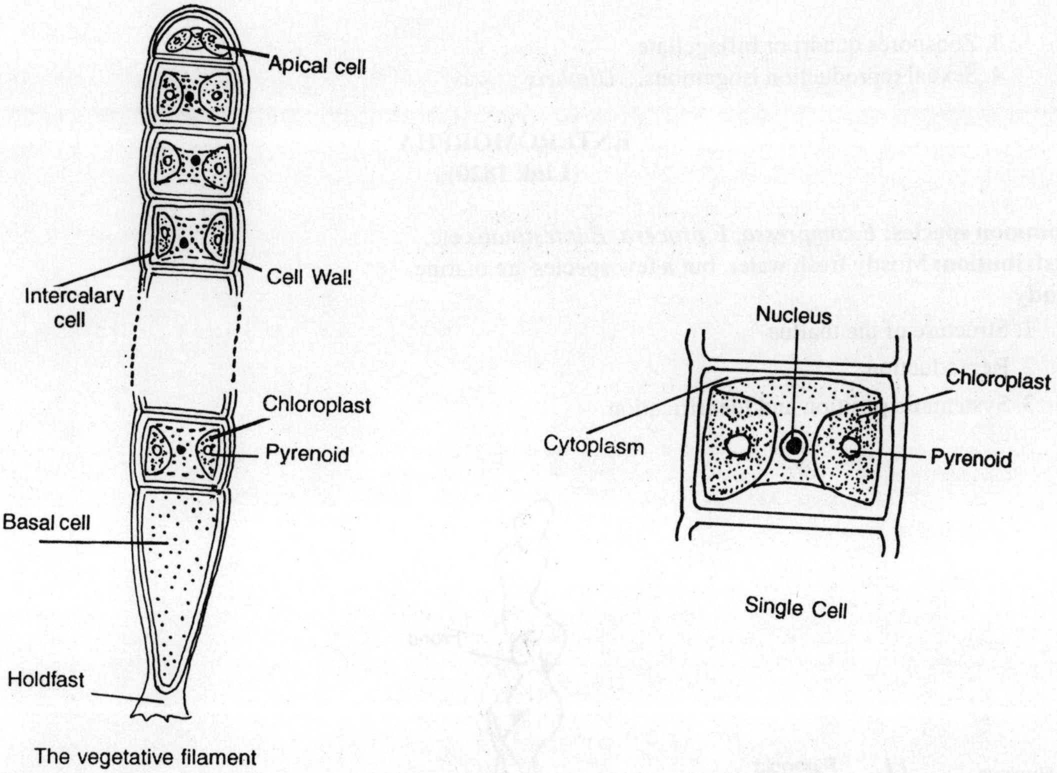

Fig. 5.11 Ulothrix

Systematic position and Identification

A 1. Chlorophyll present
2. Plant body a thallus
3. Cell wall cellulosic...*Algae*

B 1. Plastids present
2.Chlorophyll is the chief pigment
3. True nucleus present
4. Photosynthetic product is starch...*Chlorophyceae*

C 1.Plant body, unbranched filament
2. Cells uninucleate with a girdle shaped chloroplast
3.Zoospore bi or quadriflagellate...*Uliotrichales*

D 1. Cells uninucleate
2. Cells are broader than long...*Uliotrichaceae*

E 1. Chloroplast Girdle shaped
2. Hold fast present, to help in attachment

3. Zoospores quadri or biflagellate
4. Sexual reproduction isogamous... *Ulothrix*

ENTEROMORPHA
(Link 1820)

Common species: *E.compressa, E.procera, E.intestinalis* etc.
Distribution: Mostly fresh water, but a few species are marine.
Study

1. Structure of the thallus
2. Reproduction
3. Systematic position and Identification

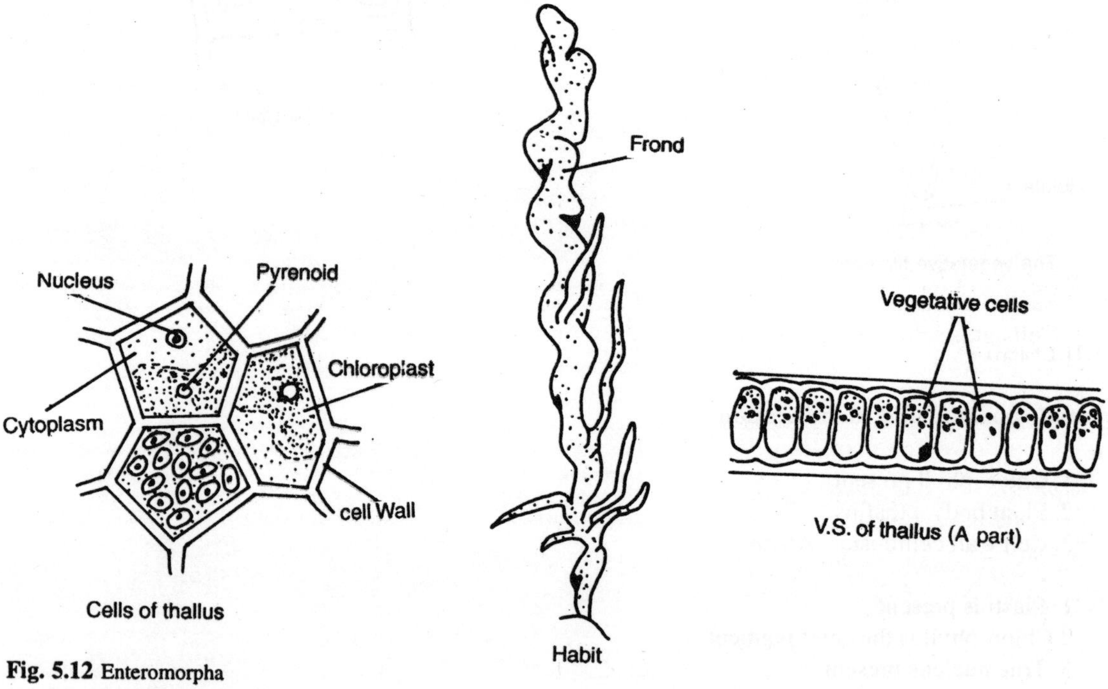

Fig. 5.12 Enteromorpha

Structure of the thallus

1. The plant body is a cylindrical or tube like thallus.
2. It is attached to the substratum with the help of a disc.
3. The thallus is one celled thick.
4. The thallus is branched, the branching being brought about by the longitudinal division of the cells.
5. The cells are many angled in shape and typically ulotrichean in structure
6. The cell wall is made up of cellulose and is striated.
7. Cells are uninucleate with a parietal chloroplast. There is a single pyrenoid embedded in the chloroplast.
8. The chloroplast faces the outside of the cell, while the nucleus is inside.

Reproduction

1. Vegetative propagation is brought about by the fragmentation of the thallus.
2. Asexual reproduction takes place by the formation of quadriflagellate zoospores.
3. Any cell of the thallus my form zoospores.
4. Sexual reproduction is iso or anisogamous.
5. The plants are dioecious, when sexual reproduction is anisogamous.

Diagrams to be drawn

1. Habit sketch
2. Section of the thallus
3. Hold fast

Systematic position and Identification

A 1. Chlorophyll present
2. Plantbody a thallus
3. Cell wall cellulosic...*Algae*

B 1. Plastids present
2. Chlorophyll is the chief pigment
3. True nucleus present
4. Photosynthetic product is starch...*Chlorophyceae*

C 1. Cells divide in more than one plane to form a sheet
2. Sexual reproduction iso or anisogamous
3. Cells uninucleate with a parietal chloroplast...*Ulotrichales*

D 1. Thallus one or two celled thick
2. Chloroplast parietal...*Ulvaceae*

E 1. Thallus is cylindrical (tubular) and branched
2. Construction is present at the region of branching
3. Hold fast disc present in attachment...*Enteromorpha.*

ULVA-Sea lettuce
(Agardh 1833)

Common species: *U.lactuca, U.lobata, U.reticulata, U.stenophyylla* etc.

Distiribution: Most of the species are marine , found atached to rocks and other substrata in the tidal zone. A few may be fesh water also.

Study

1. External features
2. Internal features
3. Reproduction
4. Systematic position and Identification

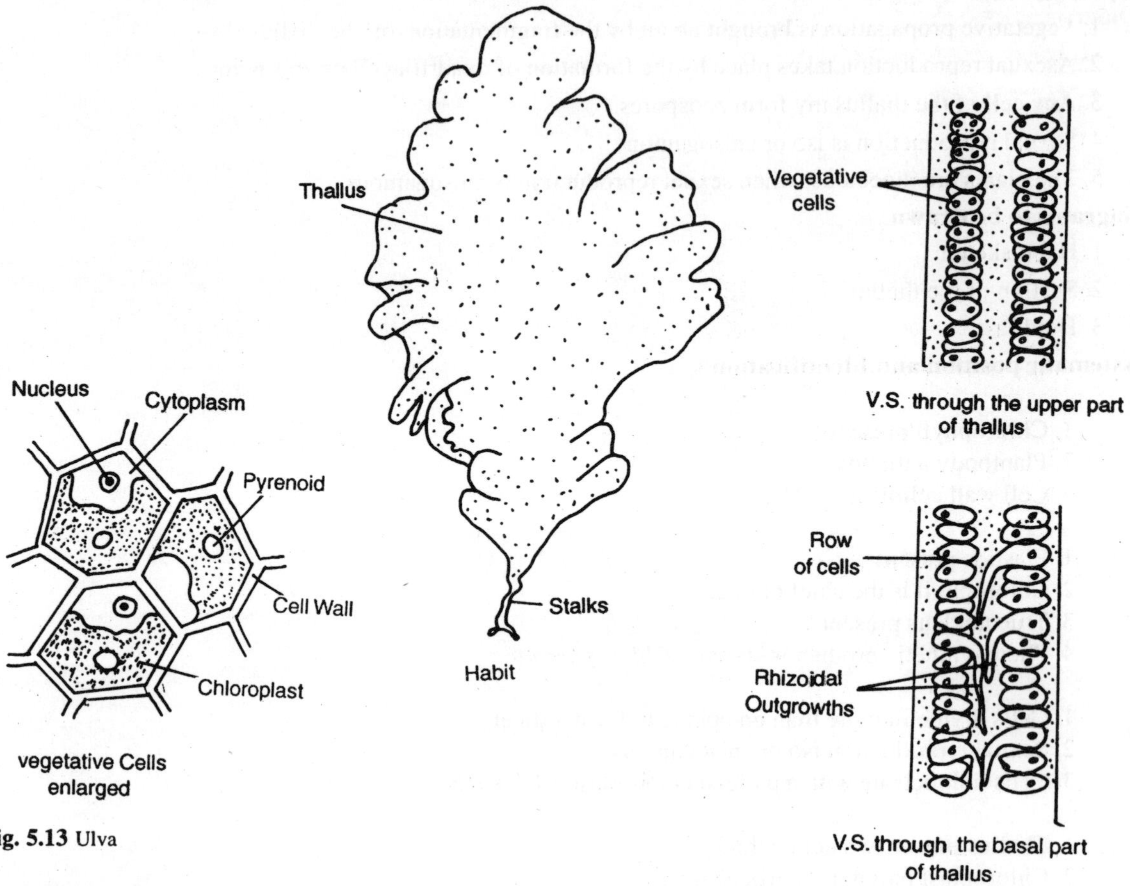

Fig. 5.13 Ulva

External features

1. The thalloid plant body is a sheet 2-3 celled in thickness.
2. The plant body is anchored to the soil with the help of a holdfast.
3. The holdfast produces many rhizoidal outgrowths that grow down into the soil.
4. The rhiziods are chlorophyllous, septate, and multinucleate.
5. The cells are uninucleate, the nucleus being present towards the inside and the chloroplast being parietal in position.

Internal structure: A long section of the thallus shows the following features.

1. The cells of the thallus are arranged in two vertical rows.
2. Cells at the basal part of the thallus give out rhizoidal out growths, they combine together to form the hold fast.

Reproduction

1. Vegetative propagation of the thallus takes place by fragmentation.
2. Asexual reporduction takes place by the formation of quadri flagellate zoospores.
3. Any cell of the thallus can form zoospores.
4. Most of the species of *Ulva* are heterothallic.
5. Sexual reproduction is mostly by isogamy but occasionally anisogamy has also been reported.

Diagrams to be drawn

1. Habit sketch.
2. L.S. of the thallus.
3. A few cells enlarged.
4. Hold fast enlarged.

Systematic position and Identification

A. 1. Chlorophyll present
2. Plant body a thallus
3. Cell wall cellulosic…*Algae*

B 1. Plastids present
2. Chlorophyll is the chief pigment
3. True nucleus present
4. Photosynthetic product is starch…*Chlorophyceae*

C. 1. Cells divide in more than one plane to form a sheet of cells.
2. Cells uninucleate with a parietal chloroplast.
3. Sexual reproduction iso or anisogamous…*Ulvales*

D. 1. Thallus one or two celled thick.
2. Choloroplast parietal…*Ulvaceae*

E. 1. Plant body flattened having two vertical rows of cells.
2. Rhizoidal outgrowths combine to form a hold fast…*Ulva*

CLADOPHORA
(Kuetzing 1843)

Common species: *C. glomerata, C.crispata, C.fracata* etc.
Distribution: Found in fresh water as well as marine seen attached to shells, stones or weeds. Some species are aegogrophilous forming small floating balls.

Study

1. Structure of the plant body
2. Reproduction
3. Systematic position and Identification.

Structure of plant body

1. The plant body is a branched filament found attached to certain substrata in water.

2. The filament is attached to the substratum with the help of rhizoidal outgrowths that are produced from the basal region.

3. The branching of the filament is lateral, a branch appearing just below the cross wall.

4. The branching later becomes pseudodichotomous due to the process of evection (Pushing)

5. Cells are longer than broad and have a stratified three layered cell wall.

7. The cytoplasm is periphereal, there being a large central vacuole.

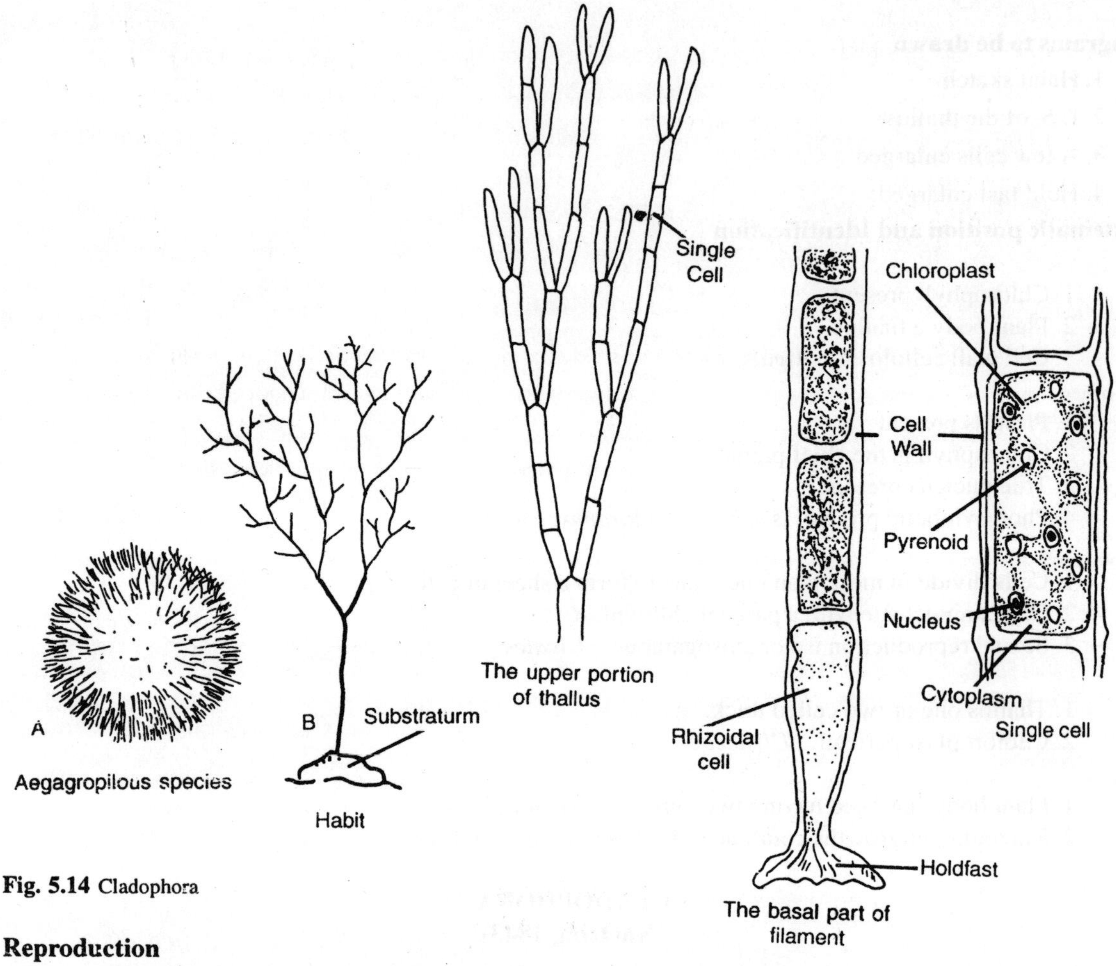

Fig. 5.14 Cladophora

Reproduction

1. Asexual reproduction is by the formation of quadri flagellate zoospores.
2. Sexual reproduction is isogamous. The gametes are biflagellae.
3. Alternation of generations seen in the life cycle.

Diagrams to be drawn

1. Habit sketch
2. A filament enlarged
3. A cell enlarged

Systematic position and Identification

A 1. Chlorophyll present
2. Plant body is thallus
3. Cell wall cellulosic...*Algae*

B 1. Plastids present
2. Chlorophyll is the chief pigment
3. True nucleus present
4. Photosynthetic product is starch...*Chlorophyceae*

C 1. Cells multinucleate.
2. Filaments branched...*Cladophorales*

D 1. Cells many times longer than borad
2. Branching is pseudodictomous...*Cladophoraceae*

E. 1. Filaments branched
2. Cells multinucleate with a reticulate chloroplast...*Cladophora*

PITHOPHORA
(Wittrock 1877)

Common species: *P.affains, P.kewensis* etc.
Distribution: Fresh water pools and ponds. Some times found attached to molluscan shells.

Study

1. External features
2. Reproduction
3. Systematic position and Identification

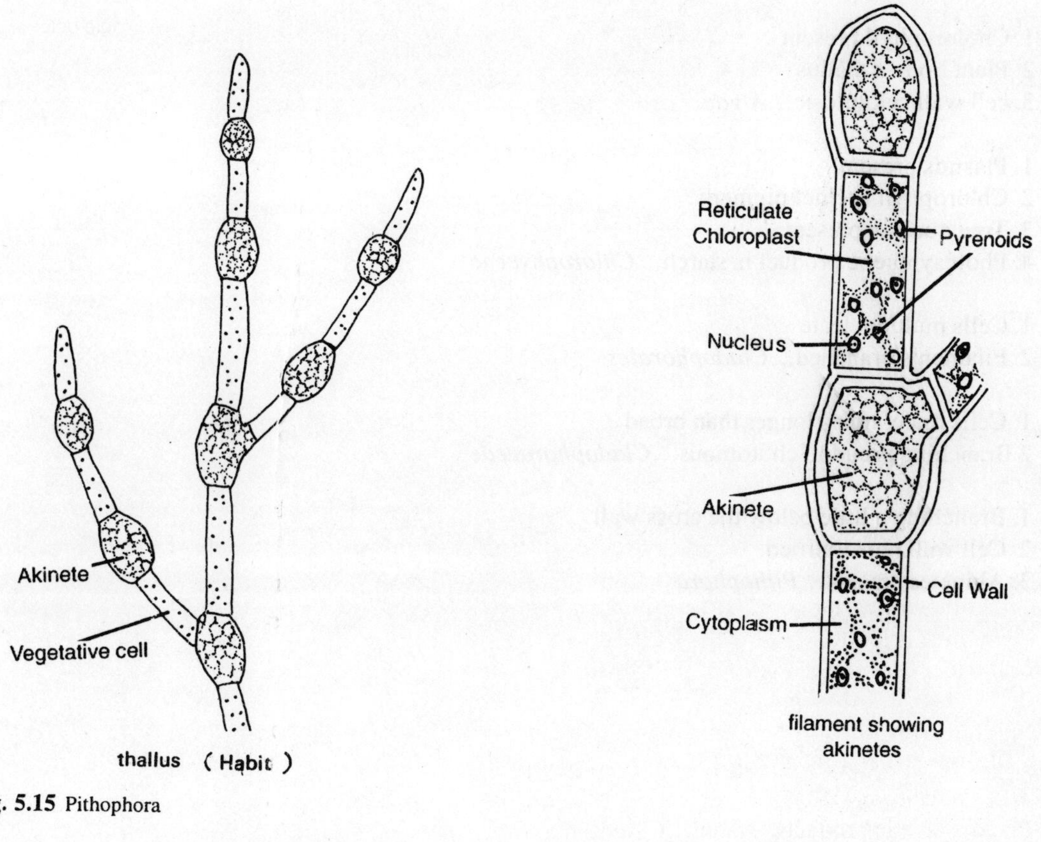

Fig. 5.15 Pithophora

External features

1. The plant body is a branched filament.
2. The cells are joined end to end to form the filament.
3. The cells have a thin wall which is not lamellated.
4. The plant body is anchored to the substratum with the help of rhizoidal outgrowths.
5. Branching of the filament is same, as seen in Cladophora, i.e., pseudodichotomous
6. Cells are longer than broad, multinucleate with a reticulate parietal chloroplast. Embedded in the chloroplast are found a number of pyrenoids.

Reproduction

1. Fragmenation of the filament results in bits, which can develop into new individuals.
2. Akinetes are formed either in intercalary or terminal cells.
3. Akinetes are thick walled dark brown in color and are rich in reserve food materials.
4. On germination akinetes develop into new individuals.

Diagrams to be drawn

1. Habit sketch
2. A part of the filament enlarged to show branching.
3. A cell enlarged

Systematic position and Identification

A 1. Cholorophyll present
2. Plant body a thallus
3. cell wall celolulosic...*Algae*

B 1. Plastids present
2. Chlorophyll is chief pigment
3. True nucleus present
4. Photosynthetic product is starch...*Chlorophyceae*

C 1. Cells multinucleate
2. Filaments branched...*Cladophorales*

D 1. Cells many times longer than broad
2.Branching pseudodichotomous...*Cladophoraceae*

E 1. Branching a little below the cross wall
2. Cell wall not stratified.
3. Akinetes present...*Pithophora*

OEDOGONIUM
(Link 1820)

Common species: *O.reliana, O.americanum, O.intermedium* etc.
Distribution: The plant is found in ponds, lakes and pools. It is usually submerged underwater attached to twigs or weeds.

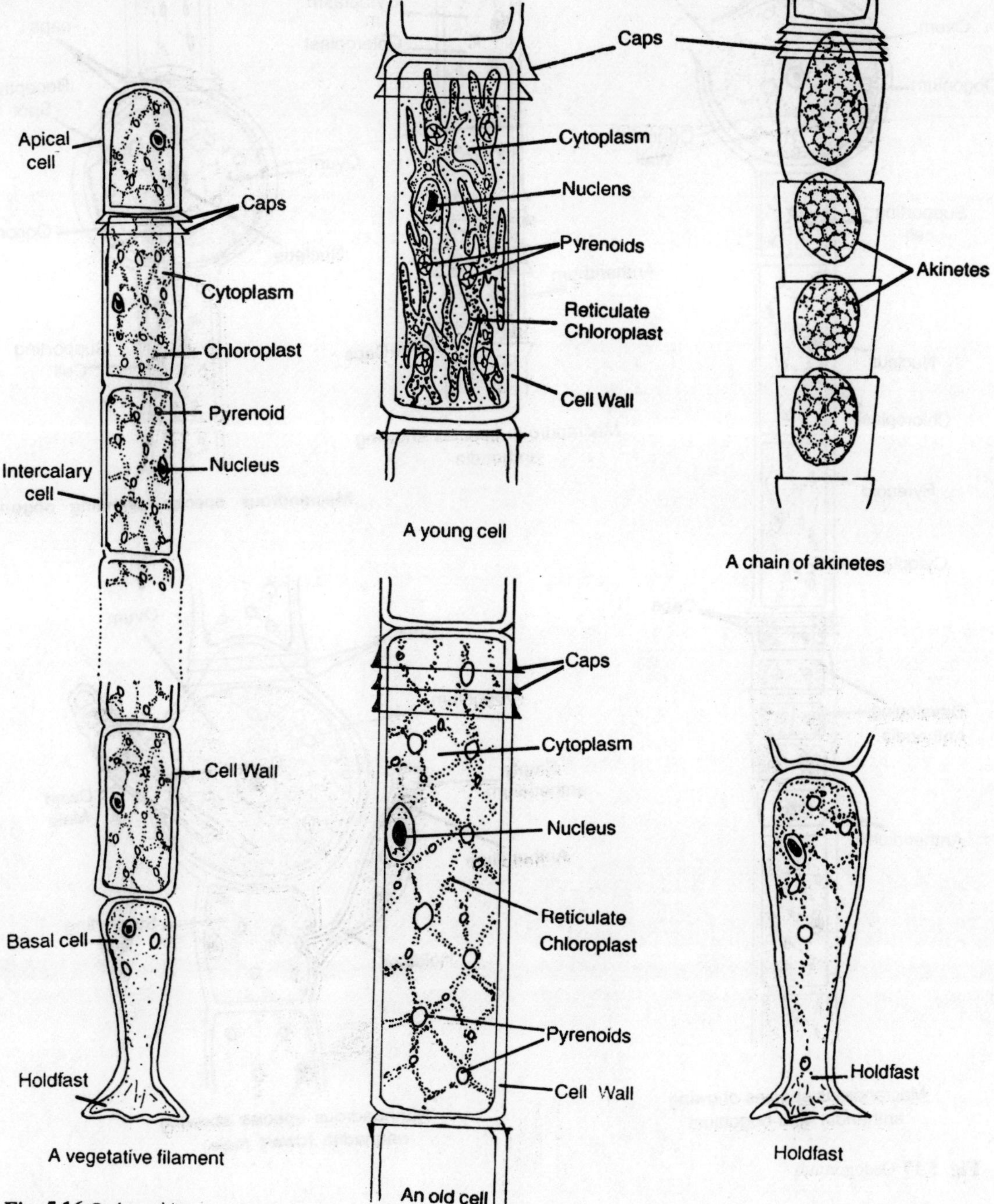

Fig. 5.16 Oedogonium

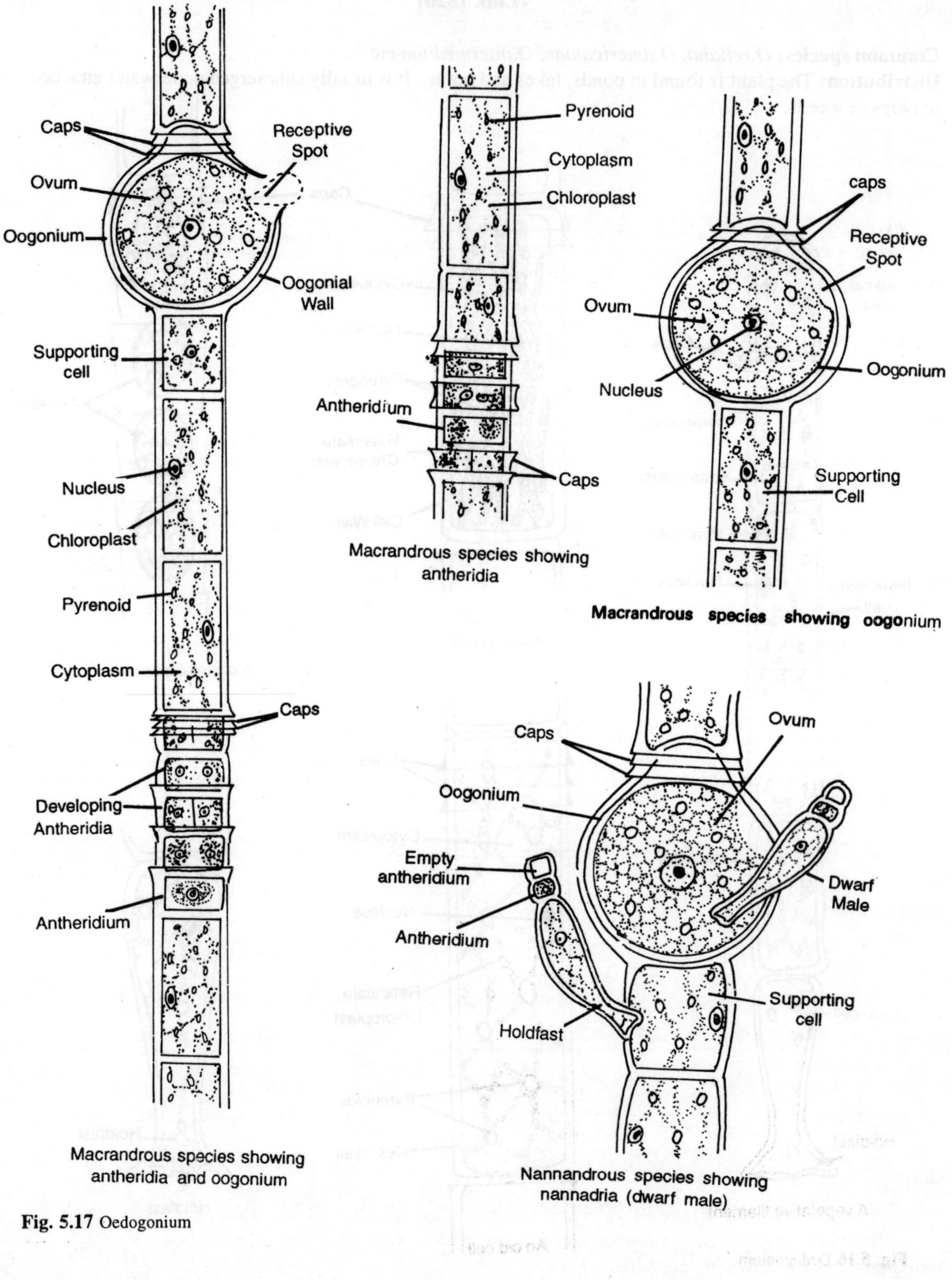

Macrandrous species showing antheridia and oogonium

Macrandrous species showing antheridia

Macrandrous species showing oogonium

Nannandrous species showing nannadria (dwarf male)

Fig. 5.17 Oedogonium

Study

1. External morphology
2. Cell division and growth
3. Reproduction
4. Systematic position and Identification

External morphology

1. The plant body is an unbranched filament, it is deep green in colour.

2. The filament consits of cylindrical cells joined end to end.

3. The basal cell is modified into a hold fast to help in attachment. The hold fast consists of a poorly developed chloroplast and has its free end modified into a disc.

4. The terminal cell usually has a rounded tip.

5. The cell has a thick stratified cell wall divided into three layers - outer chitinous, middle pectic and inner cellulosic.

There is no mucilage sheath.

Cells are uninucleate with a reticulate chloroplast having a number of pyrenoids.

Cell division and growth

1. Growth is intercalary and brought about by the addition of cells in between.

2. During cell division transverse striations called 'caps' are formed as new cell are added in between. These apical caps are characteristic of *Oedogonium*.

Reproduction-Asexual

1. Some of the free floating species undergo fragmentation.

2. Zoospores are formed in all cells except the terminal cell and the hold fast cell. Each cell forms a single zoospore.

3. Zoospores are pyriform, multiciliate, the cilia attached at anterior end.

4. The zoospore is usually released into a vesicle and then set free.

5. After a short period of free swimming, the zoospore germinates to develop into a new individual.

Reproduction-Sexual

1. Sexual reproduction is oogamous. Sex organs are antheridia and oogonia.

2. Two categories can be identified in *Oedogonium* with reference to the development of antheridia. In *Macrandrous* forms antheridia are formed on normal filaments while in *Nannandrous* forms, antheridia are formed on short or dwarf male filaments called *Nannandria.*

3. Oogonia are formed on normal filaments in both macrandrous and nannandrous forms.

4. Any intercallary cell can develop into an oogonium. The oogonium is globose, rich in cytoplasm and has a hyaline receptive sopt at one end. The oogonium is usually supported by a Suffultory cell.

5. In macrandrous forms antheridia are formed in an intercalary cell.

6. A series of transverse divisions take place in an antheridial mother cell resulting in a row of antheridia.

7. Each antheridium forms two sperms or male gametes which are exactly like the zoospore but much smaller in size.

8. In Nannandrous forms an Intercalary cell in a normal filament forms a number of *androsporangia* by undergoing repeated transverse divisions.

9. Antherozooids (sperms) are released into water.

10. Each androsporangium produces a single androspore. An androspore is exactly like a zoospore but intermediate in size between zoospore and antheozooid.

11. On release from the androsporangium, the androspore attaches itself to the oogonium or the suffultory cell. It germinates there and develops a short 3-4 celled filament called Nannandrium.

12. The terminal cell of the Nannandriumm forms the antheridia which produce antherozoids.

13. Antherozoids are similar in structure irrespective of the fact whether they are formed on normal filaments or nannandria.

14. On fertilisation, the egg develops into a zygote. Zygote is thick walled and somewhat reddish in colour.

15. Zygote undergoes a period of rest and divides to form four zoospores.

16. These zoospore develop into new individuals.

Diagrams to be drawn

1. Habit sketch
2. A few filaments enlarged
3. A cell enlarged to show structure and apical caps.
4. A cell with a zoospore.
5. A filament with oogonium and *Nannandrium.*
6. A filament with antheridia
7. A filament with androsporangia.

Systematic position and Identification

A
1. Chlorophyll present
2. Plant body is thallus
3. Cell wall cellulosic...*Algae*

B
1. Plastids present
2. Cholorophyll is the chief pigment
3. True nucleus is present
4. Photosynthetic product is starch...*Chlorophyceae*

C
1. Zoospores, antherozoids multiciliate
2. Sexual reproduction by oogamy
3. Filaments unbranched...*Oedogoniales* and *Oedoganiaceae*

D
1. Growth intercalary
2. Cap cells present
3. Nannandrium present...*Oedogonium*

BULBOCHAETAE

Common species: *B.setigera, B. intermedia* etc.
Distribution: Fresh water ponds, pools etc.

Study

1. Habit and structure
2. Reproduction
3. Systematic position and Identification

Structure of the plant body

1. The plant body is a filament found usually attached with the help of the basal hold fast cell.

2. The filament is made up of cylindrical cells and is branched unlike in Oedogonuim.

3. The branches are profuse and usually successive branches alternate.

4. Each cell base a long colorless hair with a bulbous base. (hence the name *Bulbochactae*).

5. The cells are uninucleate and have an elaborate reticulate chloroplast.

6. The chloroplast has a number of pyrenoids embedded in the meshes of the reticulum.

7. The cell wall has a thin mucilage envelope.

8. Growth is intercalary and is brought about by the cell divisions of intercalary cells.

9. The details of the cell divisions and cap formation are similar to what is seen in *Oedogonium*.

10. In most species of *Bulbochaetae* however, growth is mostly basal and not more than one cap is formed per cell.

11. Each cell after its formation grows out slightly at its apex towards one side and this protuberance is cut off as a colourless cell.

12. The cell lengthens into a hair, with the ruptured wall forming an irregular sheath around the base.

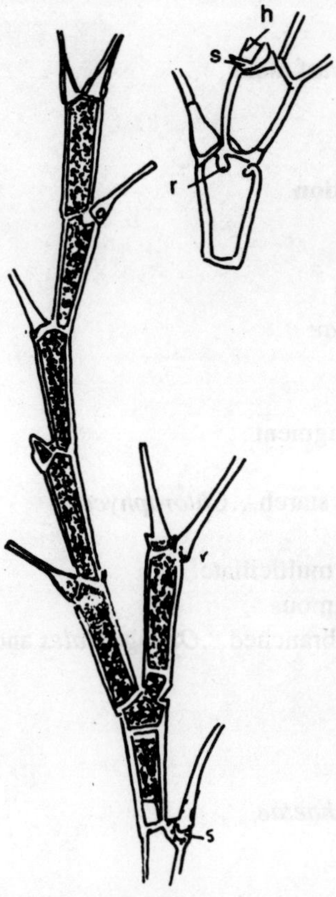

Fig. 5.18 Bulbochaetae

Reproduction-Asexual

1. Vegetative propagation is brought about by mechanical fragmentation.

2. Asexual reproduction takes place by means of zoospores. Zoospores are multi-flagellate and pyriform in shape. A circlet of flagella is found at the anterior end.

3. Each cell forms a single zoospore. Rest of the details are same as in *Oedogonium*.

Reproduction-Sexual

1. Sexual reproduction does not occur quite frequently. It is of the oogamous type.

Oogonia and antheridia are produced on the same plant in *macrandrous* species, while in *nannandrous* species, antheridia are formed on dwarf male filaments.

3. The nannandrous filaments arise from special types of spores called androspores produced in androsporangia.

4. Rest of the details of the formation antheridium, and androsporangium and oogonium are same as in *Oedogonium*.

5. Nannandrous filaments attach themselves either to an oogonium or to a cell supporting the oogonuim.

6. The dwarf males are 3-4 celled in length and produce antheridia at their tip.

7. Fertilization is similar to what is seen in *Oedogonium.*

8. Oospore (zygote) develops a thick three layered wall and the contents assume a red colour.

9. The oospore after undergoing a period of rest develops into a new individual by undergoing reduction division and forming four zoospores.

Diagrams to be drawn

1.Habit sketch
2. A filament enlarged
3. A single cell with the hair
4. Filament with oogonia and dwarf male
5. Filament with antheridia
6. Filament with androsporangia

Systematic position and Identification

A 1. Chlorophyll present
2. Plant body a thallus
3. Cell wall cellulosic...*Algae*

B 1. Plastids present
2. Chlorophyll is the chief pigment
3.True nucleus present
4. Photosynthetic product is starch...*Chlorophyceae*

C 1. Zoospores, Autherozoids multiciliate
2. Sexual reproduction oogamous
3. Filaments branched or unbranched...*Oedogoniales* and *Oedogoniaceae*

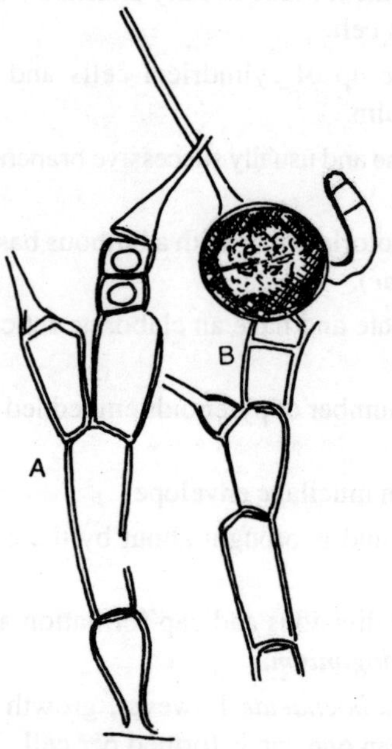

Fig. 5.19 Draparnoldia

D 1. Growth is basal or intercalary
2. Filaments branched
3. Bullous hairs present at the tip of each terminal cell...*Bulbochaetae.*

DRAPARNALDIA

Common species: *D.glomerat, D.villosa* etc
Distribution: Exclusively fresh-water. Found growing attahced to sand, twigs or as epiphytes.

Study

1. Structure of the plant body
2. Reproduction
3. Systematic position and Identification

Structure of the plant body

1. The plant body is a filament. It is anchored to the substratum with the help of a holdfast. The holdfast has rhizoidal outgrowths.
2. The plant body is heterotrichous, but the prostrate system is mostly absent.
3. Thin mucilage envelope surrounds the filament.
4. The plant body has a central axis made up of a uniseriate row of cells.
5. At regular intervals, the axis produces laterals or branches; these branches are in tufts.
6. The profuse development of the laterals, sometimes masks the main axis.
7. The apices of the branches usally terminate in hair like processes called setae.

8. The cells of the main axis are large, barrel shaped and uninucleate.
9. The chloroplast is plate like and parietal with a single pyrenoid.
10. The cells of the laterals are much smaller in size, but are similar in structure to those of the main axis.

Reproduction

1. Asexual reproduction is by quadriflagellate zoospores and hypnospores.

2. Sexual reproduction is isogamous.

3. Gametes are quadri flagellate but are much smaller in size than the zoospores.

4. The zygoe, formed by the fusion of the gametes develops directly into a new individual.

Diagrams to be drazwn

1. Habit sketch

2. A portion of the plant enlarged to show the branching .

Systematic position and Identification

A. 1. Chlorophyll present
2. Plant body a thallus
3. Cell wall cellulosis...*Algae*

B 1. Plastids present
2. Chlorophyll is the chief pigment
3. True nucleus presents
4. Photosynthetic product is starch...*Chlorophyceae*

C 1. Plant body exhibits heterotrichous habit
2. Branches end in setae...*Chaetophorales*

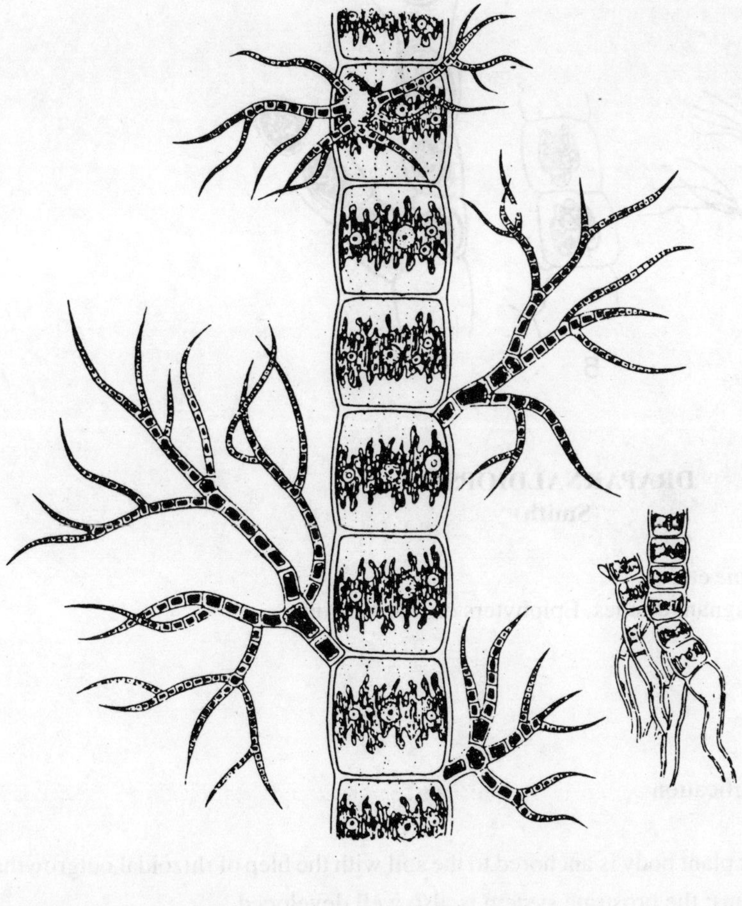

Fig. 5.20 Draparnaldia

D 1. Cells uninucleate with the parietal chloroplast
2. Terminal cells of laterals end in setae...*Chaetophoraceae*

E 1. Prostrate system almost absent
2. Cells are barrel shaped...*Draparnaldia*

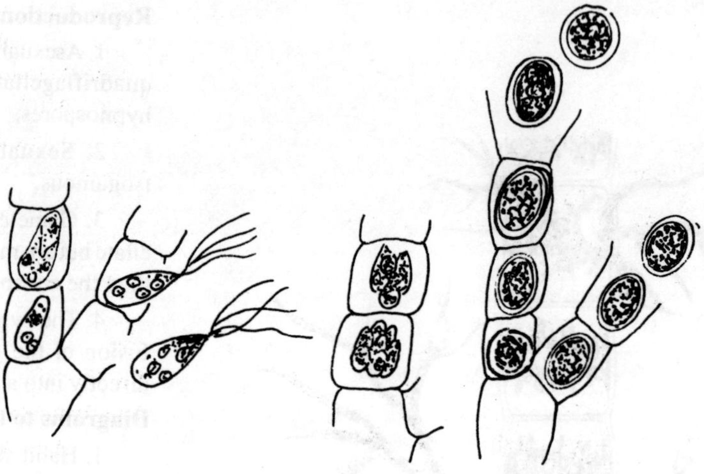

Fig. 5.21 Draparnaldia

DRAPARNALDIOPSIS
(Smith)

Common species: *D.indicia, D.alpine* etc.
Distribution: Fresh water ponds, stagnant ditches. Epiphyters or free floating

Study

1. Habit and structure
2. Reproduction
3. Systematic position and Identification

Structure of the plant body

1. The filmaentous, multicellular plant body is anchored to the soil with the hlep of rhizoidal outgrowths.

2. The plant body is heterotrichous; the prostrate system is also well developed.

3. The rhizoids are profusely branched forming a cortical covering at the basal region.

4. The erect system or the main axis consists of a uniseriate row of cells.

5. The cells in the main axix are of two types- long, barrel shaped internodal cells and short discoid nodal cells. The two types of cells alternate regularly.

6. The nodal cells bear a system of branched laterals. The branches are of two types, i.e., the branches of unlimited growth and branches of limited growth.

7. The long laterals or branches of unlimited growth arising at nodal cells (2-4 in number) have the same structure as the main axis.

8. The branches of limited growth or short laterals arise either on the nodal cell of the main axis or on long laterals. They are smaller in size, profusley branched and the cells terminate in hair like processes.

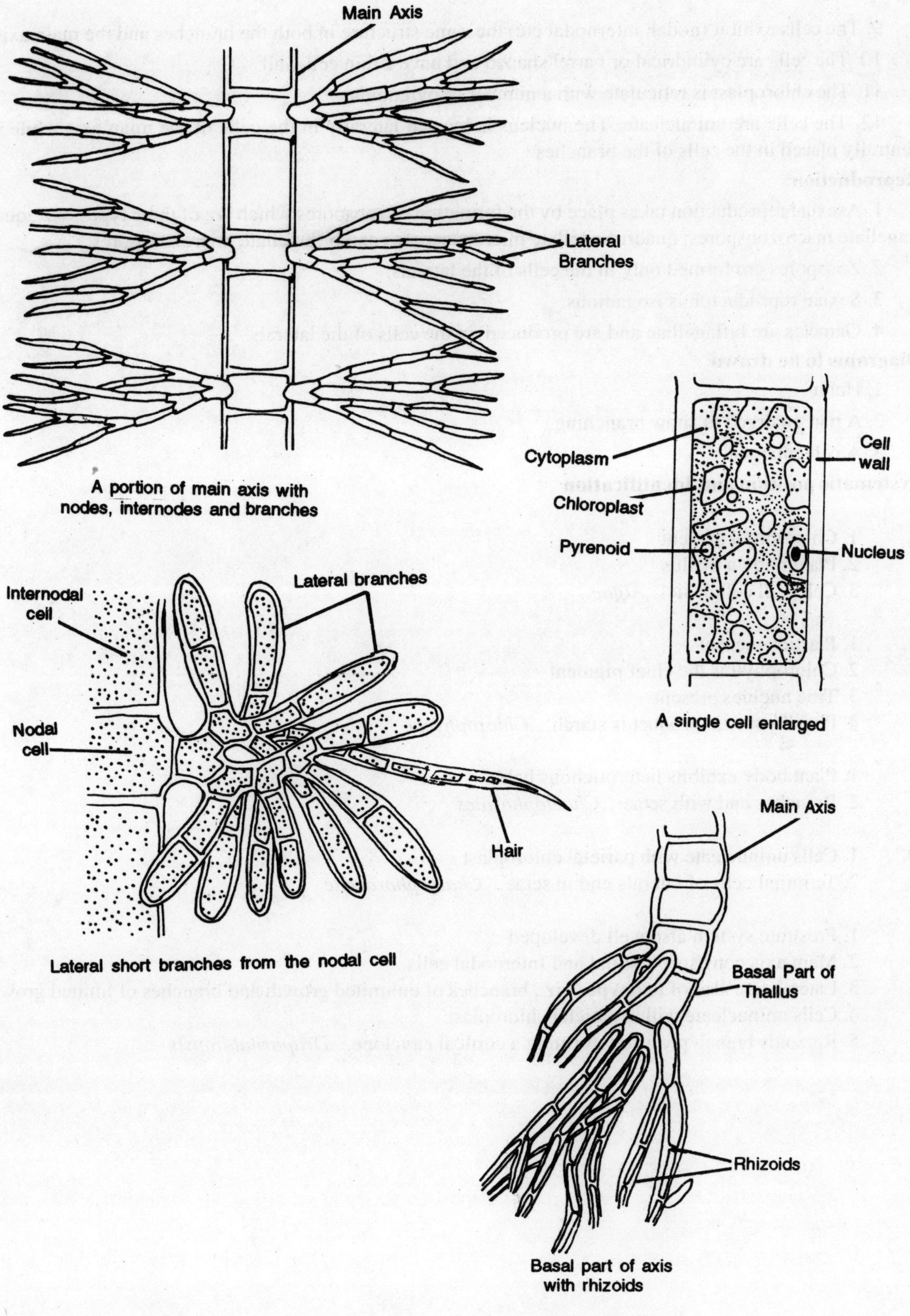

Fig. 5.22 Draparnaldiopsis

9. The cells exhibit (nodal, internodal etc) the same structure in both the branches and the main axis.

10. The cells are cylindrical or barrel shaped and have a thin cell wall.

11. The chloroplast is reticulate with a number of pyrenoids.

12. The cells are uninucleate. The nucleus is located laterally in the cells of the mian axis while it is centrally placed in the cells of the branches.

Reproduction:

1. Asexual reproduction takes place by the formation of zoospores which are of three types, viz., quadri flagellate macrozoospores, quadri flagellate microzoospores and biflagellate microzoospores.
2. Zoospores are formed only in the cells of the laterals.
3. Sexual reproducton is isogamous.
4. Gametes are biflagellate and are produced in the cells of the laterals.

Diagrams to be drawn

1. Habit
2. A part enalrged to show branching
3. A cell enlarged

Systematic position and Identification

A
1. Chlorophyll present
2. Plant body a thallus
3. Cell wall cellulosic...*Algae*

B.
1. Plastids present
2. Chlorophyll is the chief pigment
3. True nuclues present
4. Photosynthetic product is starch...*Chlorophycae*

C.
1. Plant body exhibits heterotichous habit
2. Branches end with setae...*Chaetophorales*

D.
1. Cells uninucleate with parietal chloroplast
2. Terminal cells of laterals end in setae...*Chaetophoraceae*

E
1. Prostrate system also well developed
2. Main axis consists of Nodal and Internodal cells.
3. Laterals are also of two types viz., branches of unlimited growth and branches of limited growth
4. Cells uninucleate with a parietal chloroplast.
5. Rhizoids branch profusely forming a cortical envelope...*Draparnaldiopsis*

FRITCHIELLA
(Iyengar 1932)

Common species: *F.tuberosa* is the only species recorded for the genus.
Distribution: Terrestrial alga found on damp soil, pond beds and soil rich in organic matter.

Study

1. External structure
2. Systematic position and Identification

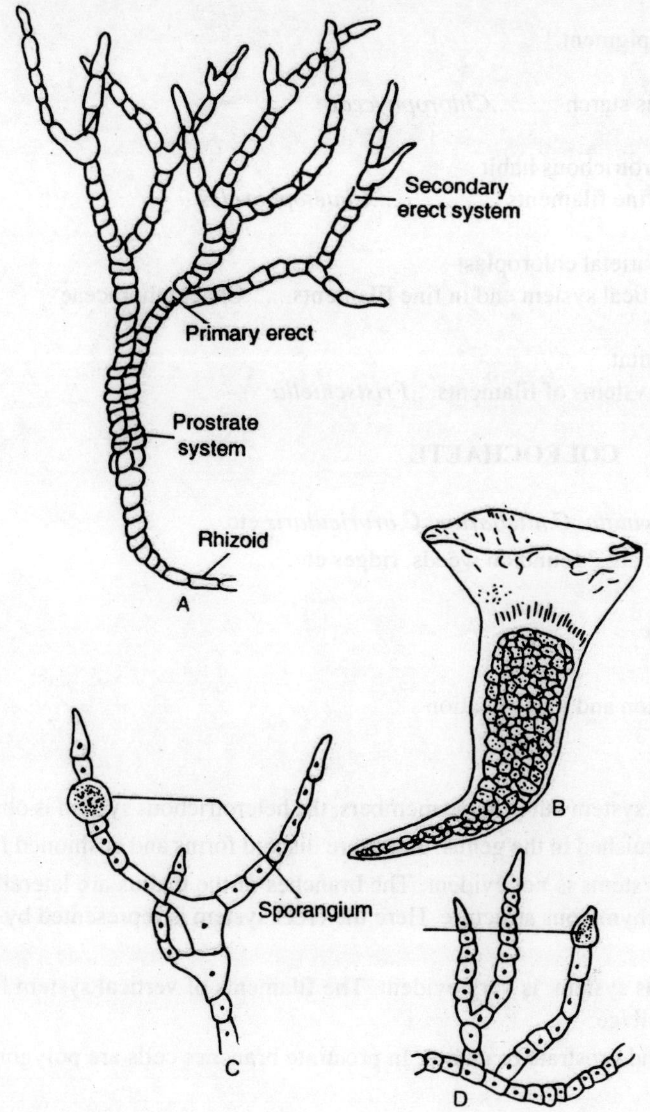

Fig. 5.23 Fritschiella

External structure

1. The microscopic plant body exhibits a clear heterotrichous habit.

2. Four different systems of thallus can be identified on microscopic examination. These are rhizoidal system, prostrate system, primary vertical system and secondary vertical system.

3. The rhizoidal system is sub-terranean, colourless, branched and septate. It helps in the anchorage.

4. The prostrate system consists of an accumulation of cells. The chlorophyllous cells form branched filaments. The cells of the filament occasionally may be distinguished into nodal and internodal cells. The nodal cells give rise to rhizoids from their lower face and primary vertical system from their upper face.

5. The prostrate system gives rise to the primary vertical system consisting of photosynthetic branched filaments. The primary vertical system is subaerial.

6. Many delicate, chlorophyllous branches arise from the primary vertical system constituting the secondary vertical system.

7. All the cells, except those in the rhizoidal system have a paretial chloroplast. The cells are uninucleate.

Diagrams to be drawn

1. Habit sketch
2. The systems of filaments (enlarged)

Systematic position and Identification

A. 1. Chlorophyll present
2. Plant body a thallus
3. Cell wall cellulosic ....*Algae*

B. 1. Plastids present
2. Chlorophyll is the cheif pigment
3. True nucleus present
4. Photosynthetic product is starch ....*Chlorophyceae*

C. 1. Plant body exhibits heterotrichous habit
2. Vertical system ends in fine filaments ....*Chaetophorales*

D. 1. Cells uninucleate with parietal chloroplast
2. Terminal cells of the vertical system end in fine filaments.......Chaetophoraceae

E. 1. Plant is terrestrial in habitat
2. Plant body having four systems of filaments....*Fristschiella*

COLEOCHAETE

Common species : *C.scutata, C.pulvinata, C.nitellarum, C.orbricularis* etc.

Distribution : Fresh water epiphytic alga, found on weeds, ridges etc.

Study 1. External structure
2. Reproduction
3. Systematic position and identification

External structure

1. The plant exhibits a heterotrichous system but in some members, the heterotrichous system is obscure.

2. Two types of forms may be distinguished in the genus. These are discoid forms and cushioned forms.

3. In discoid forms heterotrichous systems is not evident. The branches of the thallus are laterally opposed and form a compact pseudoparenchymatous structure. Here the erect system is represented by a few setae.

4. In cushioned forms, heterotrichous system is very evident. The filaments of vertical system form a cushion which will be enveloped by mucilage.

5. Shape of the cell varies in erect and prostrate branches. In prostrate brancies cells are polygonal. In erect branches, the cells are cylindrical.

6. Cells are thin walled, uninucleate and have a parietal chloroplast with one or two pyrenoids.

7. The vegetative cells have setae or bristles. These are found only in younger cells.

Reproduction

1. Asexual reproduction by zoospores.

2. Sexual reproduction is oogamous.

3. Antheridia are borne on erect branches. Each antheridium produces a single biflagellate antherozoid. Antheridia are usually found in clusters.

4. Oogonia are flask shaped and are formed terminally on the lateral branches of the erect system. The oogonium is also called carpogonium because of its flask shaped structure.

5. Fertilization results in the formation of zygote which remains within the oogonium and becomes enveloped by the surrounding cells. This becomes reddish brown in colour and is called spermocarp.

6. Spermocarp has a long period of dormancy and germinates later giving rise to zoospores.

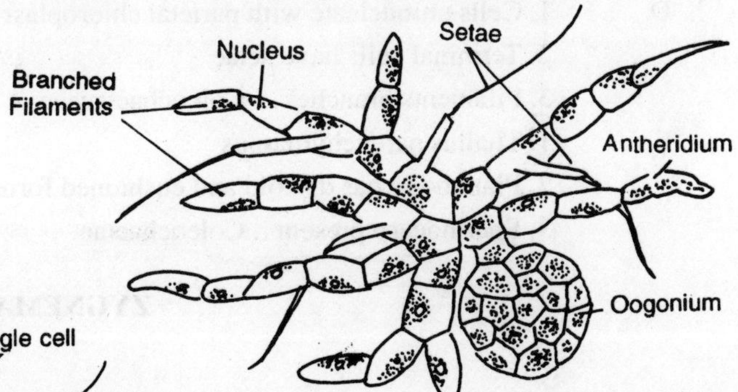

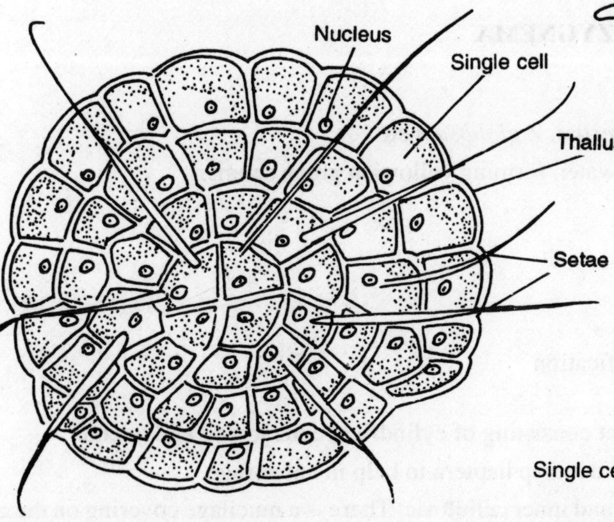

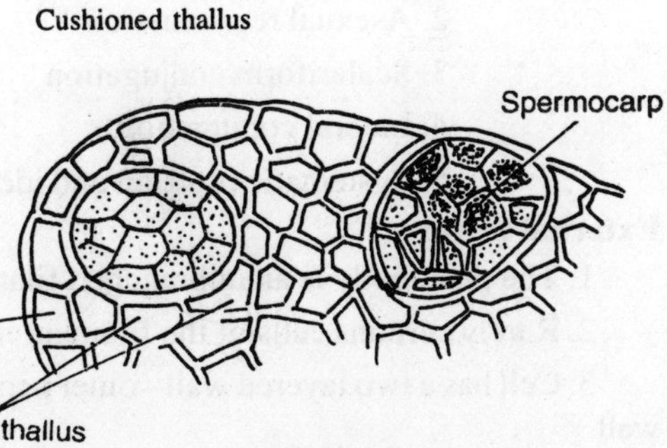

Fig. 5.24 Coleochaete; Discoid thallus (habit) Discoid thallus with spermocarp

Diagrams to be drawn

1. Habit sketch

   a. Cushioned thallus

   b. Discoid thallus

2. Antheridia

3. Oogonia and spermocarp

Systematic position and identification

A. 1. Chlorophyll present
2. Plant body a thallus
3. Cell wall cellulosic ... *Algae*

B. 1. Plastids present
2. Chlorophyll is the chief pigment
3. True nucleus present
4. Photosynthetic product is starch ... Chlorophyceae

C. 1. Plant body exhibits heterotrichous habit
2. Branches end in setae ... Chaetophorales

D. 1. Cells uninucleate with parietal chloroplast
2. Terminal cells have setae
3. Filaments branches ... Coleochaetaceae

E. 1. Thallus parenchymatous
2. Plant body has discoid and cushioned forms
3. Spermocarp present ...Coleochaetae.

ZYGNEMA

Common species : *Z.himalayensis, Z.tetrrestris, Z.globosum* etc.

Distribution : Widely distributed in fresh water, forming yellowish green scums.

Study : 1. External features
2. Asexual reproduction
3. Scalariform conjugation
4. Lateral conjugation
5. Systematic position and identification

External features

1. The plant body is an unbranched filament consisting of cylindrical cells joined end to end.

2. Rarely certains cells of the filament may develop haptera to help in anchorage.

3. Cell has a two layered wall - outer pectic and inner cellulosic. There is a mucilage covering on the cell wall.

4. Cells are uninucleate and have two stellate chloroplasts having a pyrenoid each.

Asexual reproduction

1. Fragmentation of the filament results in the breaking up of the filament into bits, each bit developing into a new individual.

2. Akinetes and Aplanospores are also formed to help in asexual reproduction.

Reproduction : sexual

1. This takes place by means of conjugation

2. Conjugation is of two types - scalarifomr and lateral.

Scalariform conjugation

1. This type of conjugation takes place between the cells of two different filaments.

2. Two filaments lie parellely and small protruberance from the two filaments grow in opposite directions and meet in the centre forming a passage called the conjugation tube.

3. Conjugation tube formed in successive cells of the two filaments gives the appearance of a ladder, hence the name Scalariform.

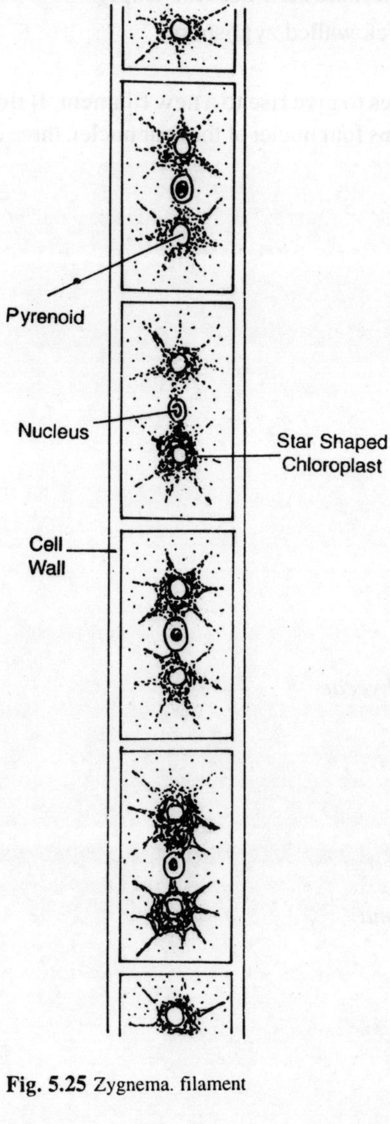

Fig. 5.25 Zygnema. filament

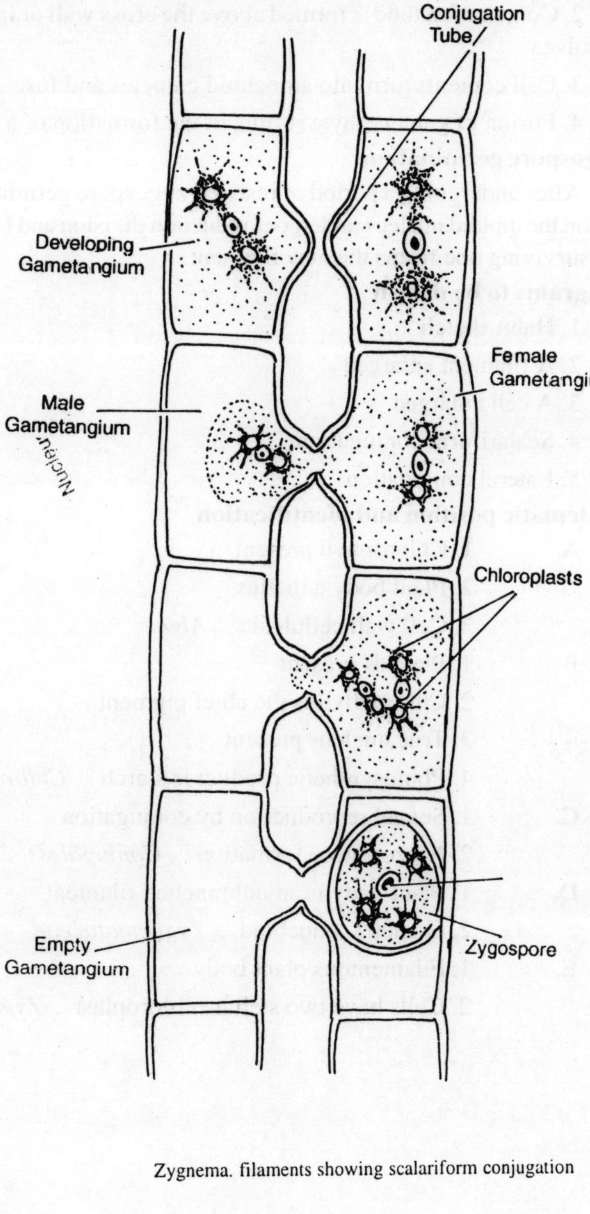

Zygnema. filaments showing scalariform conjugation

4. The contents of the cell round off and form amoeboidal gametes. The gametes are isogamous.

5. Gamete from one cell migrates to another of the opposite filament through the conjugation tube.

6. Even though gametes are morphologically isogamous the one that migrates can be called male and the other female.

7. All the cells of a filament become empty after the migrations of the gametes.

8. The gametes fuse and form a diploid zygote. It develops a very thick wall and is now called the zygospore.

Lateral conjugation

1. This type of conjugation takes place between the adjacent cells of the same filament.

2. Conjugation tube is formed above the cross wall or in some cases, the cross wall between the two cells dissolves.

3. Cell contents turn into amoeboid gametes and fuse. Alternate cells become empty.

4. Fusion of gametophyte results in the formation of a thick walled zygospore.

Zygospore germination

After undergoing a period of rest, the zygospore germinates to give rise to a new filament. Before germination the diploid nucleus undergoes reduction division and forms four nuclei of the four nuclei, three degenerat, the surviving one forms the new filament.

Diagrams to be drawn

1. Habit sketch
2. A filament enlarged
3. A cell enlarged
4. Scalariform conjugation
5. Lateral conjugation

Systematic position and identification

A. 1. Chlorophyll present
2. Plant body a thallus
3. Cell wall cellulosic ... *Algae*

B. 1. Plastids present
2. Chlorophyll is the chief pigment
3. True nucleus present
4. Photosynthetic product is starch ... *Chlorophyceae*

C. 1. Sexual reproduction by conjugation
2. No zoospore formation ... *Conjugales*

D. 1. Plant body is an unbranched filament
2. Gametes amoeboid ... *Zygnemataceae*

E. 1. Filamentous plant body
2. Cells have two stellate chloroplast ... *Zygnema*

SPIROGYRA

Common species : *S.condensata, S.jogensis, S.singularis* etc.

Distribution : Found floating in fresh water ponds and ditches, commonly called pond silk or pond scum, because of its slippery nature.

Study : 1. Study of the structure of the plant body
2. Reproduction
3. Systematic position and identification

Structure of the plant body

1. The plant body is an unbranched filament. There is no differentiation into base and apex.
2. The filament has a thick mucilaginous envelope.
3. The cells are long cylindrical and joined end to end to form the filament.
4. The cell wall is three layered - outer chitinous, middle pectic and inner cellulosic. The cross wall between the cells is replicate (pectose middle lamella having cellulose on either side).
5. The cytoplasm is peripheral with a large central vacuole. Thin cytoplasmic strands traverse across the vacuole.
6. There is a single nucleus suspended in the cytoplasm.
7. The chloroplast is ribbon shaped and is in a spiral with wavy edges. Embedded in the chloroplast are found a number of pyrenoids. Usually there will be only one spiral helix of the chloroplast. In some species two helices are found spirally in opposite direction.

Reproduction

1. Asexual reproduction by zoospore is not seen.
2. Vegetative propagation is brought about by means of fragmentation.
3. Sexual reproduction is called conjugation. The details are same as seen in *Zygnema.*
4. In *Spirogyra* also, two types of conjugation viz., scalariform and lateral are seen (for details, see Zygnema).
5. Structure and germination of zygospore are similar to what is seen in *Spirogyra.*
6. Zygospores are dark blue in colour and oval in shape. They have a thick sculputred wall.

Diagrams to be drawn

1. Habit sketch
2. A few filamentous enlarged.
3. A cell enlarged
4. Scalariform conjugation
5. Lateral conjugation
6. Zygospore germination

Systematic position and identification

A. 1. Chlorophyll present
2. Plant body a thallus
3. Cellwall collulose ... *Algae*

B. 1. Plastids present
2. Chlorophyll is the chief pigment
3. True nucleus is present
4. Photosynthetic product is starch ... *Chlorophyceae*

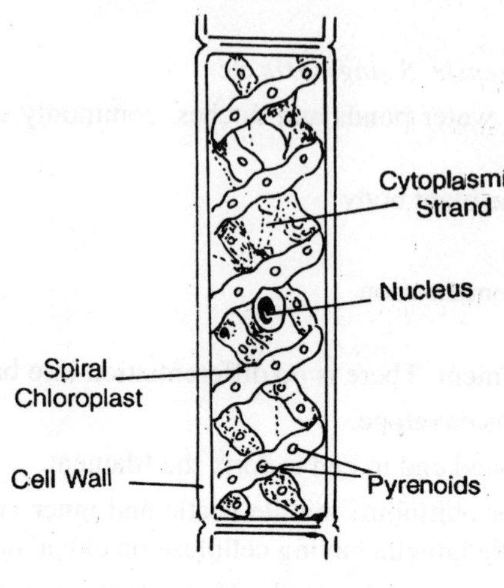

Single cell with two chloroplasts

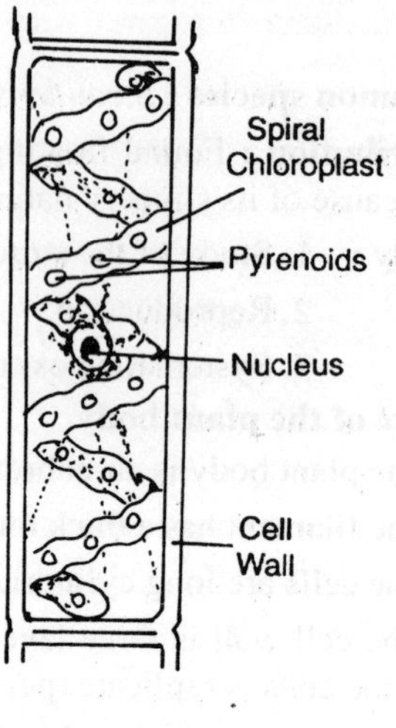

Single cell with single chloroplast

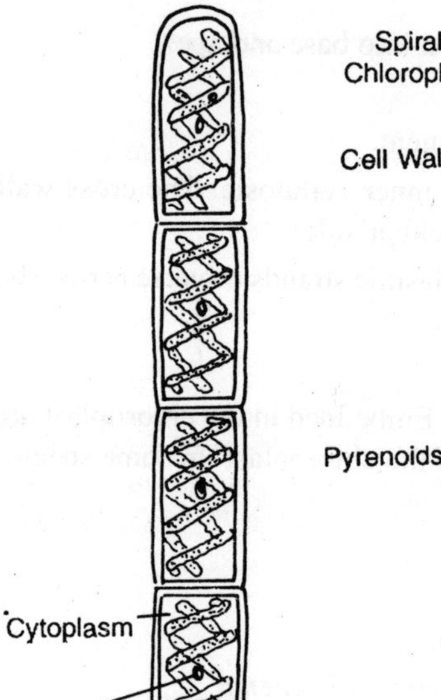

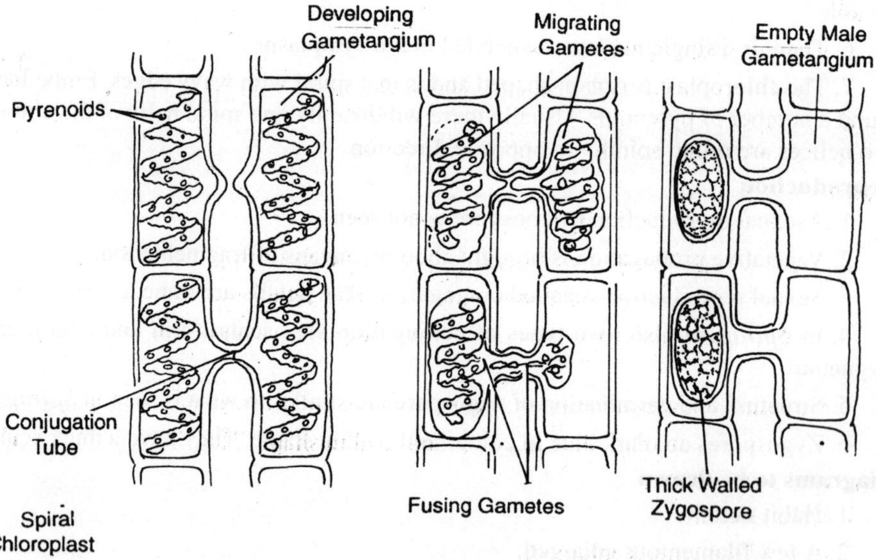

Spirogyra. Filaments showing scalariform conjugation

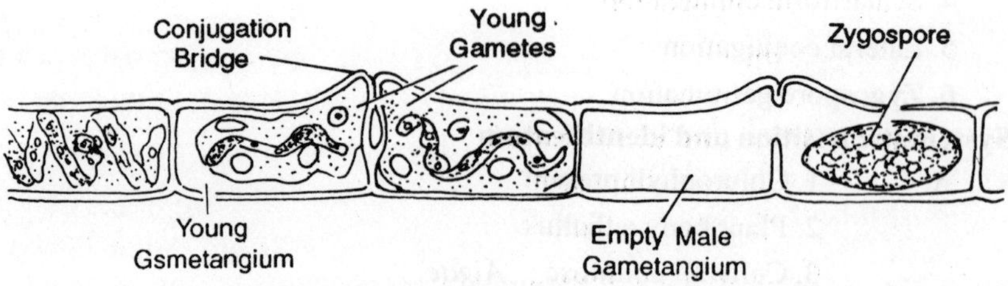

Fig. 5.26 Spirogyra. filament

Spirogyra filament showing lateral conjugation

C. 1. Sexual reproduction by conjugation
2. There is no zoospore formation ... *Conjugales*

D. 1. Plant body is an unbranched filament.
2. Gametes amoeboid ... *Zygnemataceae*

E. 1. Cell wall has thick mucilage envelope.
2. Cross walls are replicate
3. Chloroplast spiral ... *Spirogyra.*

MOUGEOTIA
(Agardh 1824)

Common species : *M.capucina, M.boodlai* etc.

Distribution : Found in ponds, semipermanent pools, lakes, slow flowing streams as bright green floating masses.

Study : 1. Structure of the plant body
2. Reproduction
3. Systematic position and identification

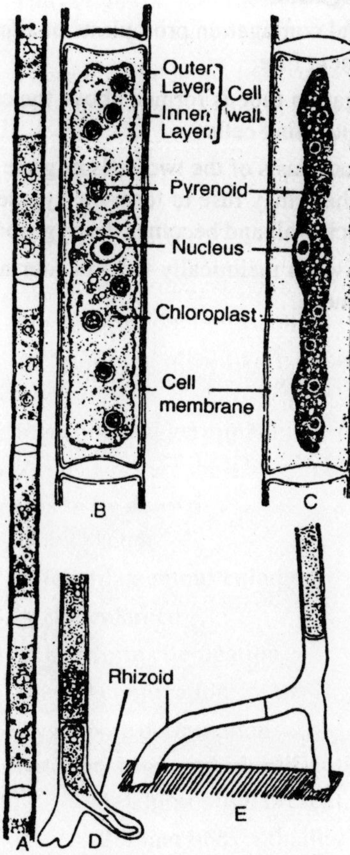

Fig. 5.27 Mougeotia. A, portion of a filament; B, face view of the cell showing structure, and the chloroplast in surface view; C, chloroplast in profile position; D-E, basal cell showing rhizoid formation (D-E, after Salisbury, B-C, based on Morris)

Structure of the plant body

1. The plant body is an unbranched filament consisting of cylindrical cells joined end to end.

2. Very rarely the basal cell may develop rhizoidal outgrowths to help in anchorage.

3. Slime layer is found deposited on the cell wall. The cell wall is two layered. It has an outer pectic and inner cellulosic layer.

4. The cross wall between the cell splits into two circular discs at a very early stage.

5. The chloroplast is axile and plate like, almost filling the cell cavity. It may have 3-4 pyrenoids.

6. The cytoplasm is peripheral with a central vacuole.

7. The single nucleus is found in the centre of the cell.

Reproduction

1. Vegetative propagation takes place by fragmentation. The filament breaks up into few celled segments. Each segment develops into a new individual.

2. Akinetes are found in some species to help in asexual reproduction. The akenites have a thick wall and rich in neutrients.

3. Occasionally aplanospores may also be formed as in some species.

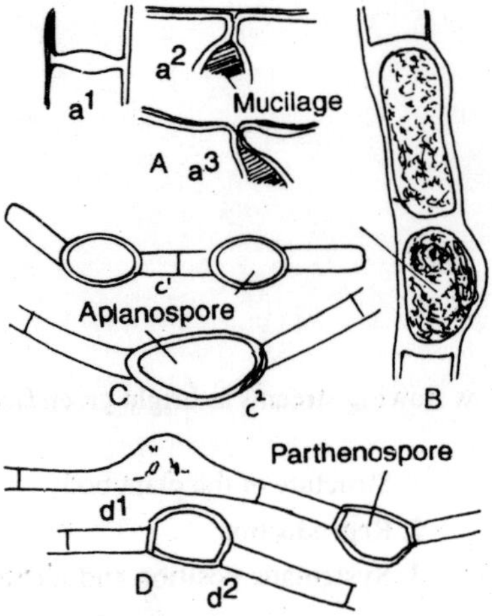

Fig. 5.28 Mougeotia. Asexual reproduction. A. mechanism of cell disjunction; B. akinetes of M.capucina; C.aplanospore formation in M.boodlai; D. parthenospore formation in Gonatonema ventricosum (A. after Fritsch; B. after Borge; C and D.

4. Sexual reproduction is by means of conjugation. It is of two types viz., scalariform and lateral.

Scalariform conjugation

1. This type of conjugation takes place between the opposite cells of two different filaments.
2. Two papillae arise from the opposite cells at corresponding points of conjugating filaments.
3. The two papillae growing in opposite directions meet in the centre establishing a conjugation tube or canal.
4. The contents of two cells form amoeboid gametes. The two gametes migrate into the conjugation tube which enlarges in diameter.
5. Gametes are isogamous. They fuse to form a zygote in the conjugation tube. The zygote secretes a thick wall around itself and will now be called zygospore.

Lateral conjugation

1. In lateral conjugation protoplasts of adjacent cells of the same filament fuse.
2. Conjugation tube is formed above the cross wall between the conjugating cells.
3. The protoplasts of the two cells migrate to the conjugation tube where they fuse to form the zygote. The zygote develops a thick wall and becomes the zygospore.
4. The zygospore undergoes a period of rest and then divides meiotically to form four nuclei. Out of these three degenerate. The surviving one forms the new filament.

Diagrams to be drawn

1. Habit sketch
2. A filament enlarged (showing Akinetes)
3. A cell enlarged
4. Stages in scariform conjugation
5. Stages in lateral conjugation

Systematic position and identification

A. 1. Chlorophyll present
2. Plant body a thallus
3. Cell wall cellulosic ... *Algae*

B. 1. Plastids present
2. Chlorophyll is the chief pigment
3. True nucleus present ... *Chlorophyceae*
4. Photosynthetic product is starch.

C. 1. No motile reproductive cells
2. Sexual reproduction by conjugation... *Conjugales*

D. 1. Plant body an unbranched filament
2. Gametes amoeboid ... *Zygnemataceae*

E. 1. Gametes strictly isogamous

2. Fusion of gametes takes place in the conjugation tube ... *Mougeotia*

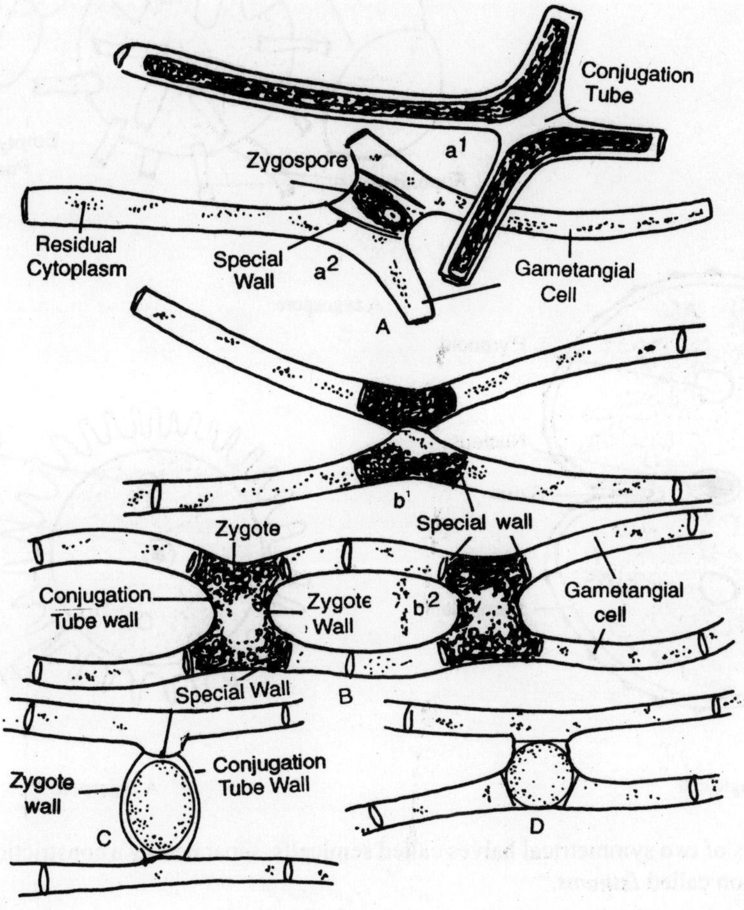

Fig. 5.29 Mougeotia.

COSMARIUM

Common species : *C.biretum, C.leotytis, C.turgidum* etc.

Distribution : Found in pools, ponds and stagnant waters rich in decaying organic matter.

Study : 1. Structure of the plant body

2. Asexual reproduction

3. Sexual reproduction

4. Systematic position and identification

Structure of the plant body

1. It is a fresh water free floating alga, commonly called Placoderm desmid or constricted desmid.

2. The plant body is unicellular, the cell one and half times longer than broad. Sometimes the cells join together to form chains or amorphous colonies.

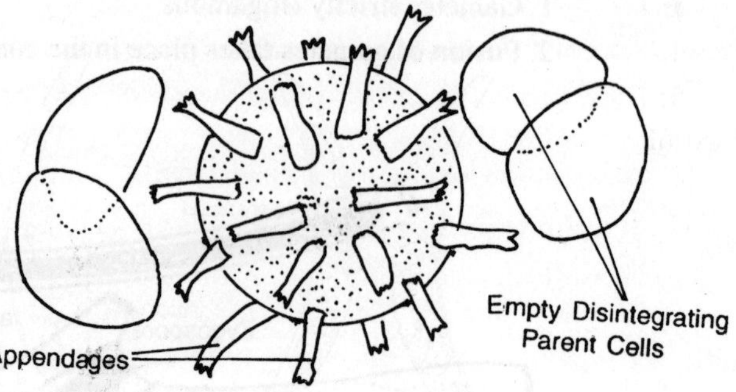

A zygospore

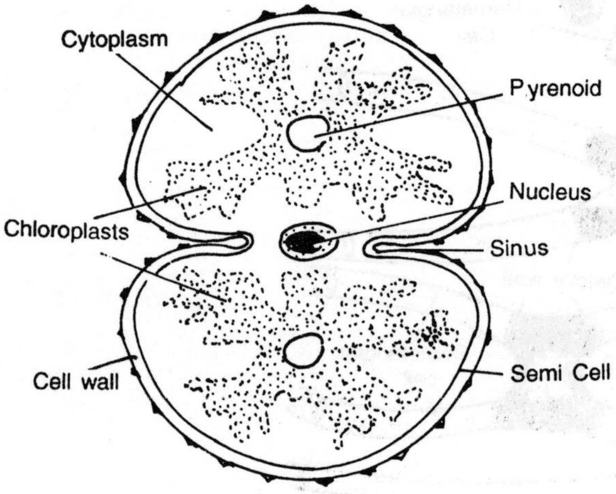

Fig. 5.30 Cosmarium. A single cell

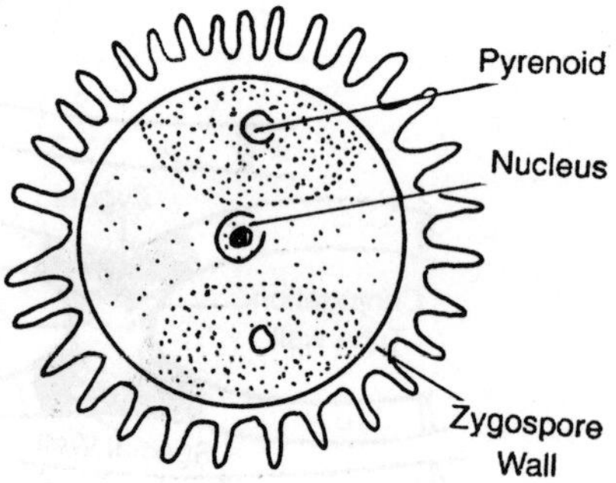

A zygospore

3. Each cell consists of two symmetrical halves called semicells, separated by a constriction (*sinus*) and joined by a narrow region called *Isthums*.

4. Each semicell may be oval, circular or elliptical in outline.

5. The cell wall may be smooth without any architectural pattern, or the ornamental wall may have three concentric layers.

6. The outer layer is gelatinous, middle layer is cellulosic impregnated with pectic compounds and innermost is cellulosic.

7. The two inner layers of the wall are porous, through which the mucilage exudes out resulting in the jerking movement.

8. Each semicell has an axial chloroplast with radiating arms. In some each semicell may have four parietal chloroplasts.

9. Each chloroplast may have one or two pyrenoids.

10. The cells are uninucleate, the nucleus being situated at the isthums region.

Asexual reproduction

1. Asexual reproduction is by cell division.

2. At the time of cell division, the nucleus mitotically divides into two. The isthums enlarges and divided into two by means of a cross wall.

3. The two daughter cells get separated and each one develops a new semicell at the isthmus region.

4. The chloroplast also divides, but the pyrenoid may be divided or a new pyrenoid may be synthesised.

5. Thus, each newly formed cell consists of one parental semicell and one new semicell.

Sexual reproduction

1. Sexual reproduction is very rare and isogamous when it takes place. It is called conjugation.

2. At the time of conjugation two cells come nearer and get encased by a common mucilage envelope.

3. The cell contents of each cell get metamorphosed into an amoeboid gamete.

4. The cell wall splits at the isthmus. The gametes come out meet at the centre freely or in the conjugation tube.

5. They fuse to form a diploid zygospore. The zygospore has a thick warty wall.

6. The remnants of the four semicells remain attached to the zygospore for some time.

7. The zygospore is brown in colour and germinates after a brief period of dormancy.

8. Nuclear fusion is effected before germination. The zygospore wall ruptures releasing two new individuals of *Cosmarium.*

Diagram to be drawn

1. Structure of the cell

2. Conjugation stages

Systematic position and identification

A. 1. Chlorophyll present
2. Plant body a thallus
3. Cell wall cellulosic ... *Algae*

B. 1. Plastids present
2. Chlorophyll is the chief pigment
3. True nucleus present
4. Photosynthetic produce is starch ... *Chlorophyceae*

C. 1. Motile reproductive cells absent
2. Sexual reproduction by conjugation ... *Conjugales*

D. 1. Plant body unicellular.
2. The cell consists of two halves called semicells ... *Desmidiaceae.*

E. 1. Plant body is unicellular having two semi cells.
2. Asexual reproduction by cell division.
3. A deep construction is present between the two semi cells .. *Cosmarium.*

CLOSTERIUM

Common species : *C.parvulum, C.colosporum* etc.

Distribution : Found widely distributed in stagnant water pools, ponds rich in decaying organic matter.

Study : 1. Cell structure
2. Reproduction
3. Systematic position and identification

Cell structure

1. It is commonly called Placeoderm desmid or unconstricted desmid.

2. The plant body is made up of a single cell. The cell is elongated and spindle like with tapering apices.

3. Majority of the species have a straight plant body, but some of them are crescent shaped.

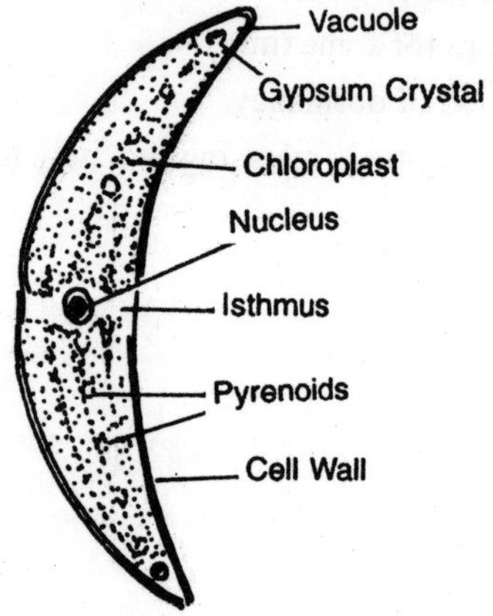

Fig. 5.31 Closterium. A single cell

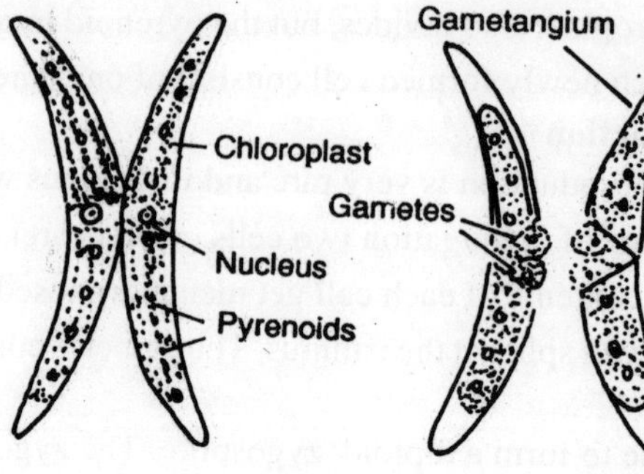

Reproductive cells showing conjugation

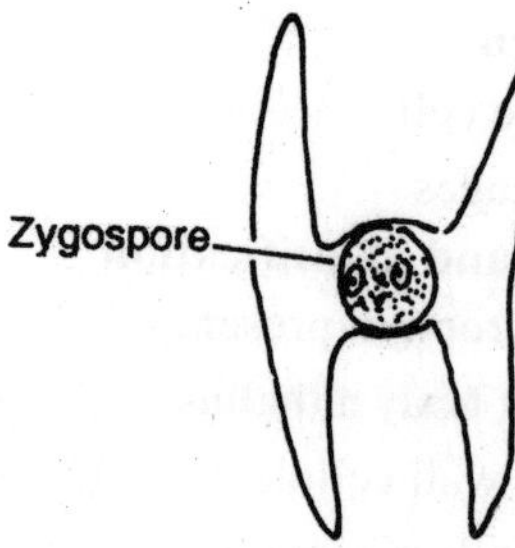

Single zygospore

4. The cell is circular in a cross section.

5. The plant body consists of two perfectly symmetrical halves called semicells.

6. A narrow protoplasmic bridge called isthmus connects the two semicells. The isthums is not constricted as in Cosmarium.

7. The cell is uninucleate, the nucleus being placed in the isthmus region. The nucleus has a conspicuous nucleolus.

8. At either end of the cell is a conspicuous vacuole containing calcium sulphate crystals showing Brownian movement.

9. The crystals do not perform any physiological function and are probably products of excretion.

10. The cell wall is three layered (see *Cosmarium*).

11. Grooves are present on the cell wall having minute pores arranged in rows. Mucilage excretes through these pores causing movement of the plant body.

12. Each semicell has an axial curved cone like chloroplast with ribs. The chloroplast has a number of pyrenoids.

Reproduction

1. Asexual reproduction is brought about by means of cell division.

2. At the time of cell division a constriction appears in the centre.

3. The nucleus divides followed by the cell. The two semicells get separated.

4. These separate and each semicell forms a new half at the broad end.

5. Each newly formed desmid has an old semicell of the parent and a new semicell formed by it.

6. Sexual reproduction takes place by conjugation and is of the isogamous type except in *C.parvulum.*

7. During conjugation two desmid cells approach each other laterally and become encased by a common mucilage envelope.

8. After arranging themselves parallely, the cell wall of the desmids splits near the middle while the inner slender layers grow in opposite directions as protruberances, meet in the centre forming a conjugation tube.

9. The protoplast of each cell gets transformed into an amoeboid gamete.

10. The gametes move into the canal and fuse forming a diploid zygospore.

11. In many species, no conjugation tube is formed. The gametes come out of the cell and fuse to form a zygospore.

12. The wall of the zygospore is thick and smooth or ornamented.

13. The zygospore after a period of rest, germinates to give rise to two individuals of *Closterium.*

Diagrams to be drawn

1. Cell structure
2. Cell division
3. Stages in conjugation.

Systematic position and identification

A. 1. Chlorophyll present
   2. Plant body a thallus
   3. Cell wall cellulosic ... *Algae*

B. 1. Plastids present
   2. Chlorophyll is the chief pigment
   3. True nucleus present
   4. Photosynthetic product is starch ... *Chlorophyceae*

C. 1. Motile reproductive cells absent
   2. Sexual reproduction by means of conjugation ... *Conjugales*

D. 1. Plant body unicellular
   2. The cell consists of two halves called semicells ... *Desmidiaceae*

E. 1. The unicellular plant body has two halves called semicells
   2. The cells elongated with no constriction at the isthmus region ... *Closterium*

CAULERPA

Common species : *C.prolifera, C.olivieri, C.fastigata, C.taxifolia, C.racemosa,C.scalpelliformis, C.peltata, C.cupressoides, C.chemntizia* etc.

Distribution : All the species are exclusively marine being distributed in the quiet shallow waters of tropical seas. Some like *C.prolifera* and *C.ollivieri* occur in Mediterranean seas.

Study : 1. External features of the thallus
2. Transverse section of the stolon
3. Reproduction

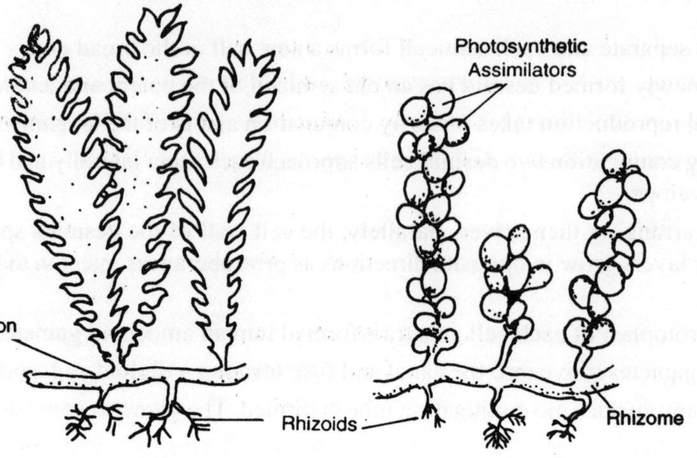

A. Caulerpa scalpelliformis B. Caulerpa racemosa

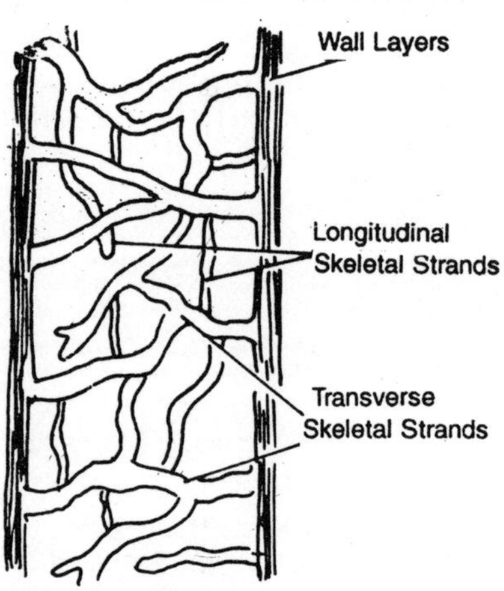

L.S. of assimilator (diagrammatic)

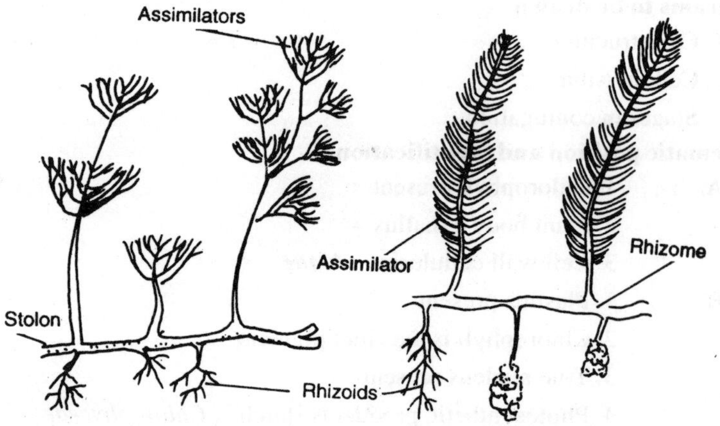

C. Caulerpa verticellata D. Caulerpa taxifolia

Habit of Plants

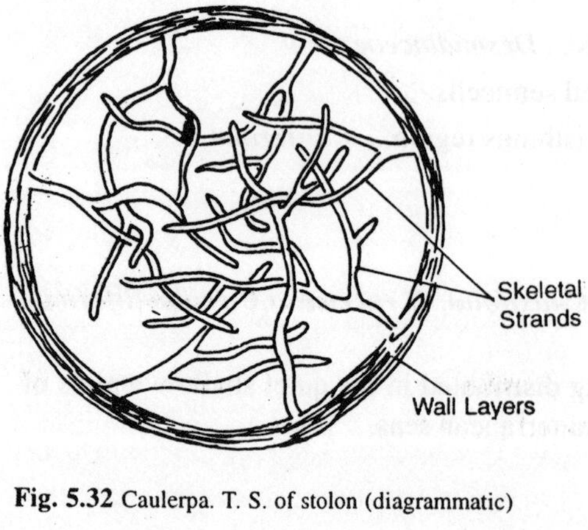

Fig. 5.32 Caulerpa. T. S. of stolon (diagrammatic)

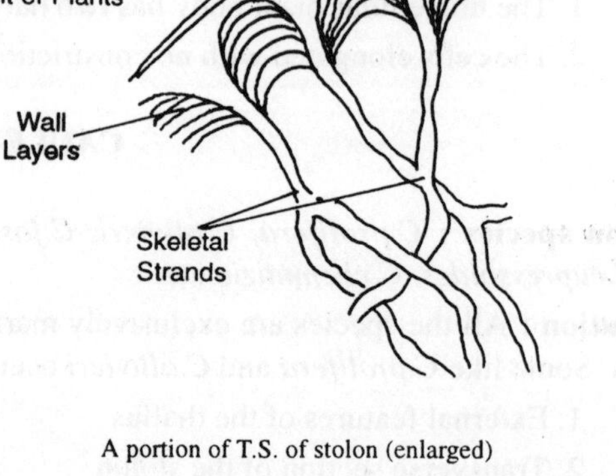

A portion of T.S. of stolon (enlarged)

External features of the thallus

1. The thallus is a very elaborate with a great amount of external complexity resembling a vascular plant.
2. The plant body however is a coenocyte; it is made up of a single cell.
3. The plant body may be distinguished into three parts viz.,
   a. Horizontal stolon
   b. Rhizoidal structures
   c. Photosynthetic assimilators.
4. The stolon is horizontal. It is a branched rhizone like structure. It may be as long as one metre in some species.
5. The stolon on its under surface bears a number of branched rhizoids. These help in anchorage and absorption.
6. On the upper surface the stolon bears a number of green leaf like photosynthetic assimilators.
7. The photosynthetic assimilators show a great variety in their shape structure etc. In fact the species of *Caulerpa* are identified on the basis of assimilators, as the stolon and rhizome are same in all the species.
8. The species of *Caulerpa* are named after the plants and their structures which they resemble.
9. *C.scalpelliformis*, has flat assimilatiors with margins dissected (not deeply) into tooth like projections.
10. *C.peltala* has assimilators bearing a number of peltate discs.
11. *C.taxifolia* has assimilatiors much like in *C.scalpelliformis*, but dissections are very deep and the tooth like projections are narrow.
12. *C.verticellata* has assimilators in the form of cylindrical whorled structures.
13. *C.*chemnitzia has assimilators with small tubular, trumpet like lateral appendages.
14. *C.hypnoides* has moss like bushy assimilators.
15. *C.fastigata* has irregularly branched thread like assimilators.

Anatomy of the stolon

1. A cross section of the thallus is somewhat circular in outline.
2. There are no cross walls. The cell wall is made up of a mixture of pectin and cellulose.
3. Internal to the cell wall is the peripheral cytoplasm containing numerous discoid chloroplasts (no pyrenoids) and nuclei.
4. There is a large vacuole in the centre. It conists of colloids and is trasversed by a number of skeletal strands called trabeculae.
5. The trabeculae are found in all parts of the plant body; they are exceptionally well developed in stolon.
6.. The trabeculae develop from the wall and are at right angles to the surface.
7. The trabeculae in the stolon are closely interwoven and offer mechanical support.

Reproduction

1. Vegetative propagation takes place by means of death and decay of the stolon. The branches get separated and develop into individual plants.
2. Sexual reproduction is anisogamous and species are heterothallic.
3. The assimilators act as gametangia and produce gametes which come out through the extrusion papillae.
4. The gametes fuse to form the zygote which develops into a new individual.

Diagrams to be drawn

1. Habit sketch of the species available in the laboratory.
2. Transverse section of the stolon.
3. Assimilators with extrusion papillae.

Systematic position and identification

A. 1. Chlorophyll present
2. Plant body a thallus
3. Cell wall cellulosic ... *Algae*

B. 1. Plastids present
2. Chlorophyll is the chief pigment
3. True nucleus present
4. Photosynthetic product is starch ...*Chlorophyceae*

C. 1. Thallus coenocytic, tubular with no cross walls
2. Thallus is macroscopic and looks complex
3. Chloroplasts discoid.
4. Cell has a large central vacuole...*Siphonales*

D. 1. Plants are marine in habitat.
2. Thallus has a horizontal stolon, with rhizoids below and assimilations above ...*Caulerpaceae.*

E. 1. Plant body coenocytic
2. Assimilatiors show great variety.
3. Extrusion papillae develop on assimilatiors...*Caulerpa*

The various speices of *Caulerpa* given have to be identified on the basis of the nature of the assimilators.

CODIUM

Common species : *C.fragile*

Distribution : Widely distributed in warm tropical seas; grows on rocks between the tide lines.

Study : 1. Structure of the thallus
2. Internal structure
3. Reproduction
4. Systematic position and identification

Structure of the thallus

1. The plant body is a much branched, soft thallus, tubular or cylindrical in appearane.
2. The branches are interwoven into a prostrate cushion like or globose mass.
3. The prostrate system bears a number of erect cylindrical shoots which are dichotomously branched.
4. The plant body is spongy to touch and macroscopic.
5. The plant body is anchored with the help of discs or rhizoids.

Internal structure

1. A transverse section of the branch of a thallus reveals an aerial core of densely interwoven colourless filaments.
2. Lateral branchlets called utricles arise from the aerial core of filaments.
3. The utricles lie in a palisade like layer surrounding the central aerial core.
4. Each utricle has a thick peripheral layer of cytoplasm internal to the cell wall surrounding a central vacuole.

5. The cytoplasm has many nuclei and discoid chloroplasts without pyrenoids; most of the chloroplasts are found at the tip of the utricles.

6. Filaments of the axial core develop annular ingrowths here and there especially at the bases of the utricles.

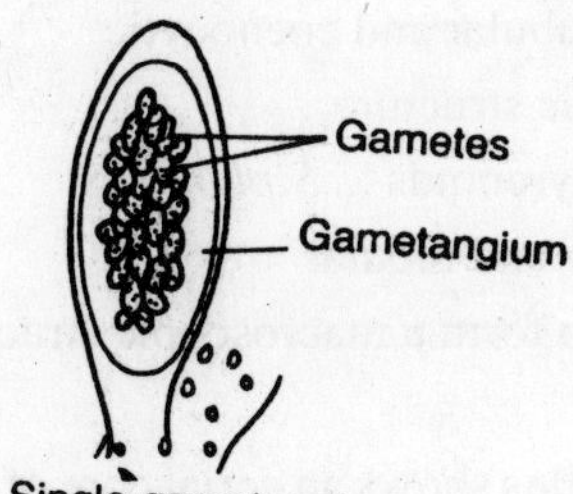

Single gametangium

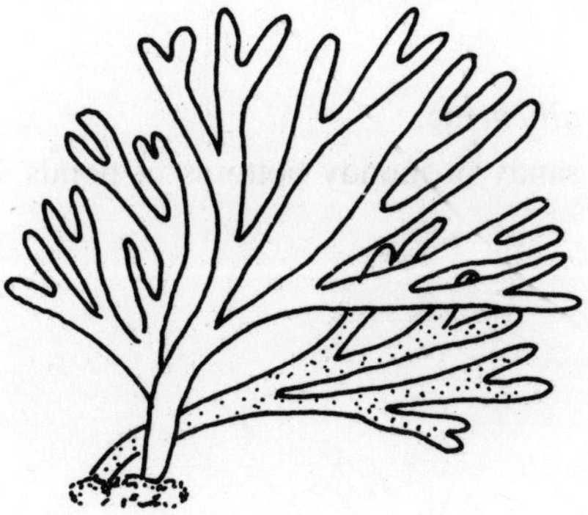

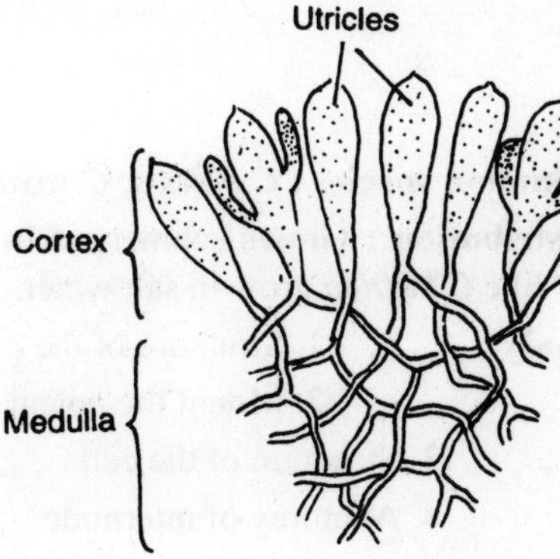

Fig. 5.33 Codium. Thallus showing habit

T. S. of thallus (A portion)

Introduction

1. Asexual reproduction is by fragmentation of the plant body.

2. Sexual reproduction is anisogamous. The plants are heterothallic.

3. Gametangia are formed on one side of the utricles.

4. Both micro (male) and macro (female) gametes are biflagellate.

5. Male and female gametangia may be distinguished by their distinct colouration. Male gametangia are golden brown in colour while female gametngia are dark green.

6. The gametes fuse to form a zygote, which directly give rise to a new plant.

Diagrams to be drawn

1. Habit sketch

2. Cross section of the thallus

3. Gametangia arising from the utricle

Systematic position and identification

A. 1. Chlorophyll present

2. Plant body a thallus

3. Cell wall cellulosic ... *Algae*

B. 1. Plastids present

2. Chlorophyll is the chief pigment

3. True nucleus present

4. Photosynthetic product is starch ... *Chlorophyceae*

C. 1. Thallus single celled, tubular and coenocytic.

2. Thallus is a macroscopic structure

3. Chloroplasts have no pyrenoids ... *Siphonales*

D. 1. Plant body soft, spongy and tubular

2. Branches interwoven to form a macroscopic structure ... *Codiaceae.*

E. 1. Tubular plant body

2. Cross section of the thallus shows an aerial core of filaments and a palisade layer of utricles.

3. Gametangia arise from utricles ... *Codium.*

CHARA

Common species : *C.baltica, C.roxallina, C.zeylanica, C.wallichi* etc.

Distribution : Grows submerged in fresh water and upon sandy or muddy bottoms of ponds. Some species like *C.baltica* grow in salt water.

Study : 1. Structure of the plant body

2. Mount the apical cell

3. Structure of the cell

4. Anatomy of internode

5. Asexual reproduction

6. Structure of Globule (tease open and study)

7. Structure of Nucule (tease open and study)

8. Systematic position and identification

Structure of plant body

1. The plant body is commonly called stone wort or aquatic horse tail because of its superficial resemblances to *Equisetum.*

2. The plant body has a central axis, 20-40 cms long and differentiated into nodes and internodes.

3. The axis is attached to the substratum with the help of rhizoids, which arise either at the base or from lower nodes.

4. The rhizoids are uniseriate and branched.

5. The main axis is branched profusely and the branches are of two types viz., branches of limited growth and branches of unlimited fruit.

6. The branches of limited growth, also called 'leaves' arise in whrols of 3-15 at the nodal region. They are small 5-15 internodes long and green in colour. Very often short, spiny, secondary laterals are produced from the leaves.

7. The branches of unlimited growth resemble the main axis in having nodes and internodes. They are produced at the nodes. The branches of unlimited growth at their nodal region bear the leaves.

Structure of the apical cell

1. The growth of the main axis as well as the branches takes place with the help of a single dome shaped apical cell.

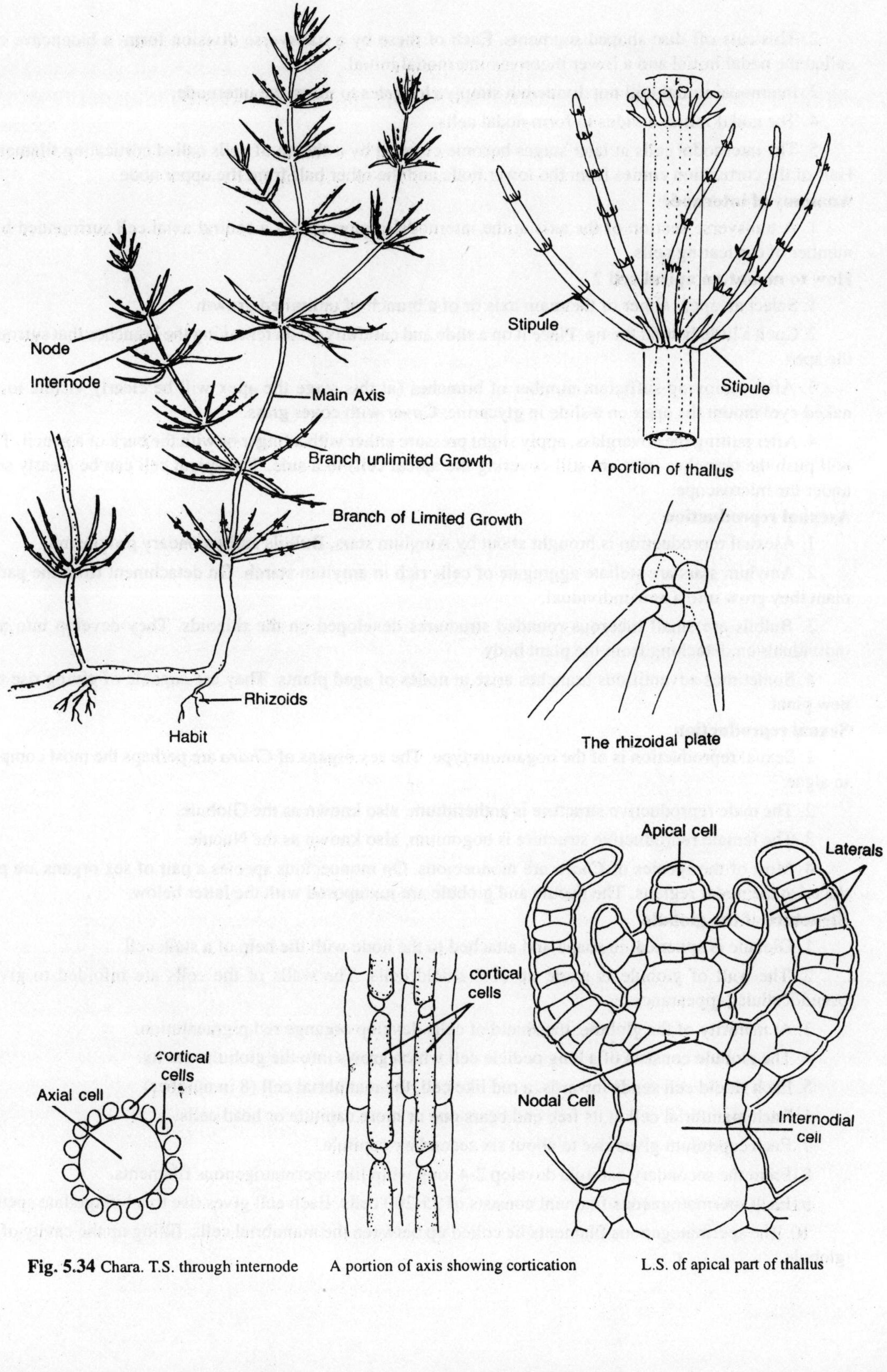

Fig. 5.34 Chara. T.S. through internode   A portion of axis showing cortication   L.S. of apical part of thallus

2. This cuts off disc shaped segments. Each of these by a transverse division form a biconcave cell called the nodal initial and a lower biconvex internodal initial.

3. Internodal initial will not divide but simply elongates to form the internode.

4. The nodal initial divides to form nodal cells.

5. The internodal cells at later stages become covered by a sheath of cells called corticating filaments. Half of the cortication comes from the lower node and the other half from the upper node.

Anatomy of internode

1. A transverse section of the axis at the internodal region shows a central axial cell surrounded by a number of corticating cells.

How to mount an apical cell ?

1. Select an apex, either of the main axis or of a branch of unlimited growth.

2. Cut it a little behind the tip. Place it on a slide and carefully go on removing the branches that surround the apex.

3. After removing sufficient number of branches (at this stage the apex will be clearly visible to the naked eye) mount the apex on a slide in glycerine. Cover with cover glass.

4. After putting the coverglass, apply slight pressure either with a finger or with the back of a pencil. This will push the branches (that are still covering the apical cell) to a side. The apical cell can be clearly seen under the microscope.

Asexual reproduction

1. Asexual reproduction is brought about by Amylum stars, Bulbils and secondary protonema.

2. Amylum stars are stellate aggregate of cells rich in amylum starch. On detachment from the parent plant they grow into a new individual.

3. Bulbils are small tuberous rounded structures developed on the rhizoids. They develop into new individuals on detaching from the plant body.

4. Sometimes adventitious branches arise at nodes of aged plants. They are capable of giving rise to a new plant.

Sexual reproduction

1. Sexual reproduction is of the oogamous type. The sex organs of *Chara* are perhaps the most complex in algae.

2. The male reproductive structure is antheridium, also known as the Globule.

3. The female reproductive structure is oogonium, also known as the Nucule.

4. Most of the species of *Chara* are monoecious. On monoecious species a pair of sex organs are produced at the nodal regions. The nucule and globule are juxtaposed with the latter below.

Structure of the globule

1. Globule is spherical in shape and attached to the node with the help of a stalk cell.

2. The wall of globule is made up of 8 shield cells. The walls of the cells are infolded to give a pseudocellular appearance.

3. At maturity of the globule, the shield of cells develop organge red pigmentation.

4. The globule consists of a long pedicle cell which grows into the globular cavity.

5. Each shield cell sends inwards, a rod like cell, the manubrial cell (8 in number)

6. Each manubrial cell at its free end bears one or more capitula or head cells.

7. Each capitulum gives rise to about six secondary capitula.

8. From the secondary capitula develop 2-4 long whip like spermatogenous filaments.

9. Each spermatogneous filament consists of 25-200 cells. Each cell gives rise to a biflagellate sperm.

10. The spermatogenous filaments lie colied up between the manubrial cells, filling up the cavity of the globule.

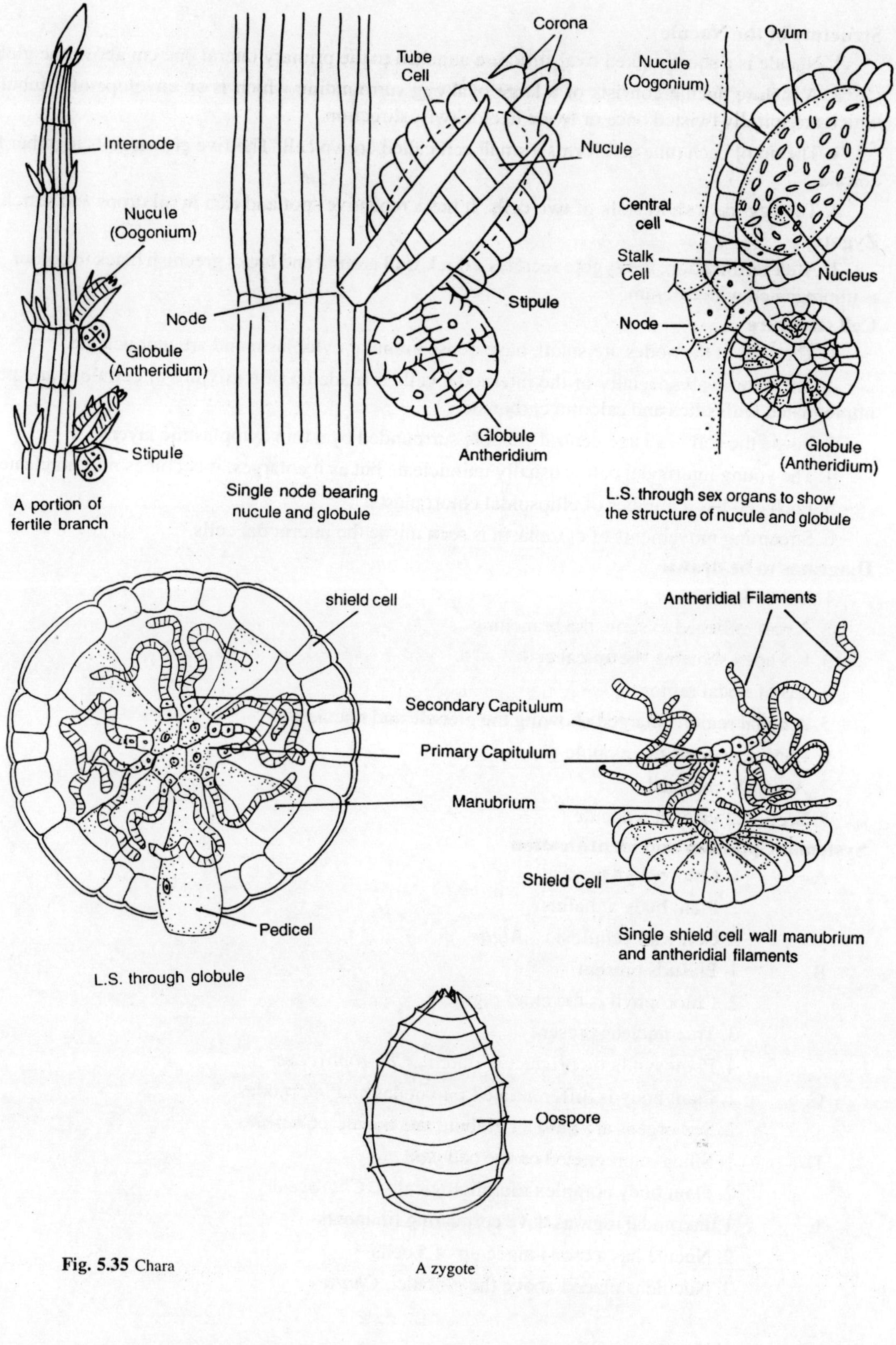

Fig. 5.35 Chara A zygote

Structure of the Nucule

1. Nucule is a short stalked oval structure attached to the primary lateral one cm above the globule.

2. A mature nucule consists of a large oval egg surrounding which is an envelope of 5 tubular cells, which are spirally twisted once or twice in clockwise direction.

3. The tip of each tube cell forms a small cell called crown cell. The five crown cells together form the corona.

4. The egg has a short stalk of two cells. It has a receptive spot and rich in oil drops and starch grains.

Zygote

1. After fertilization, the zygote secretes a thick wall around and looks greenish black in colour. The wall is impregnated with calcium.

Cell structure

1. The cells at the nodes are small, have dense granular cytoplasm and are uninucleate.

2. The outer wall specially of the internodal cells is made up of a mixture of cellulose and pectin and impregnated with silica and calcium carbonate.

3. Inside the cell is a large central vacuole surrounded by a thin cytoplasmic layer.

4. The young internodal cell is usually uninucleate but as it enlarges, it becomes multinucleate.

5. Each cell has a number of ellipsoidal chloroplasts.

6. Streaming movements of cytoplasm is seen inisde the internodal cells.

Diagrams to be drawn

1. Habit sketch
2. A part enlarged to show the branching
3. L.S apex showing the apical cell
4. T.S of nodal region
5. A nodal region enlarged showing the globule and nucule.
6. Long section of the globule
7. A nucule enlarged
8. Sectional view of a nucule.

Systematic position and identification

A. 1. Chlorophyll present
2. Plant body a thallus
3. Cell wall cellulosic ...*Algae*

B. 1. Plastids present
2. Chlorophyll is the chief pigment
3. True nucleus present
4. Photosynthetic product is starch...*Chlorophyceae*

C. 1. Plant body is differentiated into nodes and internodes.
2. Sex organs are unique; globule and nucule...*Charales*

D. 1. Silica impregnated on the cell wall
2. Plant body complex morphologically...*Characeae*

E. 1.Internodal regions have corticating filaments
2. Nucule has a crona made up of 5 cells
3. Nucule is placed above the globule...*Chara*

NITELLA
(Agardh 1824)

Common species : *N.cornia, N.opara, N.gracilis*

Distribution : Found in fresh water pools, ponds etc., in comparatively deep waters.

Study : 1. External morphology
2. Structure of a cell
3. Structure of sex organs
4. Systematic position and identification

External morphology

1. The plant body looks morphologically complex. It is macroscopic.

2. The plant looks like the pteridophyte, *Equisetum* plant hence called aquatic horse tail.

3. The plant has basically 3 parts:
   a. Main axis
   b. Rhizoids
   c. Branches

4. The main axis is divided into nodes and internodes. Nodes are multicellular while internodes are unicellular being made up of an elongated internodal cell. There is no cortication (see *Chara*) of the internodal cell.

5. Rhizoids help in the anchorage of the plant body to the substratum. They arise from the lowest node, branched and multicellular with oblique septa. A quadrant rhizoidal plate is present, from which arise the rhizoids.

6. The branhes arise at the nodal region of the main axis. They are of two types viz., branches of limited growth and branches of unlimited growth.

7. The branches of limited growth or short laterals are 5-10 internodes long and are borne in a whorl. At their nodal regions, the short laterals, bear the secondary short laterals, which are unicellular and hair like. The secondary short laterals are also called stipules.

8. The branches of unlimited growth, also known as long laterals, are produced at the nodal regions in pseudo-axillary positions. They resemble the main axis in all respects.

Structure of the cell

1. The cell at the internode which can be considered a typical cell has a cellulose wall.

2. Within the cell, there is a large central vacuole.

3. The cytoplasm is peripheral and encloses a single nucleus and many discoid chloroplasts. The chloroplasts do not have pyrenoids.

Reproduction

1. Sexual reproduction is oogamous. The species of *Nitella* are either monoecious or dioecious. Just as in *Chara*, here also, the antheridium is known as the globule and, the oogonium is called the nucule.

2. In monoecious species, the nucule and globule are produced at the nodal regions of secondary short laterals.

3. At every node, usually the nucule and globule are placed one above the other with the nucule below and globule above.

Structure of the globule

1. The globule is orange red in colour and is spherical in shape.

2. It is attached to the plant at the nodal region with the help of a long stalk cell or pedicel cell.

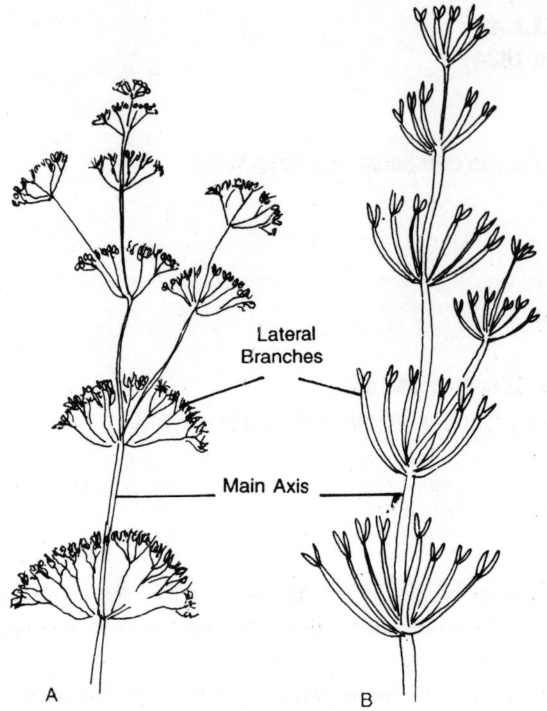

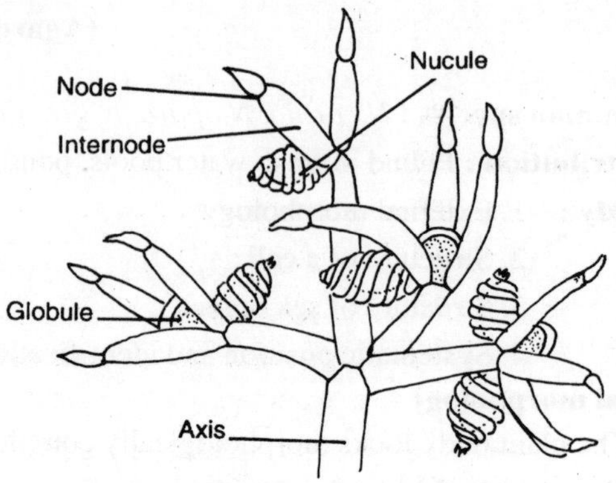

A portion of thallus showing sex organs

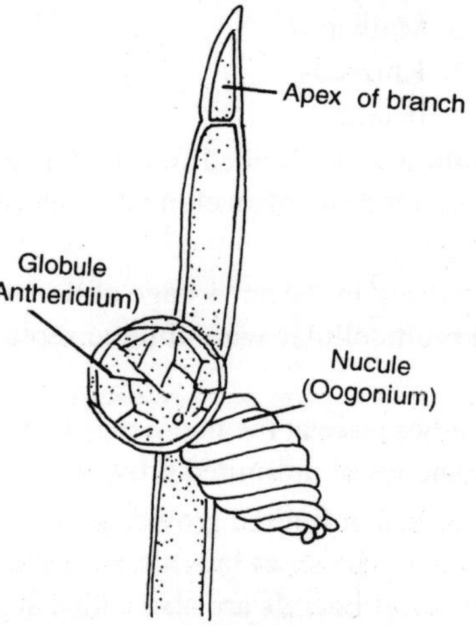

A node with Globule and nucule

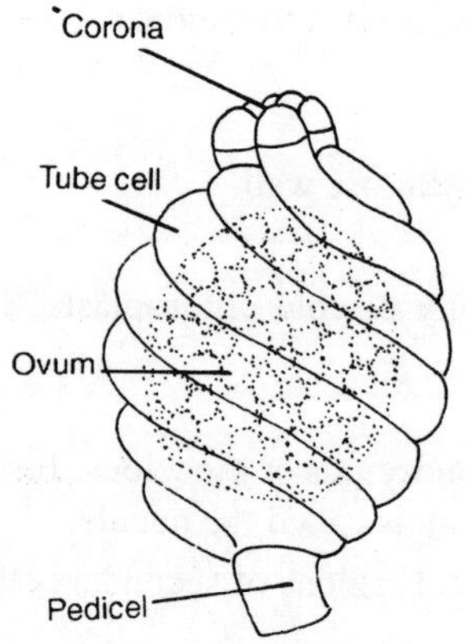

A nucule

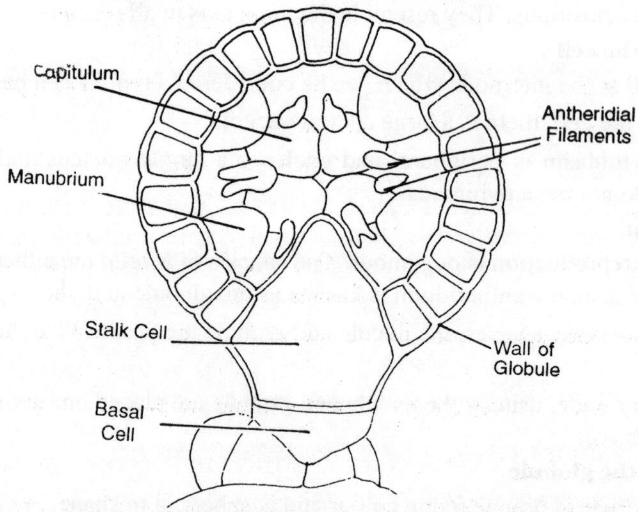

Fig. 5.36 Nitella. A nucule

Globule

3. The wall of the globule is made up of eight shield cells.

4. Each shield cell produces into the cavity of the globule a rod shaped manubrial cell. There are totally 8 manubrial cells.

5. Each manubrial cell has 4-6 primary capitulum cells.

6. Each primary capitulum in turn bears 2-4 secondary capitula.

7. From each cell of the secondary capitulum arise 2-4 spermatogenous filaments. Each filament is 25-100 cells long.

8. The cells of the spermatogenous filaments, each one of them produce a biflagellate antherozoid or sperm.

Structure of Nucule

1. The nucule is an oval structure deep green to greenish black in colour and is found in the nodal region below the globule. It has a short pedicel.

2. The nucule has a single oosphere or egg rich in granular cytoplasm. The cytoplasm also has starch and oil as reserve food.

3. The envelope of the ovule is made up of 5 tubes which are coiled around the nucule.

4. The tip of the nucule is covered by 10 crown cells constituting the corona. The cells are arranged in two superposed tiers.

Fertilization

1. The cells of the corona open apart to facilitate the entry of antherozoids.

2. The zygote or oosphore develops a thick wall around it and becomes silicified.

3. The zygospore after undergoing a period of rest germinates to develop into a new individual.

Diagrams to be drawn

1. Habit sketch

2. Nodal region enlarged to show the branching

3. Nucule and globule enlarged

4. Sectional views of nucule and globule

Systematic position and identification

A. 1. Chlorophyll present
2. Cell wall cellulosic
3. Plant body a thallus ... *Algae*

B. 1. Plastids present
2. Chlorophyll is the chief pigment
3. True nucleus present
4. Photosynthetic produce is starch ... *Chlorophyceae*

C. 1. Thallus is differentiated into nodes and internodes
2. Sex organs are globule and nucule... *Charales*

D. 1. Silica deposit is found on the plant body
2. Rhizoids are multicellular... *Characeae*

E. 1. At the nodal region globule is above and nucule below
2. There is no cortication of the internode
3. The corona of the nucule has 10 cells ... *Nitella*

CHAPTER 6

XANTHOPHYCEAE

The members of the class have yellowish green chromoplasts containing chlorophyll beta carotene and xanthophyll. Food reserve is leucosin and oil. There is no formation of starch. Plant body may be unicellular or multicellular. The motile reproductive cells are biflagellate, which are unequal in length (Heterokontae)

Asexual reproduction is by zoospores or aplanospores. Sexual reproduction is mostly anisogamous and occasionally isogamous. In *Vaucheria* however sexual reproduction is oogamous.

The class includes 15 genera and 375 species (Smith) majority of which are freshwater in distribution.

Smith classifies Xanthophyta into six orders of which Heterosiphonales is one. This order includes *Vaucheria* and *Botrydium*.

Fritsch considered Xanthophyceae as a separate class in algae. He classified the class into 4 orders, of which Heterosiphonales is one. But Fritsch classified *Vaucheria* under Siphonales of the class Chlorophyceae. In this book , classification of Smith is followed with reference to *Vaucheria*.

VAUCHERIA
(Vaucheria de Candolle)

Common species : *V. sessilis, V. polysperma, V. amphibia, V. geminata, V. terrestris* etc.

Distribution: About half a dozen species are marine among the 40 odd species. The remainder are fresh water inhabiting well aerated streams or terrestrial regions. Terrestrial species grow on damp soil forming extensive green mats.

Study: 1. Structure of the plant body

2. Asexual Reproduction

3. Systematic position and classification

Structure of the plant body:

1. The thallus is a sparingly branched tubular cell frequently attaining a length of several centimetrs.

2. In terrestrial forms the thallus is attached to the soil with the help of hapteron.

3. There is a centrally running continuous vacuole present throughout the length of the plant body.

4. The cell wall is thin and made up of two layers; outer pectic and inner cellulosic.

5. The cytoplasm is peripheral. It has numerous discoid chromatophores on its outer face (towards the cell wall) and many nuclei (coenocytic) towards the inner face.

6. The chromatophores are small, circular or elliptical in outline and have no pyrenoids.

7. Minute droplets of oil are found in cytoplasm.

8. There are no cross walls. They are formed only during the formation of reproductive structures or to separate off a dead portion.

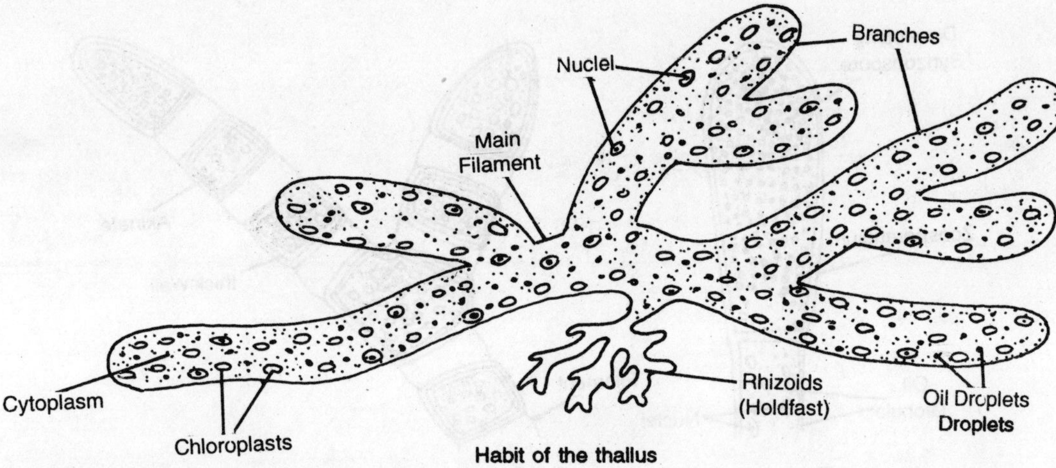

Fig. 6.1 *Vaucheria*

Asexual reproduction

1. Asexual reproduction takes place by:
   (a) Fragmentation
   (b) Synzoospore formation
   (c) Aplanospores and Akinetes
   (d) Hypnospores

2. Any mechanical breakage will result in fragments which can develop into new individuals.

3. Zoospores are mostly formed in aquatic species.

4. Zoospores are formed singly in club shaped sporangia at the tip of the branches.

5. The sporangium is separated from the rest of the plant body by means of a cross wall.

6. At the time of formation of a zoospore, the nuclei and chromatophores reverse their position, in the zoosporangium, so that the nuclei lie outside.

7. The zoospore is multinucleate and multiflagellate. It is called a *compound zoospore or synzoospore*

8. The distal wall of the sporangium opens to form a pore through which, the zoospore squeezes itself out.

9. On being liberated out of the sporangium, the zoospore loses its flagella and germinates to give rise to a new individual.

10. Akinetes or Aplanospores or non-motile spores are usually formed in terrestrial species.

11. In each sporangium only one aplanospore or akinete is formed.

12. In some terrestrial species, the entire thallus forms a number of thick walled *hypnospores* by transverse segmentation.

13. Each hypnospore may directly develop into a new individual.

14. The contents of the hypnospores may sometimes divide to form a number of cysts.

15. The plant of *Vaucheria*, with transverse septation having a number of cysts, appears like an alga by name *Gongrosira*, hence this stage is also called the *Gongrosira stage.*

16. On liberation from the plant, each cyst develops into a new individual.

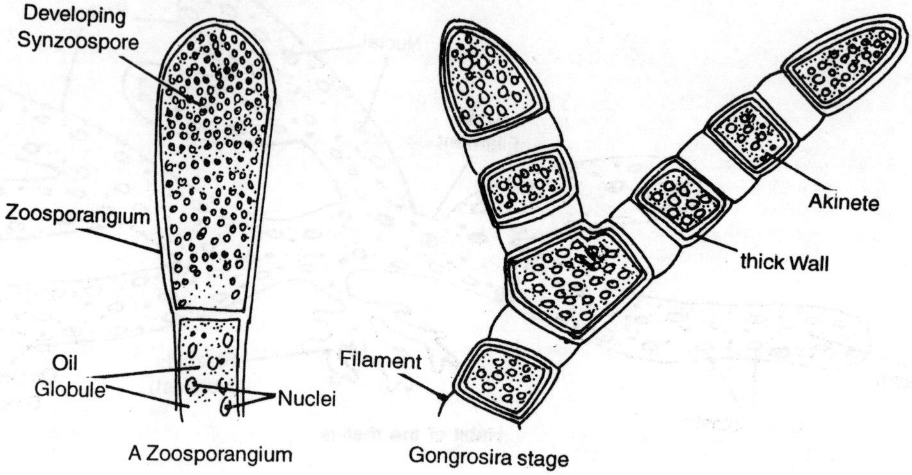

Fig. 6.2 Vaucheria (Gongrosira stage)

Sexual reproduction

1. Sexual reproduction is of the oogamous type. Sex organs are antheridia and oogonia.

2. Species of *Vaucheria* may be monoecious or dioecious (mostly monoecious).

3. Antheridia and oogonia arise as outgrowths either on the main thallus adjacent to each other or on special lateral branches.

4. In some species, a lateral branch terminates with an oogonium, with an antheridium below.

5. In certain instances, a single antheridium will be associated with many oogonia.

6. An antheridium is a tubular structure, sometimes assuming a hook shape.

7. At maturity the antheridium bends and the terminal part is cut off from the remainder by the development of a cross wall.

8. The protoplasm of the antheridium gets divided into a number of biflagelate antherozooids.

9. The mature antherozooids are liberated out of the antheridium through a terminal pore.

10. The oogonium is a globose or oval structure. It is sessile.

11. At the time of development of an oogonium the cytoplasm flows into the bulged portion of the filament.

12. The oogonium is formed by the lateral bulging of the filament.

13. Into this bulge migrates a mass of cytoplasm called wanderplasm.

14. At maturity, the oogonium gets separated from the filament by means of a cross wall.

15. An oogamium, at the time of fertilization consists of an uninucleate egg.

16. The oogonium develops a papilla or beak on one side to form a hyaline receptive spot.

17. It is at this hyaline spot that the sperms enter the oogonium. The fusion of the sperm and egg results in the formation of zygote. The zygote has a thick wall and develops into a new individual.

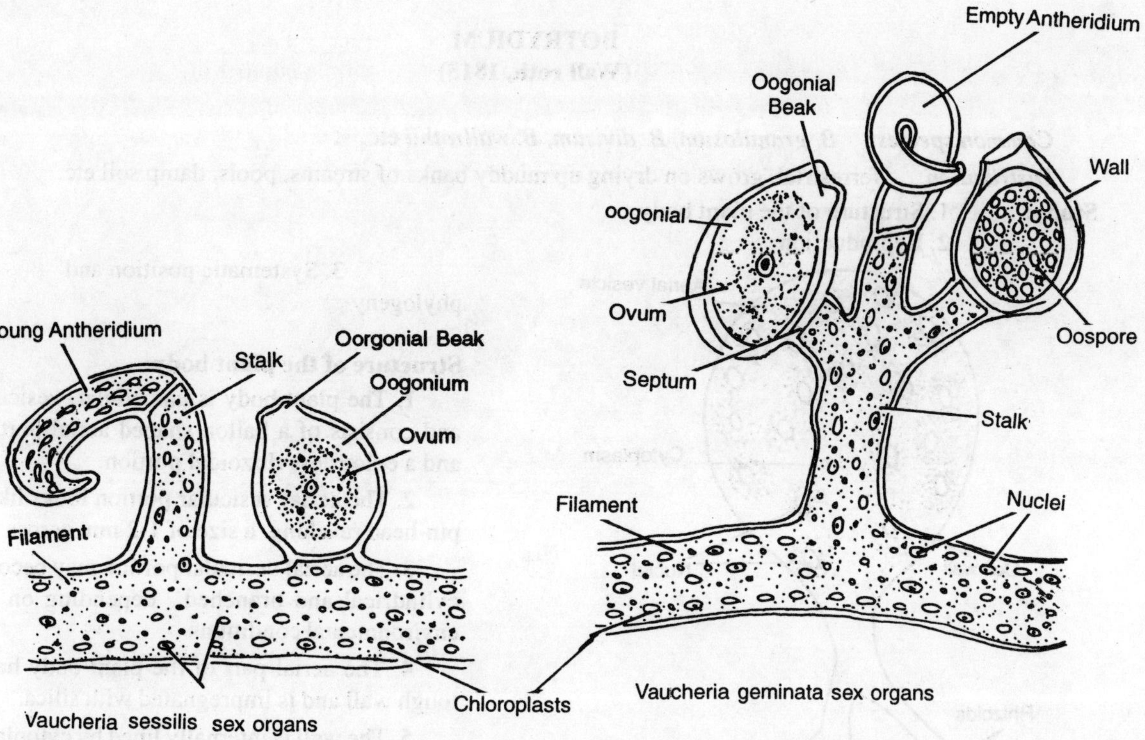

Fig. 6.3 *Vaucheria*

Diagrams to be drawn

1. Habit sketch
2. A part of the plant enlarged.
3. Zoosporangium with zoospore.
4. Aplanospore.
5. Antheridium and oogonium

Systematic position and Identification

A. 1. Chlorophyll present
2. Plant body a thallus
3. Cell wall cellulosic…*Algae*

B. 1. Reserve food material is soil
2. Motile reproductive cells have unequal flagella … *Xanthophyceae*

C. 1. Plant body siphonous with no cross walls.
2. Nuclei many in the cell … *Heterosiphonales*

D. 1. Chloroplasts lack pyrenoids.
2. Siphonein and Siphonoxanthin pigments absent …*Vaucheriaceae*

E. 1. Plant body a coenocyte
2. Thallus branched
3. Cross walls are formed only during formation of reproductive structures
…*Vaucheria*

BOTRYDIUM
(Wall roth, 1815)

Common species: *B. granulosum, B. divisum, B. wallrothii* etc.

Distribution: Terrestrial, grows on drying up muddy banks of streams, pools, damp soil etc.

Study: 1. Structure of the plant body
2. Reproduction
3. Systematic position and phylogeny

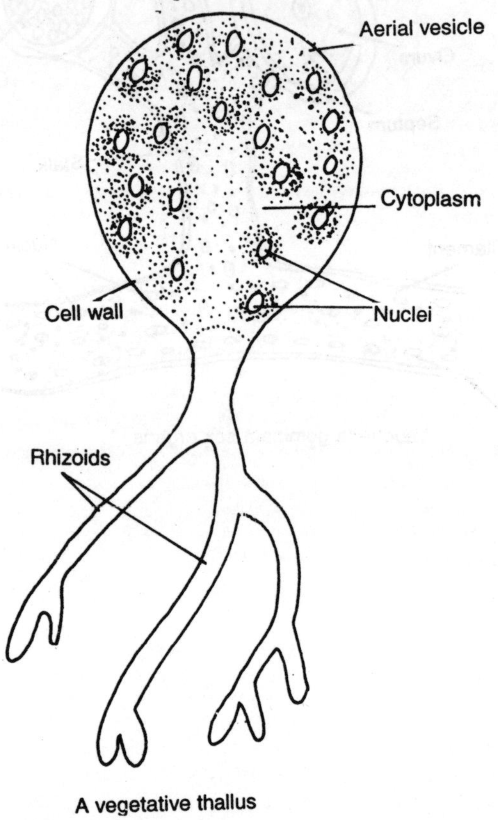

Fig. 6.4 Botrydium

Structure of the plant body:

1. The plant body is unicellular, vesicular and consists of a ballon shaped aerial portion and a colourless rhizoidal portion.

2. The aerial vesicular portion looks like a pin-head reaching a size of 1-2 mm across.

3. The aerial vesicular portion may become cylindrical and branched, depending on the environmental conditions.

4. The aerial part of the plant body has a tough wall and is impregnated with silica.

5. The wall is internally lined by cytoplasm in which are embedded numerous nuclei and chromatophores.

6. The chromatophores are discoid and often connected to one another by dense strands of cytoplasm.

7. In young plants pyrenoid like bodies are found in chromatophores, but the reserve food is never starch.

8. Reserve food is oil or leucosin.

9. There is a large central vacuole, but the cytoplasmic strands traverse across the vacuole.

10. The underground rhizoidal portion is colourless and is either sparingly or profusely branched. It has no chromatophores, but nuclie are present.

11. Rhizoidal portion helps in anchorage and absorption.

Reproduction:

1. Asexual reproduction takes place by a variety of methods.

These are : (a) Buds (b) Aplanospores (c) Zoospores (d) Hypnospores or Cysts.

(a) In some species buds are formed from the aerial part. These develop into new individuals.

(b) Large number of small, uninucleate, pyriform, biflagellate zoospores are formed. These come out of the vesicle, swim for some time in water and develop into new individuals.

(c) When moisture is not sufficient to form the zoospores, the protoplast of the cell gets divided into a number of non-motile, uni or multinucleate aplanospores.

(d) The aplanospores during dry season develop a thick wall and are called *hypnospores*. In some instances cysts are also formed. These are of 3 types.

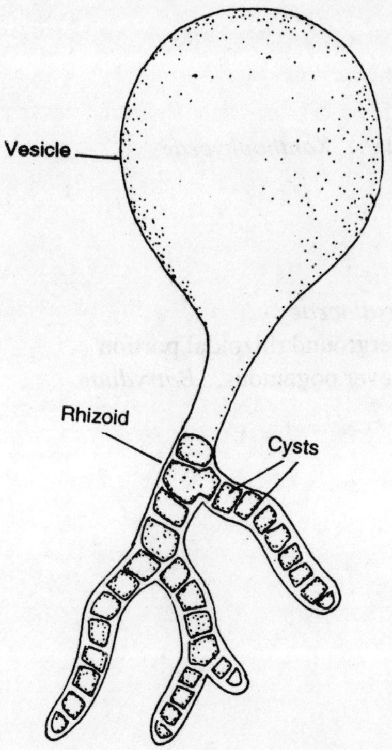

Fig. 6.5 A plant body bearing rhizoidal cysts (hypnospores)

(i) *Macrocysts:* One large cyst for the entire vesicle.

(ii) *Sporocysts:* Many cysts are formed in the vesicle.

(iii) *Rhizocysts :* Many cysts are formed in the rhizoidal part.

1. Sexual reproduction is either isogamous or anisogamous.

2. Gamete formation is seen mostly during rainy season.

3. The species of *Botrydium* may be monoecious or dioecious.

4. During gamete formation, the protoplast in the vesicle gets divided into a number of uninucleate gametes.

5. The gametes are pyriform, have one to four chromatophores, no eye spots and have unequal flagella.

6. Gametic fusion results in the formation of zygote which develops into a new individual.

Diagrams to be drawn

1. Habit sketch
2. A plant enlarged
3. Plant showing hypospores.
4. Plant showing rhizocysts.

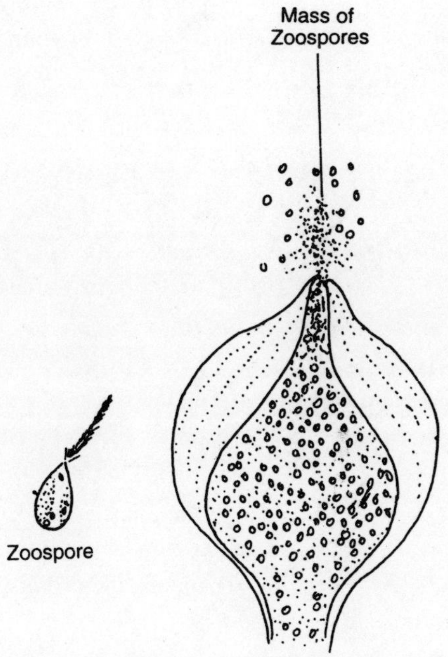

Fig. 6.6 Discharge of swarmers from vesicle

Systematic position and Identification

A. 1. Chlorophyll present
2. Plant body is a thallus
3. Cell wall cellulosic ...*Algae*

B. 1. Reserve food material is oil
2. Motile reproductive cells have unequal flagella ... *Xanthophyceae*

C. 1. Plant body siphonous with no cross walls.
2. Nuclei many in the cell...*Heterosiphonales*

D. 1. Coenocytic plant body
2. Central vacuole present
3. Plant body always terrestrial in habitat ...*Botrydiaceae*

E. 1. Plant body divided into aerial vesicle and underground rhizoidal portion
2. Sexual reproduction iso or anisogamous but never oogamous...*Botrydium*

CHAPTER 7

Bacillariophyceae

The Bacillariophyceae are commonly called diatoms. They include a large number of unicellular and colonial algae, which vary greatly in form and size. They are most widely distributed among algae. The primary distinguishing feature of Bacillariophyceae is that the cell wall is silicified and consists of two halves, which fit together as do the lid and box. The cell has one to many chromatophores.

Reproduction is by cell division and auxospore formation. Auxospores are usually zygotic in nature, produced by the union of gametes, but may not be always so.

There are about 170 genera, and 55,000 species majority of which are living but some known only in fossil conditions.

Certain genera are found only in fresh water and certain others only in salt water. Many are found in both the habitats. The marine forms of diatoms constitute a predominant part of the planktonic algae. They are so small, that as many as 35,500,000 diatoms may be present in one cubic meter of water. Many of the diatoms grow as epiphytes attached to the surface of higher plants and other algae.

The marine planktonic diatoms together with dinoflagellates are of fundamental biological importance, as they form the first link in the food chain of all animals in the ocean.

Diatoms usually exists as single cells, but sometimes cells may be joined by mucilage to form filamentous structures. The cells have a beautiful symmetry. They may be in the form of circles, squares, boat shaped, triangles and various other types of curves.

Diatomaceous Earth: Even after the death of diatom cell, the silicified wall remains unaltered, and millions of these accumulate at the bottom of the oceans and become hardened. This fossil deposit is known as diatomaceous earth and is of great importance economically.

Daitomaceous earth is of great commercial value. It is used as mild abrasive in silver polish and tooth paste. At one time it was used as the absorbent for nitroglycerine in dynamites.

Diatomaceous earth also finds extensive use as a filter in sugar refineries, as an insulator in blast furnaces etc.

Classification: The class Bacillariophyceae is divided into two orders ,viz., Pennicae or Pennales and Centricae or centrales.

Pennales: These are characterised by:

1. Cells are elongated or boat shaped in outline
2. They are predominantly fresh water.
3. They exhibit locomotion due to streaming movement of cytoplasm.
4. These cells are bilaterally symmetrical.

Centrales:

1. The shape of the cell is like a circular pill box.
2. They are predominantly marine.
3. There is no locomotion.
4. There is no raphe.

In this book a member of the order Pennales: *Pinnularia* which is studied as a typical example is given.

PINNULARIA
(Ehrenberg)

Common species: *P. major, P. viridis, etc.*

Distribution: Commonly found in fresh water ponds, and ditches; forms an oily layer on top of water.

Study: 1. Cell structure

(a) Girdle view

(b) Valve view

2. Reproduction
3. Systematic position and Identification

Cell structure:

1. Structurally the diatom cell consists of two parts or halves which fit in like a lid and box.
2. Of the two halves, the smaller one is called *Hypotheca* and the bigger one is called *Epitheca.*
3. The cell has a size of 140 μ x 20 μ
4. Each valve or half consists of a flat top (valve face) and the margin or girdle representing the overlapping face.
5. A diatom cell can be viewed from two different angles. The view from the side is called *girdle view,* and the view from the top is called the *valve view.*

Valve view:

1. Only one valve can be seen.
2. The wall can be clearly seen and it is called frustule
3. The cell wall is made up of pectin and it is silicified.
4. On the wall are seen prominent markings called costae. The costae represent rows (striae) of small pores or cavities in the frustule.
5. Along the centre of the valve is present a long slit in the wall. This is called Raphe, it is sigmoid in shape.
6. At the centre of the raphe is a swelling called the *Central nodule.*
7. At either end of the raphe are also present two nodules called *Polar nodules*
8. The mucilage extends out of the wall through openings or pores and is supoosed to help in locomotion.
9. Inside the silicified wall is the cytoplasmic lining, surrounding a vacuole.

Girdle view:

1. In the girdle or side view both the valves are seen (only one valve is seen in the valve view)
2. Costae can be seen in the frustule.
3. Inside the wall is the cytoplasm having a single nucleus.
4. One or two chromatophores may be seen distributed in the cytoplasm.
5. Oil and volution globules are present in the cytoplasm

Reproduction

1. The commonest method of reproduction seen in diatoms is cell division.

2. During cell division, the two halves get separated after the nuclear division. Each daughter cell gets only one valve from the parent. It has to develop another valve by itself. The newly formed valve is always a hyoptheca. This is irrespective of the fact, whether it gets a hypotheca or epitheca from the parent cell.

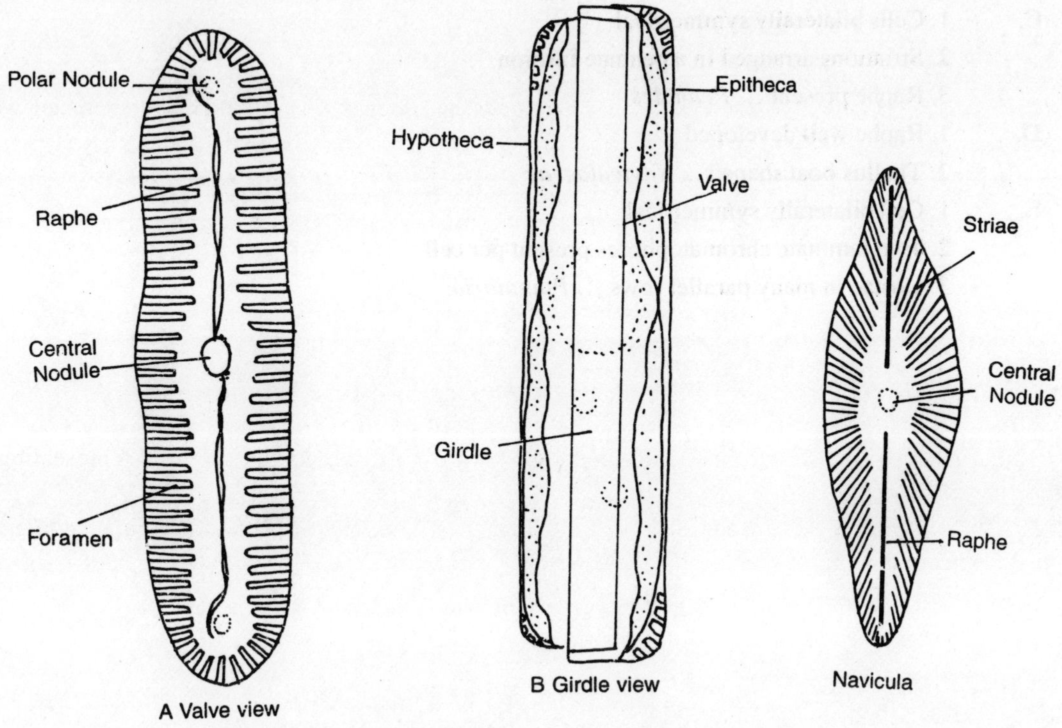

Fig. 7.1 Diatoms (Valve view and girdle view)

Auxospores:

1. Auxospores are elongated spores formed in the diatom cell to bring about reproduction.

2. Auxosopres are both asexual as well as sexual. They are formed by any one of the following five ways.

(i) Two cells conjugate to form a single auxospore.

(ii) Two cells conjugate to form two auxospores

(iii) Two cells being enveloped into a common envelop, but each giving rise to an auxospore without conjugation

(iv) One cell gives rise to one auxospore.

(v) One cell gives rise to two auxospores.

The first two methods given above are obviously sexual. The last three methods appear to be asexual.

Systematic position and Identification

A. 1. Chlorophyll present
2. Plant body thallus
3. Cell wall cellulosic … *Algae*

B. 1. Plant body unicellular
2. Cell is made up of two halves
3. Silica is deposited on the cell wall … *Bacillariophyceae*

C. 1. Cells bilaterally symmetrical
2. Striations arranged in a pennate fashion
3. Raphe present … *Pennales*

D. 1. Raphe well developed
2. Thallus boat shaped ... *Naviculaceae*

E. 1. Cell bilaterally symmetrical
2. Two laminate chromatophores present per cell
3. Costae in many parallel rows … *Pinnularia*

CHAPTER 8

PHAEOPHYCEAE

(Phaeophyta Brown algae)

The members of Phaeophyceae are commonly called "Brown Algae because of predominance of Xanthophyll a yellowish brown pigment in the chromatophores. Of the six Xanthophylls, only three are found in Phaeophyceae.

The class comprises of 195 genera and 1000 species. Of these, except three all are marine forms found mostly in cold waters than tropical seas. Two genera *Sargassum* and *Dictyota* however, are abundant even in warm waters. Many of them are sub-littoral in distribution found at a depth of 20-90 meters. They are normally attached to rocks and other objects by holdfasts. In *Fucus*, the hold fast is discoid but not like in *Laminaria*. Many of them grow as epiphytes or as endophytes. In some, as in *Sargassum* the plant body may get detached and float in large masses. Sometimes, they are found in such abundance in the sea, as in the sea waters between the coast of Florida and West Indies, to give the name Sargasso sea. Here, they form large floating masses covering an area of one quarter million square miles.

Range of vegetative Construction:

The plant body is Multicellular. It is filamentous and branched as in *Ectocarpus* or a thin flat thallus as in *Dictyota*. It may be also a tough leathery thallus as in *Fucus*. In *Ectocarpus*, the plant body exhibits heterotrichous arrangement with erect and prostrate systems. Heterotrichous organisation is absent in higher Phaeophyceae.

The largest forms of brown algae are called kelps or rock weeds, they range from one meter to 90 meters. One of the giant kelps *Macrocystis pyrifera* is as long as 90 meters.

Cell structure:

The outer most layer is the cell wall having an inner cellulose and an outer gelatinous portion. Calcium is reported to be present in the middle layer of the wall, whereas the outer layer has compounds like algin, fucosin etc. Algin chemically is calcium salt of alginic acid either whole or in part. The cytoplasm has well developed nucleus and chromotophores. Nuclear division is Mitotic.

Anatomy of thallus :

Simple phaeophyceae like *Ectocarpus* have only one layer of cells, unlike *Fucus, Laminaria* etc. which show differentiation into epidermis, cortex and medulla. The cells of medulla help in conduction.

Pigments: Chromatophores are discoid usually. They have in addition to chlorophyll (less of chlorophyll), carotene and xanthophyll. Besides these, additional brown pigments, fucoxanthin and phycophein are also present. Fucoxanthin is a carotenoid pigment (C_{40} H_{45} O_6) consisting of Fucoxanthin A and B. In addition to these, others like Violoxanthin, Flavoxanthin, Neoxanthin, Neofucoxanthin A and B are also present. Pyrenoids are absent, though pyrenoid like bodies are reported in *Ectocarpus siliculosis*.

Product of assimilation is not starch but complex carbohydrates like Laminarin, Mannitol and some fat like substance. In some like *Laminaria*, iodine is present as important reserve.

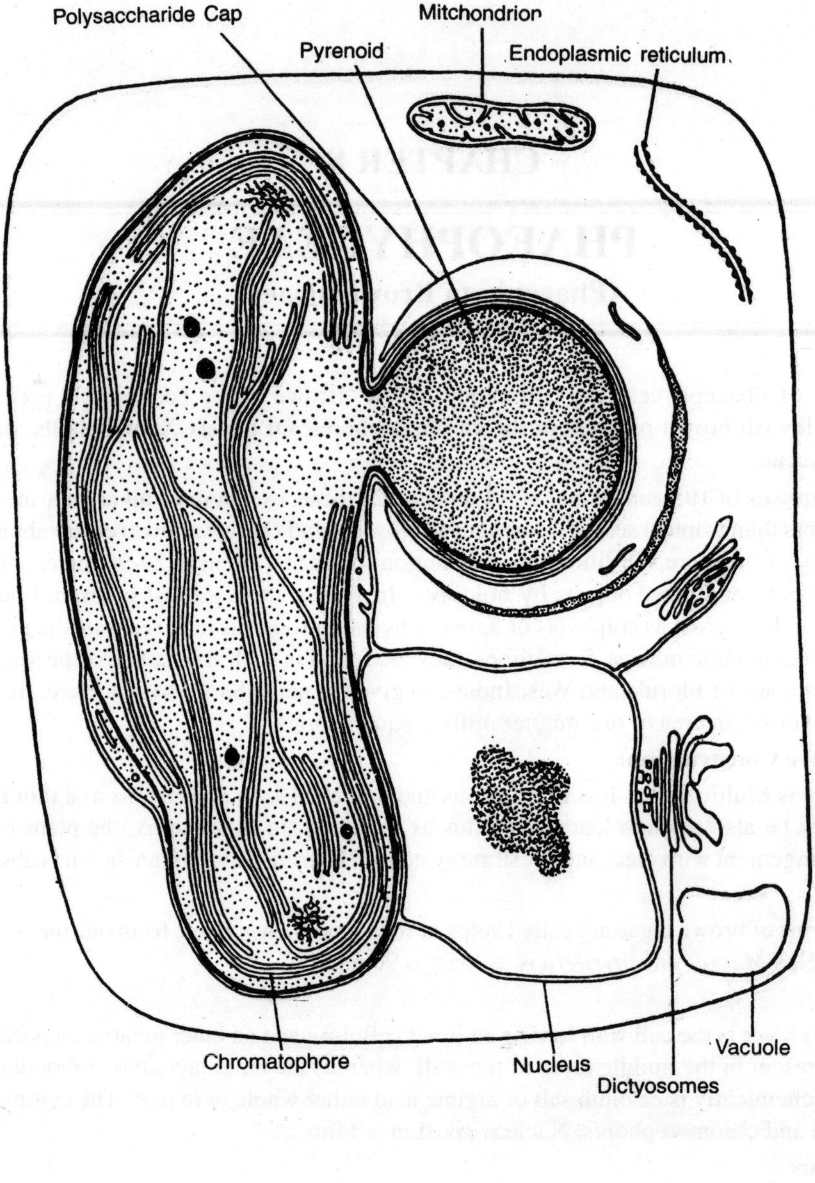

Fig. 8.1 Generalised structure of Phaeophycean cell

Reproduction:

Vegetative propagation is by fragmentation. Asexual reproduction is by motile or non-motile spores and they are formed in sporangia like unilocular sporangia or multilocualr sporangia. The zoospores formed in the former are haploid as they are formed after Meiosis. In the latter, zoospores are diploid. Consequently, they develop into sporophytes only. In some phaeophyceae, like *Dictyota,* nonmotile spores called tetraspores are formed. Tetraspores are haploid as they are formed after reduction division.

Sexual reproduction

Is isogamous, anisogamous or oogamous.

Alteration of generations: Many Phaeophyceae exhibit isomarphic alternation of generations.

Classification: Earlier phycologists divided Phaeophyceae based on structure and methods of reproduction into 2 orders.

1. *Phaeosporales*: Presence of zoospores, sexual reproduction is anisogamous or oogamous.
2. *Cyclosporales:* Zoospores absent, sexual reproduction is anisogamous or oogamous.

Kyalin (1933) divided phaeophyceae on the basis of life cycle pattêrn into 3 sub classes.

(1) Isogeneratae: Shows isomorphic alteration of generation. It is divided into 5 orders of which order Ectocarpales is one.

(2) *Cyclosporae*: Shows heteromorphic alternation of generations. It is divided into 6 orders.

(3) *Heterogeneratae:* No alternation of generations. Has only one order – Fucales.

Fritsch divides the Class Phaeophyceae into 9 orders namely:

1. Ectocarpales
2. Tilopteridales
3. Culteriales
4. Sporochnales
5. Desmarestiales
6. Laminariales
7. Sphacelariales
8. Dictyotales and
9. Fucales

In this book, *Ectocarpus, Laminaria, Dictyota, Sargassum* and *Fucus* are discussed.

ECTOCARPUS
(Lyngbye 1819)

Common species: E. coniferus, E. fariculatus, E. secundes, etc.

Distribution: Marine, distributed world wide but frequent in temperate oceans. Grows also as epiphyte on Fucales.

Study:
1. Structure of the plant
2. Reproduction
3. Systematic position and Identification

Alternation of generations:

Ecotocarpus exhibits isomorphic alternation of generations, in which the sporophyte, and gametophyte are similar morphologically. The difference between the two becomes obvious only during reproduction.

Structure of the plant (sporophyte)

1. The thallus is made up of thin, brownish, profusely or sparingly branched filaments.
2. The thallus is differentiated into an attached prostrate system and an erect system.
3. The cells are joined end to end in an uniserriate manner.
4. The postrate system is profusely branched.
5. The erect system is a branched tuft having the ultimate branches attenuated into an acute apex.
6. In some species, the older portion becomes corticated by a layer of descending rhizoidal branches.
7. Each filament is several centimeters long and monosiphonous. Cells are attached end to end.
8. The filaments are branched, the branching is lateral and the branches lie below the cross walls.
9. Due to erection, sometimes branching may appear dichotomous.
10. In some species branches taper and terminate in uniserriate hairs.
11. Growth in length of the prostrate system is apical .
12. In the vertical system growth is intercalary and is brought about by short meristematic cells found near the base of the vertical branches.

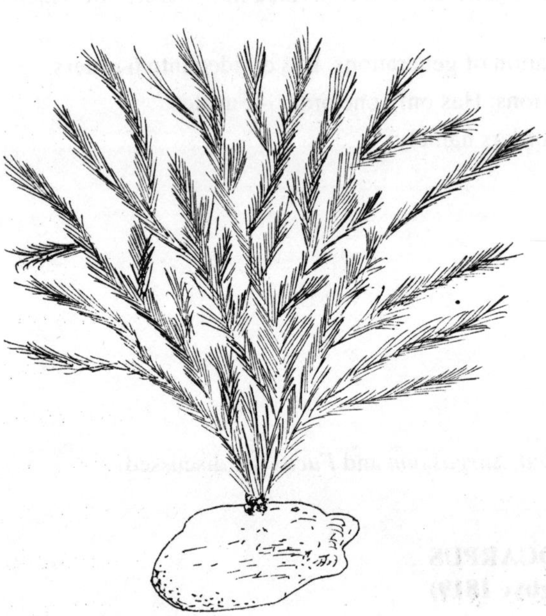

Fig. 8.2 Ectocaprus. Habit

Cell structure:

1. Cell wall consists of an outer gelatinous part and an inner cellulose part. Algin is also present on the outer layer.
2. There is a central vacuole, traversed by thin cytoplasmic threads.
3. The cell has a single nucleus and many discoid chromatophores. Sometimes the chromatophores may be band shaped..

Reproduction:

1. The life cycle exhibits isomorphic alternation of generations.
2. The sporophyte reproduces asexually, while the gametophyte reproduces sexually.
3. Asexual reproduction is brought about by means of zoospores.
4. Zoospores produced in plurilocular sporangia are diploid, while zoospores produced in unilocular sporangia are haploid.

Plurilocular Sporangia:

1. These develop on the sporophyte at the end of lateral branches.
2. The sporangia are elongated cone like structures and are made up of a number of cubical cells arranged in several vertical rows.
3. Each cubical cell develops into a biflagellate zoospore. This zoospore is diploid as it develops without a reduction division on the sporophyte.
4. The zoospores develop back into sporophyte.
5. The zoospores are called neutral spores and the sporangium neutral sporangium.

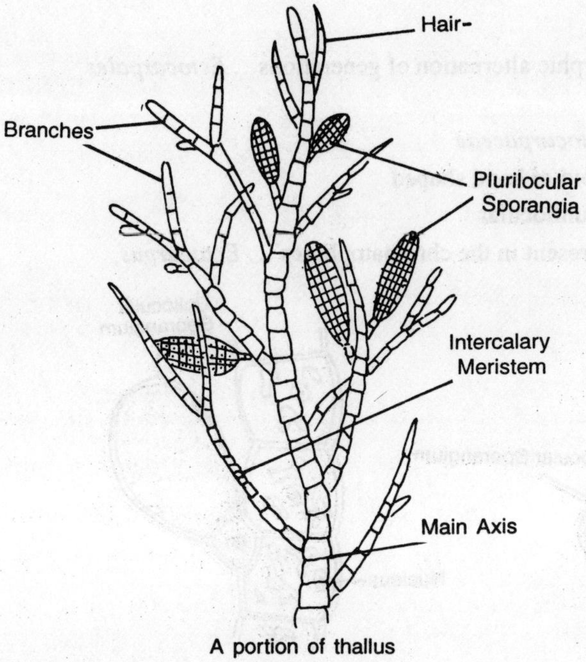

Fig. 8.3 Ectocaprus - A portion of thallus

Unilocular Sporangia:

1. These are also stalked, globose or pear shaped structures produced laterally on the lateral branches of the sporophyte.

2. The sporangium has many diploid nuclei, which undergo reduction division to produce a large number of haploid nuclei.

3. Each haploid nucleus develops into a biflagellate haploid zoospore.

4. The haploid zoospores develop into gametophytes.

Gametophyte:

1. The gametophyte resembles the sporophyte in all respects except for the reproductive organs.

Reproduction:

1. The gametophyte reproduces sexually. Sexual reproduction is isogamous.

2. Gametes are formed in plurilocular gametangia, which look exactly like plurilocular sporangia.

3. Plurilocular gametangia produce a number of haploid biflagellate gametes.

4. Gametes fuse to form a diploid zygote which directly gives rise to the sporophytic plant body of Ectocarpus.

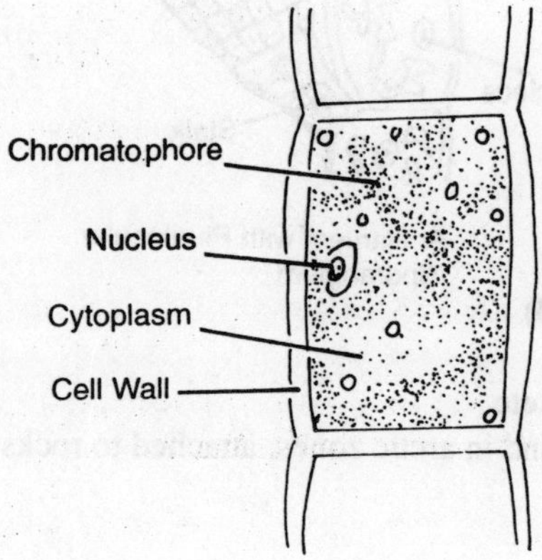

Fig. 8.4 Single cell enlarged

Diagrams to be drawn:

1. Habit sketch
2. A part of the plant to show branching
3. Plurilocular sporangia
4. Uenilocular sporangia
5. Plurilocular gametangia

Systematic position and Identification

A. 1. Chlorophyll present

2. Plant body a thallus

3. Cell wall cellulosic … *Algae*

B. 1. Chromatophores brown in colour

2. Reserve food material is mannitol and lamenarin

3. Flagella attached on the sides in reproductive cells … *Phaeophyceae*

C. 1. Plant body filamentous
2. Growth trichothallic
3. Life cycle exhibits isomorphic alternation of generations... *Ectocarpales*

D. 1. Thallus monosiphonous
2. Habit heterotrichous...*Ectocarpaceae*

E. 1. Chromatophores are discoid or band shaped
2. Sporangia plurilocular or unilocular
3. Pyrenoid like structures present in the chromatophores ... *Ectocarpus*

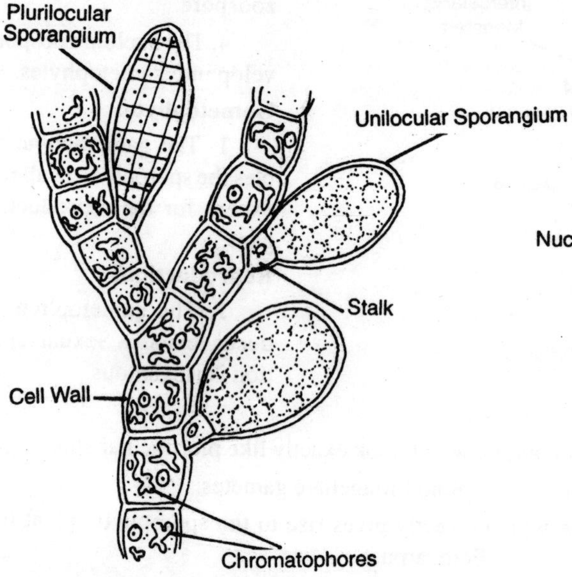

A portion of thallus showing unilocular and plurilocular Sporangia

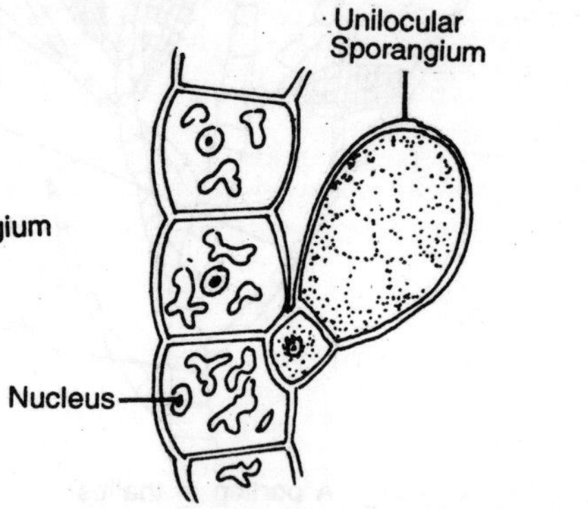

A Filament with Plurilocular Sporangium

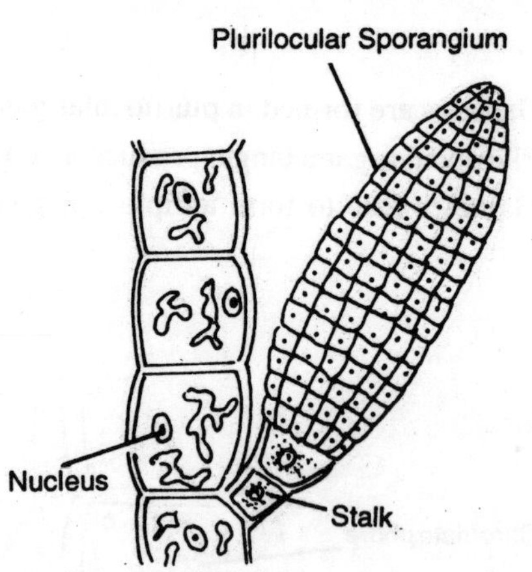

A filament with Plurilocular Sporangium

Fig. 8.5 Ectocarpus

LAMINARIA
(Lamouroux, 1813)

Common species: *L. japonica, L. flavicans, L. agardhii,* etc.

Distribution: Marine, common in temperate seas. Found in arctic zones, attached to rocks and other substrata.

Study: 1. External structure of the plant
2. Internal structure
3. Reproduction
4. Systematic position and Identification

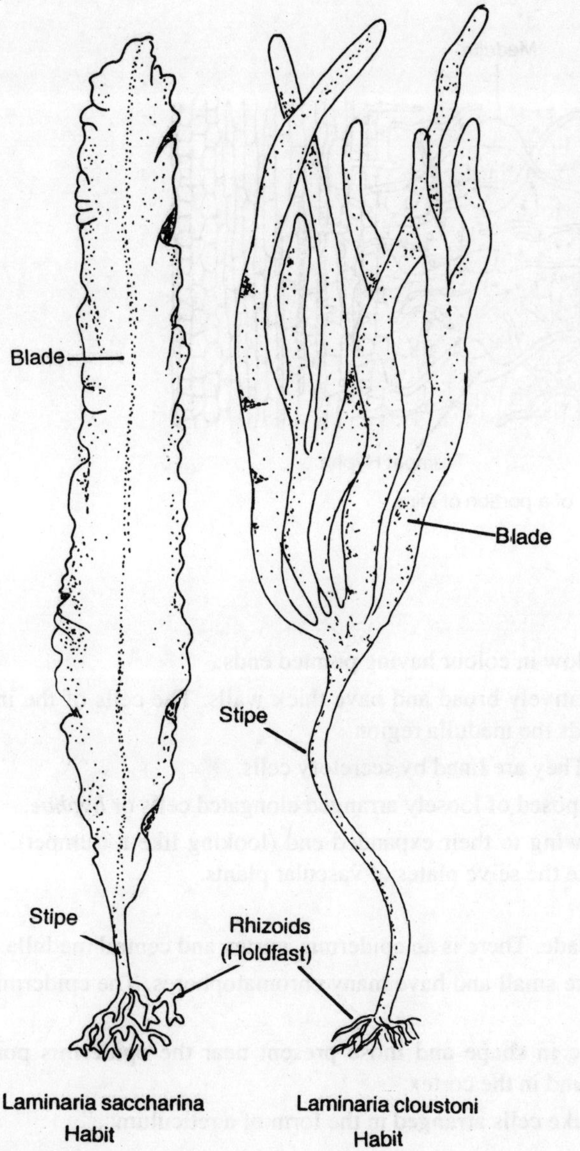

Fig. 8.6 Lamainaria

External structure:

1. The sporophytic plant body is distinguishable into three parts, namely, hold fast, stipe and blade. Altogether the plant reaches a length of 12-15 meters.

2. Holdfast is present at the basal part of the plant body and helps in anchorage. Holdfast may be in the form of a disc shaped structure. The disc may produce a number of rhizoidal out growths (Hapteron)

3. The middle stalk or axis like part of the plant is called the stipe. It may be smooth, cylindrical or flattened.

4. The lamina or blade is the terminal part of the plant. The lamina is flat with smooth or wavy margins

5. There will be one to many blades (lamina) on the stipe depending on the species. Only one blade is seen in *L. Saccharin* while in *L. cloustorii* there are many blades.

6. The blade is 2-4 meters long and has no mid rib i.e., the stipe does not continue in the lamina region.

Internal structure:

The internal structure of the stipe can be studied by having a cross as well as a long section.

Long section of the stipe:

1. A long section shows 3 regions viz, epidermis, cortex, and central medulla.

2. Epidermis has barrel shaped cells with chromatophores. These cells are meristematic and hence epidermis is also known as *Meristoderm*. A cuticle is found bordering the epidermis.

3. The cortex has vertically elongated cells. It can be distinguished into outer cortex and inner cortex.

4. In the medulla region are found loosely arranged cells.

Cross section of the stipe:

1. In the cross section also, three regions viz., epidermis, cortex and medulla are seen.

2. Epidermis is similar to what is seen in long section . Cells are small and cubical.

3. Internal to the epidermis is the cortex. This is divided into outer cortex and inner cortex.

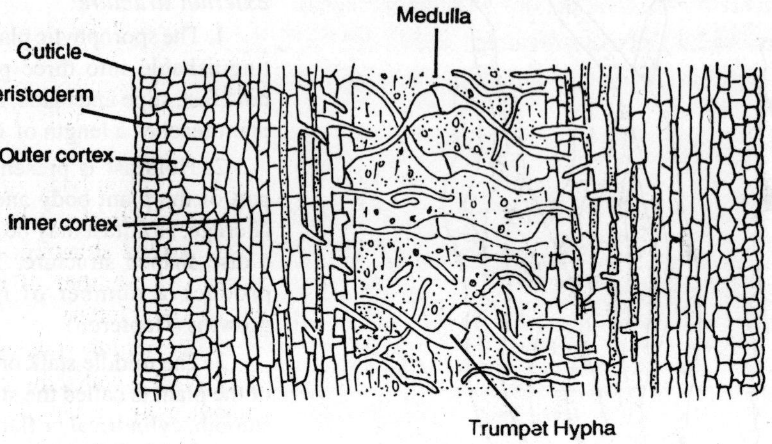

T.S. of a portion of stipe

Fig. 8.7 Lamainaria

4. Outer cortex has elongated cells, pale yellow in colour having pointed ends.

5. The cells of the inner cortex are comparatively broad and have thick walls. The cells of the inner cortex develop certain tubular outgrowths towards the medulla region.

6. Mucilage canals are found in the cortex. They are lined by secretory cells.

7. The central region called medulla, is composed of loosely arranged elongated cells or *hyphae*.

8. These cells are called *trumpet hyphae* owing to their expanded end (looking like a trumpet). The trumpet hyphae have a perforated plate much like the seive plates of vascular plants.

Lamina

1. Anatomically the lamina resembles the blade. There is an epidermis, cortex and central medulla.

2. Epidermis is the outermost layer. Cells are small and have many chromatophores. The epidermis is covered by a gelatinous layer.

3. The cells of the cortex are isodiametric in shape and those present near the epidermis posses chromatophores. Many mucilage cavities are found in the cortex.

4. Medulla is composed of vertical hyphae like cells arranged in the form of a reticulum.

Reproduction:

1. The sporophyte reproduces asexually by the development of unilocular sporangia. The unilocualr sporangia produce zoospores after reduction division.

2. The unilocular sporangia are arranged in clusters called sori on either surface of the lamina. Between the sporangia are present many hair like processes called paraphyses.

3. The sporangia undergo reduction division to produce a large number of haploid zoospores. These zoospores develop into gametophytes.

Sexual Reproduction:

1. The gametophytes are small and filamentous. They are branched. The male and female gametophytes are different.

2. Sexual reproduction is Oogamous.
3. Antheridia are produced at the tips of lateral branches.
4. Oogonia are small globose structures.
5. The male and female gametophytes are highly reduced structures.

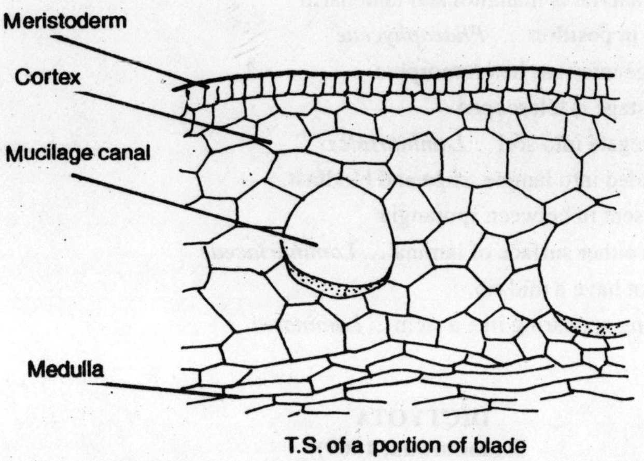

Fig. 8.8 Lamainaria

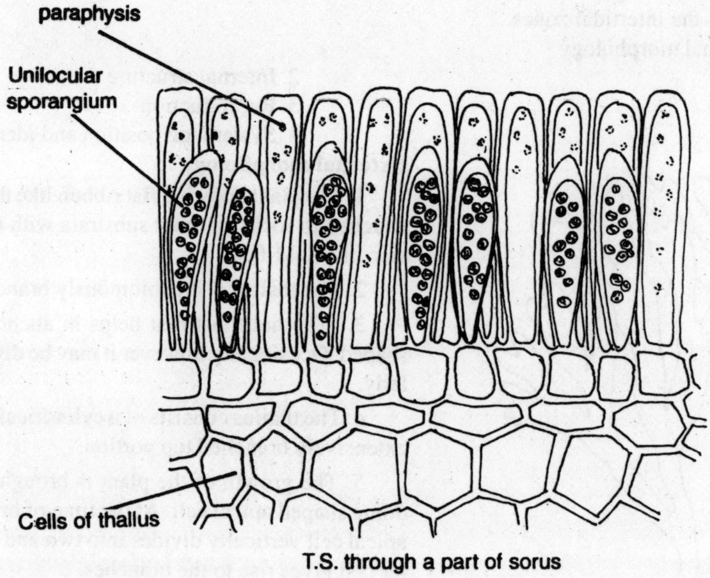

Fig. 8.9 Lamainaria

Diagrams to be drawn:

1. Habit sketch
2. T.S. and L.S. of stipe
3. T.S. of lamina

Systematic position and identification:

A. 1. Chlorophyll present
2. Plant body is a thallus
3. Cell wall cellulosic ... *Algae*

B. 1. Chromatophores brown in colour.
2. Reserve food material is mannitol and lamenarin
3. Flagella lateral in position ... *Phaeophyceae*

C. 1. Alternation of generations heteromorphic
2. Gametophytic stage is telescoped
3. Sporangia aggregate into sori ...*Laminariales*

D. 1. Plant body divided into lamina, stipe and hlodfast.
2. Paraphyses present in between sporangia
3. Sori present on either surface of lamina ... *Laminariaceae*

E. 1. Lamina does not have a mid rib.
2. Stipe is very long appearing like a stem ... *Laminaria.*

DICTYOTA
(Lamouroux, 1809)

Common species: *D. dichotoma.*

Distribution: Marine alga, found widely distributed in tropical seas, growing attached to rocks in the intertidal zones.

Study: 1. External morphology
2. Internal structure
3. Reproduction
4. Systematic position and identification

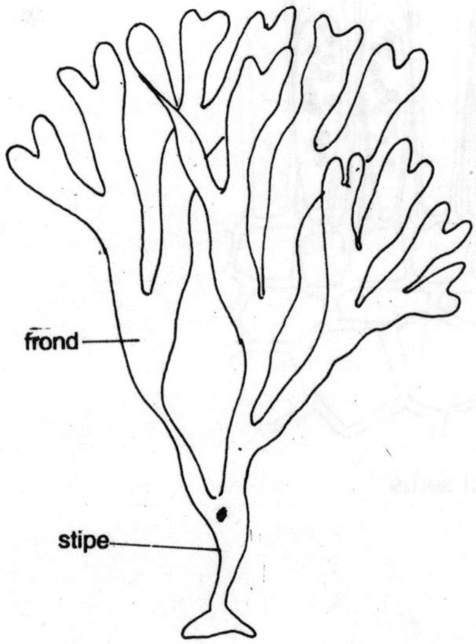

Fig. 8.10 Dictyota

External morphology :

1. The plant body is a flat ribbon like thallus, found attached to rocks or other substrata with the help of a disc like hold-fast.

2. The thallus is dichotomously branched.

3. The hold-fast that helps in anchorage is disc shaped. Occaionally however it may be divided irregularly.

4. The thallus consists of a cylindrical stalk and an extensively branched top portion.

5. The growth of the plant is brought about by a dome shaped apical cell. At the time of branching, the apical cell vertically divides into two and each daughter cell gives rise to the branches.

6. The cells of the thallus found below the apical cell are arranged in a radial manner.

Internal structure of the thallus :

1. A cross section of the thallus reveals three layers of cells viz., two layers of small cells and a middle layer of large cells.

2. The central layer of large cells also called medulla, has thin walled cells. The cells have a large vacuole. They have but few chromatophores. Sometimes the chromatophores may be entirely absent. The cells of the medulla are rich in fucosan vesicles and many refractive granules. The cells are uninucleate.

3. The upper and lower layers of peripheral cells are smaller in size and have many chromatophores.

4. Mucilage filled colourless multicellular hairs are present on both the upper and lower faces of the epidermis.

Reproduction:

1. Asexual reproduction takes place by means of tetrasporangia.

2. The tetrasporangia are spherical cells present on either surface of the thallus. They may occur singly or in clusters called sori.

3. The tetrasporangia are diploid as they are formed on the sporophytic plant. They undergo reduction division to form four tetrahedrally arranged tetraspores.

4. The tetraspores develop into gametophytes.

Sexual reproduction:

1. Sexual reproduction is oogamous. The gametophytes are dioecious. The sex organs namely antheridia and oogonia are borne on both the surfaces of the thallus.

2. Both antheridia and oogonia occur in clusters called sori.

Antheridia :

1. The antheridial sori are produced on either surface of the thallus. Each sorus has 100-300 antheridia.

2. The sorus on all sides is covered by long sterile hair like cells forming a protective sheath.

3. Each antheridium is an elongated sac and has a short stalk made up of a single cell.

4. The antheridium is multinucleate. Each nucleus surrounds itself with a little cytoplasm and develops into an uniflagellate antherozoid.

Oogonium :

1. The oogonia are also clustered into sori. Each sorus has a group of 25-50 oogonia.

2. There are no sterile cells in the oogonial sorus to form an envelope.

3. Oogonia also have a single celled stalk, sac like in appearance and have a single egg each.

4. Fertilization results in the formation of egg. Egg and antherozoids meet outside. Fusion is external.

Diagrams to be drawn

1. Habit sketch
2. T.S of the thallus
3. Surface view of tetrasporangia
4. T.S of thallus showing tetrasporangia
5. T.S. of thallus through an antheridial sorus
6. T.S. of thallus through an oogonial sorus

Systematic position and identification

A. 1. Chlorophyll present
2. Plant body is a thallus
3. Cell wall cellulosic ... *Algae*

B. 1. Chromatophores brown in colour.
2. Reserve food material is mannitol and Lamenarin ... *Phaeophyceae*

C. 1. Thallus flat and dichotomously branched.

2. Asexual reproduction by non-motile spores.

3. Growth is apical ... *Dictyotales*

D. 1. Sex organs aggregated into sori.

2. Sperms flagellate with non motile egg.

3. Fertilization external ... *Dictyotaceae.*

E. 1. Thallus dichotomously branched.

2. No midrib in the thallus.

3. Central medulla is single layered ... *Dictyota.*

SARGASSUM
(Agardh, 1821)

Common species : *S.cinerium*

Distribution : All the species are confined to tropical seas. Found attached to rocks in intertidal zones. Sometimes may from large floating masses extending over several miles.

Study : 1. Morphology of the thallus.

2. Anatomy of stem

3. Reproduction

4. Systematic position and identification

Morphology of the thallus :

1. The plant body is macroscopic more of less bushy resembling an angiosperm plant.

2. The plant body is mostly attached to rocks and other substrata, sometimes found free floating also.

3. The plant has a central axis or stem which is attached in the young condition to the substratum with the help of hold fast.

4. The axis is differentiated into nodes and internodes.

5. The axis is branched. It bears richly branched primary laterals or long shoots of unlimited growth.

6. The main axis as well as long shoots bear a number of flat, leaf like photosynthetic laterals (Branch of first order).

7. Leaves are narrow with a terrete margin. They may or may not have a midrib. They help in photosynthesis.

8. The leaves are sterile and have numerous dots called cryptostomata. These lead internally into empty chambers called cryptoblasts.

9. The axis also bears a system of secondary branches (Branches of second order) which are modified into air bladders to help in buoyancy.

10. In the axils of leaves there is a tertiary branching (Branches of third order) consisting of numerous finger shaped projections. These are called receptacles and are fertile.

11. The receptacles internally have conceptacles bearing reproductive organs.

Anatomy of stem

A cross section of the thallus shows three regions viz., epidermis or meristoderm, cortex and medulla.

1. Epidermis is the outermost layer. This is often invested with a mucilage covering.

2. The cells of the epidermis are columnar and have no intercellular spaces.

3. The cells are meristematic, and have a large member of plastids and fucosan granules.

4. Cortex is multilayered consisting of thick walled polygonal cells.

5. The cells of the cortex help in storage, besides offering mechanical support.

6. The innermost cortical cells gradually merge with the cells of medulla and have gelatinised walls.

7. Medulla is the central zone and it has loosely arranged cells.

8. The outer most region of medulla has thick walled cells but the inner region has thin walled cells.

9. The cells of meduall help in conduction.

Anatomy of leaf

1. The leaf in its internal structure resembles the stem.

2. The outline is elliptical and has a number of pores leading to cryptostomata.

3. Outermost layer is epidermis, made up columnar cells which are closely packed. There is a mucilage envelope for the epidermis.

4. Internal to the epidermis is a layer of cells rich in chromatophores. This is the chief photosynthetic zone of the leaf. This is the assimilating tissue.

5. Next to the assimilatory tissue is the storage zone consisting of thin walled isodiametric cells.

6. In the middle of the leaf is a zone of conducting tissue.

Asexual reproduction

1. This is brought about by fragmentation.

2. No zoospores are formed.

Sexual reproduction

1. Sexual reproduction is oogamous. Sex organs viz., oogonia and antheridia are borne in flask shaped structures called conceptacles, which are in turn borne in receptacles.

Structure of a receptacle :

1. Receptacles are small finger shaped branches.

2. A cross section of a receptacle is elliptical in outline.

3. Lining the margin of the receptacle internally are found a large number of flask shaped cavities called conceptacles.

4. Each conceptale is either male or female, they bear either antheridia or oogonia.

Structure of male conceptacle :

1. The conceptacle is a flask shaped cavity. It opens to the exterior through a small opening called the ostiole.

2. The wall of the conceptacle is composed of small flat cells having many chromatophores.

3. From the floor of the conceptacle arise many hair like structures called paraphyses.

4. Near the ostile also, hair like structures are found. These are called periphyses.

5. The antheridia are borne on some of the branched paraphyses.

6. The antheridia are small globose structures having a two layered wall.

7. About 64 antherozoids are produced inside the antheridium. The antherozoids are biflagellate.

Structure of a female conceptacle :

1. Female conceptacles are similar in structure to the male conceptacles, except that they posses oogonia.

2. Oogonia do not develop on the paraphyses. They arise directly on the wall of the conceptacle.

3. The oogonium has a single celled stalk which is embedded in the wall of the conceptacle.

4. The oogonium has a 3 layered wall.

5. A young oogonium has eight nuclei, but seven of them abort, so that a mature oogonium has a single egg.

Structure of a sterile conceptacle:

1. These are conceptacles that do not bear any sex organs.
2. Numerous paraphyses arise from the floor of the conceptacle.
3. Near the ostiole are hairs called periphyses whose cells are rich in chromatophores.

Fertilization:

1. The fusion of the gametes results in the formation of zygote which directly develops into a new plant of *Saragssum*

Diagrams to be drawn

1. Habit sketch
2. T.S. of stem
3. T. S. of leaf
4. Ground plan of T.S.of Receptacle
5. A sterile conceptacle enlarged
6. A male conceptacle enlarged
7. An antheridial branch enlarged
8. A female conceptacle enlarged
9. An oogonium enlarged

Systematic position and identification

A. 1. Chlorophyll is present
   2. Plant body is a thallus
   3. Cell wall cellulosic ... *Algae*

B. 1. Chromatophores brown in colour.
   2. Reserve food material is mannitol and Laminarin ... *Phaeophyceae*

C. 1. Plants have a complex morphological structure.
   2. They show a high degree of anatomical differentiation.
   3. No zoospores are formed.
   4. Sex organs in special flask shaped structures called conceptacles... *Fucales*

D. 1. Plant body bears distinct foliar organs.
   2. Thallus is branched, branching is radial.
   3. Air bladders present ... *Sargassaceae.*

E. 1. 'Leaves' narrow with a distinct midrib
   2. Air bladders present
   3. Receptacles in clusters, axillary to the leaves ... Sargassum.

FUCUS
(Linnaeus, 1753)

Common species : *F.vesiculosus, F.serratus* etc.

Distribution : Marine alga, found in temperate seas. Found attached to rocks in intertidal zones.

Study : 1. External morphology
2. Internal structure of the frond
3. Reproduction
4. Systematic position and identification

External morphology

1. The plant body is a flattened dichotomously brnached ribbon like thallus.

2. The dark brown thallus is leathery in texture and is anchored to the substratum with the help of hold fast.

3. The plant body is divided into three parts viz., holdfast, stipe and frond.

4. The stipe is a stalk like structure arising from hold fast. It continues in the region of the frond in the form of a midrib.

5. The terminal flattened ribbon like portion of the plant body is called the frond.

6. The frond is dichotomously branched and the surface is slippery due to the presence of gelatinous material.

7. The frond has a midrib, along the line of which in some speices are found air bladders. These help in buoyancy.

8. The margin of the frond may be smooth or serrate.

9. The growth is brought about by an apical cell situated in a furrow.

10. Along the length of the frond on the surface are found numerous, small dot like structures called Cryptostomata or sterile conceptacles.

11. Mature fronds, at their tips bear swollen fertile structures called receptacles. These receptacles contain many flask shaped cavities called conceptacles.

Anatomy of frond

1. A cross section of the frond shows three regions viz., meristoderm, cortex and medulla.

2. Meristoderm or epidermis consists of meristematic cells. It is made up of a single layer of closely packed cells.

3. Often the meristoderm is enveloped by a cuticle.

4. The cells of the meristoderm have numerous chromatophores.

5. Cortex consists of large, thin walled cells. The cells become broader towards the centre of the frond. They have but few chromatophores.

6. The central region called medulla, consists of loosely arranged cells with intercellular spaces filled by mucilage.

7. The cells of medulla help in conduction.

Anatomy of midrib (Stipe)

1. In its internal structure, the stipe resembles the frond. It has an outer meristoderm, middle cortex and inner most medulla.

2. Meristoderm and cortex are similar in structure to that of the frond.

3. The medulla consists of long irregular cells called hyphae which offer mechanical support.

Reproduction

1. Vegetative propagation is brought about by means of fragmentation.

2. Sexual reproduction is oogamous brought about by oogonia and antheridia.

3. The sex organs are bone in flask shaped cavities called conceptacles which in turn are borne in the terminal receptacles.

4. Each receptacle in a transverse section shows a number of conceptacles lining the margin internally.

5. Species of *Fucus* may be monoecious or dioecious.

6. Conceptacles also may be bisexual (bear both sex organs) or unisexual.

Structure of the male conceptacle

1. The conceptacle is a flask shaped structure. It has an inner basal layer of fertile cells.

2. Many paraphyses (sterile hairs) arise from the floor of the conceptacle. These are multicellular and may be branched or unbranched.

3. The conceptacle opens to the exterior through a small opening called the ostiole.

4. The antheridia arise internally on short or long filaments which terminate in a paraphysis.

5. The antheridia are small, pyriform and stalked.

6. The antheridia have a two layered wall viz., the outer Exochite and inner Endochite.

7. Each antheridium produces many biflagellate, uninucleate antherozoids.

Structure of the female conceptacle

1. The conceptacle is similar in structure to the male except that it bears oogonia.

2. Many oogonia are found in each conceptacle arising directly from the wall of the conceptacle.

3. The female conceptacle has paraphyses ostiole etc just like the male conceptacle.

4. Oogonia have a short stalk. They arise singly and are globose in shape.

5. The wall of the oogonium is very thick. It has eight functional eggs or oospheres.

6. At maturity, the eggs come out of the conceptacle through the ostiole. However they remian attached to the conceptacle base.

7. Fusion of the gametes results in a zygote which develops into a new individual.

Diagrams to be drawn :

1. Habit sketch
2. Anatomy of frond
3. T.S. of receptacle (ground plan)
4. Male and female conceptacles enlarged.

Systematic position and identification

A. 1. Chlorophyll present
 2. Plant body is a thallus
 3. Cell wall cellulosic ... *Algae*

B. 1. Chromatophores brown in colour.
 2. Reserve food is mannitol and Lmainarin... *Phaeophyceae*

C. 1. Plants have a complex morphological structure.
 2. They show a high degree of internal tissue variety.
 3. No zoosopres are formed.
 4. Sex organs in conceptacles ... *Fucales*

D. 1. Frond has a distinct midrib
 2. Air bladder intercalary (when present)
 3. Oogonia have 8 functional eggs ....*Fucaceae*

E. 1. Plants have a definite hold fast
 2. Frond leathery dark brown with prominent midrib.
 3. Branching is dichotomous
 4. Conceptacles unisexual or bisexual ... *Fucus*

CHAPTER 9

RHODOPHYCEAE – Red Algae

The red algae are primarily distinguished from other algae, by their sexual reproduction in which, aflagellated male gametes (spermatia) are passively transported to the female sex organ, carpogonium. Plastids of red algae contain several pigments among which R.phycoerythrin is usually present in large quantities to mask pigments and give red colour.

Occurrence

There are about 400 genera and 2,500 species. About 50 species belonging to about a dozen or more genera are fresh water.They are restricted to rapidly flowing streams and falls.

A majority of red algae are marine in distribution, the members may be attached or occasionally free floating. They are widely distributed in all global oceans including Arctic and Antarctic. They grow attached to rocks and other substrata. Some may be epiphytic.

Morphology of the thallus

Except for two genera, all others are multicellular. The thallus is fundamentally filamentous. The thalli are macroscopic reaching a size of 30 cms. The filaments are branched, may be monosiphonous or polysiphonous.

Pigments: The red algae owe their name to the red colour. But, some fresh water forms like *Batrachospermum* are green. They possess phycoerythrin (red) and phycocyanin (blue-green) in addition to chlorophyll. Phycoerythrin is in abundance so as to mask the green colour. Phycoerythrin is absent in fresh water forms like *Batrachospermum.*

Cell structure

Generally cells are uninucleate but some are multinucleate having 3000 to 4000 nuclei in a single cell. Except in a few there is a central vacuole in the majority of species with cytoplasm forming a thin layer internal to cell Wall. The cell wall is made up of cellulose and various other pectic compounds (Kylin 1943). Of cells of primitive red algae have a single stellate chromatophore having a pyrenoid. These however lack the starch sheath found in Chlorophyceae. In advanced members, chromatophores are discoid. Chromatophores have carotenes, xanthophylls, red pigments, blue green pigments in addition to chlorophyll. The product of assimilation is called Floridean starch. It is considered to be related to amylopectin. Many red algae accumulate soluble sugar Floridoside, a galactoside of glycerol. Cells are connected to one another, through plasmodesmata.

Reproduction

Motile reproductive cells are entirely wanting. Asexual reproduction takes place by a variety of spores like monospores, carpospores, tetraspores etc.

Sexual Reproduction

The male sex organ is spermatangium and the female, the flask shaped carpogonium having a long neck called trichogyne. Spermatia liberated from spermatangia fuse with the egg of the carpogonium to form the zygote. The zygote may or may not undergo reduction division, and forms carpospores. In the former case, the carpospores are haploid and in the latter they are diploid and develop into a diploid generation in which there is reduction division producing haploid generation through tetraspores.

Majority of red algae show alternation of generations which may be biphasic or triphasic.

Classification: The class Rhodophyceae is divided into two subclasses, viz., *Bangioideae* and *Florideae.* In Bangioideae, forms are simple and corporpores are haploid. There is only one order. Bangiales. In Florideae, the plant body is highly complex and carpospores are diploid. Members are both fresh water and marine. This sub class includes six orders, viz., Nemalionales, Gelidiales, Cryptomaniales, Gigartinales, Rhodomenales, and Ceramiales. The last two orders are discussed with an example for each order.

BATRACHOSPERMUM
(Roth 1797)

Common species: *B. moniliforme, B. densum, B. vagum, etc.*

Distribution: Found in fresh water streams and rivulets. Usually found attached to substrata like stones and twigs.

Study:
1. Structure of the plant body
2. Reproduction
3. Systematic position and Identification

Structure of the plant body:

1. The plant body is a greenish or pink bushy filamentous structure.
2. It consists of a central axis divided into nodes and internodes.
3. The main axis is uniseriate (monosiphonous). It is anchored to the substratum with the help of old shoots forming a prostrate system.
4. The main axis is corticated and at the nodal region it bears branches.
5. Branches are lateral and are of 2 types, viz., branches of unlimited growth and branches of limited growth.
6. The branches of unlimited growth resemble the main axis in all respects and in turn are divided into nodes and internodes.
7. The branches of limited growth are also known as *glomerules* are short and 10-15 cells long. They are made up of bead like cells.
8. The branches of limited growth found both on the main axis as well as on the branches of unlimited growth at the nodal regions.
9. The glomerules arise in large numbers at the nodal regions giving a bushy appearance to the plant. Occasionally however they may be also found sparsely along the internodal region.
10. The cells are rectangular in shape, uninucleate and have a wall made up of cellulose.
11. There are many discoid chromatophores in each cell. Each chromatophore has a pyrenoid.

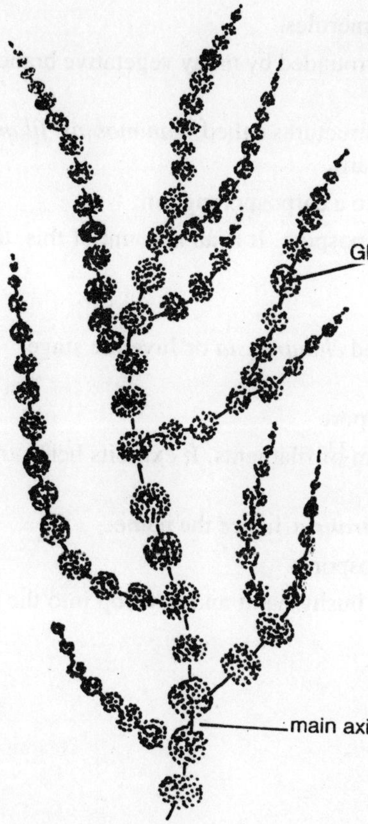

Fig. 9.1 Batrachospermum. External features (habit)

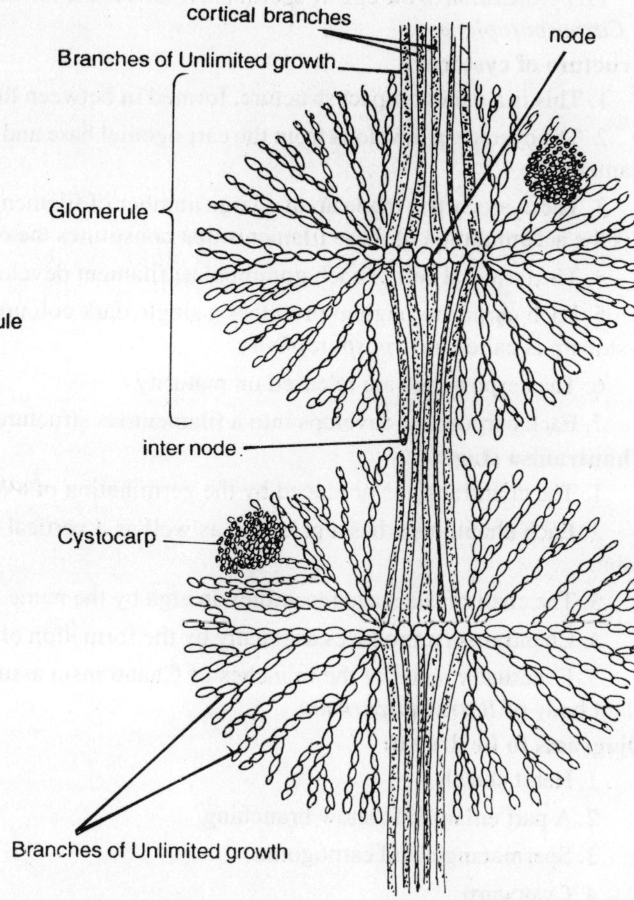

Fig. 9.2 Batrachospermum. A portion of thallus showing whorl of laterals (glomerules)

Reproduction:

1. Vegetative propagation is brought about by means of fragmentation. Each fragment can develop into a new individual.

2. Sexual reproduction is oogamous. Sex organs are *spermatangia* and *carpogonia*.

3. Species of *Batrachospermum* may be *monoecious* or *dioecious*.

4. The sex organs are usually borne at the tips of short laterals (branches of limited growth).

5. The terminal cell of a short lateral divides and forms 204 spermatangia or antheridia.

6. Each spermatangium produces an uninucleate non-motile *Spermatium* or *Male cell.*

7. At maturity the spermatia are liberated from the plant. They are dispersed by water current. But they do not have any flagella.

8. The female reproductive structure, viz., the carpogonium is also produced at the tip of a short lateral.

9. The carpogonium is a unicellular flask shaped structure. It has a bulbous basal region and an elongated neck like portion called the trichogyne.

10. There is a single egg at the basal region of the carpogonium.

11. Fertilization of the egg by spermatia results in the formation of a zygote which develops into *Cystocarp* or *Carposporophyte.*

Structure of cystocarp

1. This is a dark compact structure, formed in between the glomerules.

2. The cystocarp develops from the carpogonial base and is surrounded by many vegetative branches or filaments.

3. The cystocarp is made up of a large number of filamentous structures called '*gonimoblast filaments*'. It is the accumulation of these filaments that constitutes the cystocarp.

4. The terminal cell of each gonimoblast filament develops into a carposporangium.

5. Each carposporangium produces a single dark coloured carpospore. It is an account of this, that the cystocarp is called a *carposporophyte.*

6. The carpospores are released on maturity.

7. Each carpospore develops into a filamentous structure called *chantransia* or Juvenile stage.

Chantranisa stage:

1. The chantransia is produced by the germination of a carpospore.

2. Each chantransia has a prostrate as well as a vertical system of filaments. It exhibits heterotrichous habit.

3. The chantransia stage resembles an alga by the name *Chantransia*, hence the name.

4. Chantransia reproduces asexually by the formation of monospores.

5. Eventually some of the branches of Chantransia assume a bushy habit and develop into the typical plant body of *Batrachospermum.*

Diagrams to be drawn:

1. Habit sketch
2. A part enlarged to show branching
3. Spermatangia and carpogonia
4. Cystocarp
5. Chantransia

Systematic position and Identification:

A. 1. Chlorophyll present
2. Plant body is a thallus
3. Cell wall cellulosic ... *Algae*

B. 1. Motile reproductive cells absent
2. Chromatophores red pigmented ... *Rhodophyceae*

C. 1. Plant body has monosiphonous axis.
2. Chromatophores discoid .
3. Juvenile stage present in the life cycle... *Nemalionales*

D. 1. Plant body is found in fresh water.
2. Life cycle is diphasic and Haplobiontic ...*Batrachorpermaceae*

E. 1. Plant body bushy
2. Short laterals have bead like cells
3. Juvenile stage present in life cycle ... *Batrachospermum*

GRACILARIA
(GREVILLE, 1830)

Common species: *G. armata, G. compressa etc.*

Distribution: Marine in distribution occuring also as a lithophyte at the intertidal zones in tropical seas.

Study: 1. Morphology of the thallus.
2. Anatomy of the thallus
3. Reproduction
4. Systematic position and classification

Morphology of the thallus:

1. The thallus is a branched, cylindrical or compressed structure made up of a number of filamentous structures.

2. The thallus also called frond is attached to the substratum with the help of a discoid holdfast. The hold fast consists of a number of closely packed thread like structures.

3. The fronds are 25-40 cms, long, dark red in colour and dichotomously branched.

4. Some of the branches of the frond have attenuated apices, terminating in hair like structures.

5. The fronds are monosiphonous, have uniseriately arranged cells. The fronds have a gelatinous envelope on their surface.

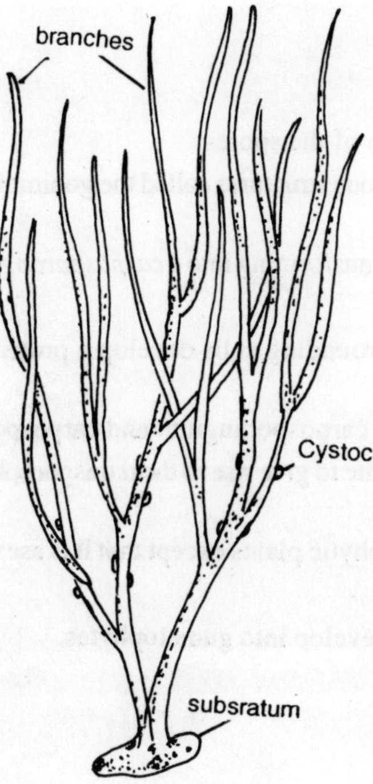

Fig. 9.3 Gracilaria. Thallus bearing cystocarps.

Anatomy of the thallus:

1. A transverse section of the thallus reveals three regions, namely, peripheral layer, cortex and the medulla.

2. The cells of the peripheral layer are small, uninucleate and rich in cytoplasmic contents.

3. The mature cells of the peripheral layer may be multinucleate.

4. Unicellular hairs are produced from some of the peripheral cells.

5. Internal to the peripheral layer is the cortex. It is made up of many small cells. The cells have many chromatophores.

Reproduction:

1. All the species of *Gracilaria* are dioecious.

2. The life cycle exhibits alternation of generations.

3. There are three types of plants, male plants, female plants and sporophytes. All of them are similar morphologically.

4. In the male plant, the reproductive structures are spermatangia.

5. The spermatangia develop from the cortical cells. Each spermatangium produces a spermatium. The spermatia are non-motile structures. They are dispersed by water current.

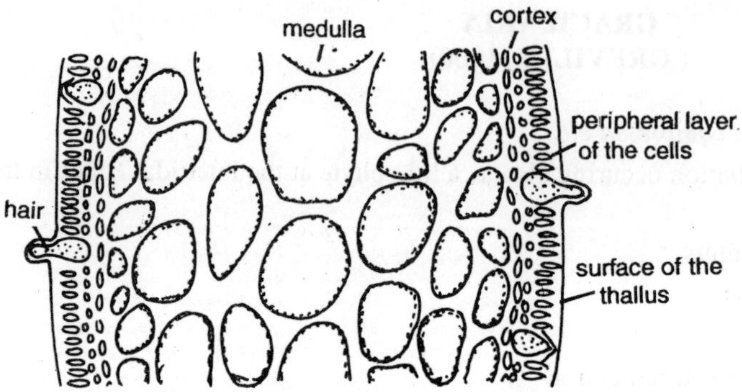

Fig. 9.4 Gracilaria. T.S. thallus.

6. In the female plant, the female reproductive organs, namely, the carpogonia develop from the cortical cells.

7. Some of the cortical cells distined to produce carpogonia, develop into supporting cells. Each supporting cells produces a hypobasal cell and a carpogonium proper.

8. The carpogonium is made up of a single flask shaped cell. It has a swollen base in which is situated the egg and an elongated trichogyne.

9. The fusion of the egg and spermatium results in the formation of the zygote which develops into the cystocarp or the carposporophyte.

Structure of the cystocarp:

1. The cystocarp is a globase structure. It has a short stalk.
2. The cystocarp has an opening called the ostiole, for the liberation of the spores.
3. At the base of the cystocarp is a placenta which gives rise to filamentous structures called the gonimoblast filaments.
4. Each gonimoblast filament is 3-4 celled long with the terminal cell transforming into a *carposporangium.*
5. Each carposporangium produces a diploid carpospore.
6. When the cystocarp is producing gonimoblast filaments, the surrounding cells develop a protective sheath.
7. The cystocarp has a wall, a placental cell, gonimoblast filaments, carposporangium and carpospores.
8. At maturity, the carpospores are liberated from the plant and germinate to give rise to the tetrasporophyte.

Structure of the tetrasporophyte:

1. Tetrasporophytes are exactly similar in morphology to the gametophytic plants except that it is asexual.
2. The sporophytes produce tetrasporangia from the cortical cells.
3. Each tetrasporangium produces four haploid tetraspores which develop into gametophytes.

Diagrams to be drawn:

1. Habit sketch
2. Anatomy of the thallus
3. Spermatangium
4. Carpogonium
5. Tetrasporophyte

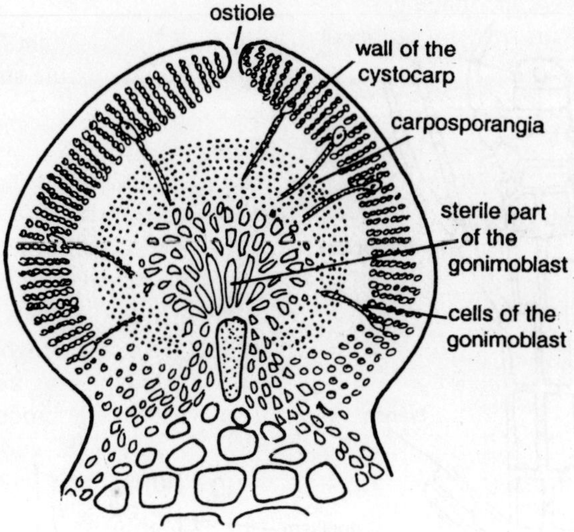

Fig. 9.5 Gracilaria. L.S. cystocarp

Systematic Position and Identification

A. 1. Chlorophyll present

2. Plant body is a thallus

3. Cell wall cellulosic ... *Algae*

B. 1. Motile reproductive cells absent

2. Chromatophores red pigmented

3. Cystocarp present ... Rhodophyceae

C. 1. Thallus dichotomously branched

2. Male and female sex organs develop from cortical cells...*Gigartinales*

D. 1. Carpogonia develop from hypobasal cells.

2. Tetraspores arranged in the form of a cross ...*Gracilariaceae*

E. 1. Thallus is monosiphonous

2. Epidermis or peripheral layered with a gelatinous layer...*Gracilaria*

PLOYSIPHONIA
(Greville, 1824)

Common species: *P. fastigata, P. platicarpa, P. variegata, P. ferulacea, P. urceolata, etc.*

Distribution: Marine alga found along the Altantic and Pacefic coasts. Sometimes grows as epiphyte on Fucales.

Study: 1. Structure of the plant body

2. Cell structure

3. Reproduction

4. Systematic position and identification

Structure of the plant body:

1. The plant body is a complex, branched, filamentous thallus.
2. It forms, dense, tufted, bushy filaments of purple red colour.
3. The thallus shows distinction into two types of filaments, viz., (a) Basal prostrate or decumbent filament and (b) vertical filaments that arise from the prostrate filaments.
4. The prostrate filament is attached to the substratum with the help of aseptate rhizoids.
5. The free ends of the rhizoids enlarge to form disc like haptera.
6. The filaments show dichotomous or lateral branching.
7. The main axis as well as the branches show a polysiphonous (multiaxial) arrangement
8. There is a central row of elongated cells called the *axial siphon.*
9. The axial siphon is surrounded by four to twenty pericentral siphons.
10. Some of the ultimate branches may be Monosiphonous (uniaxial) terminating into hair like structures called trichoblasts.

11. The trichoblasts may be sterile or fertile bearing the sex organs.

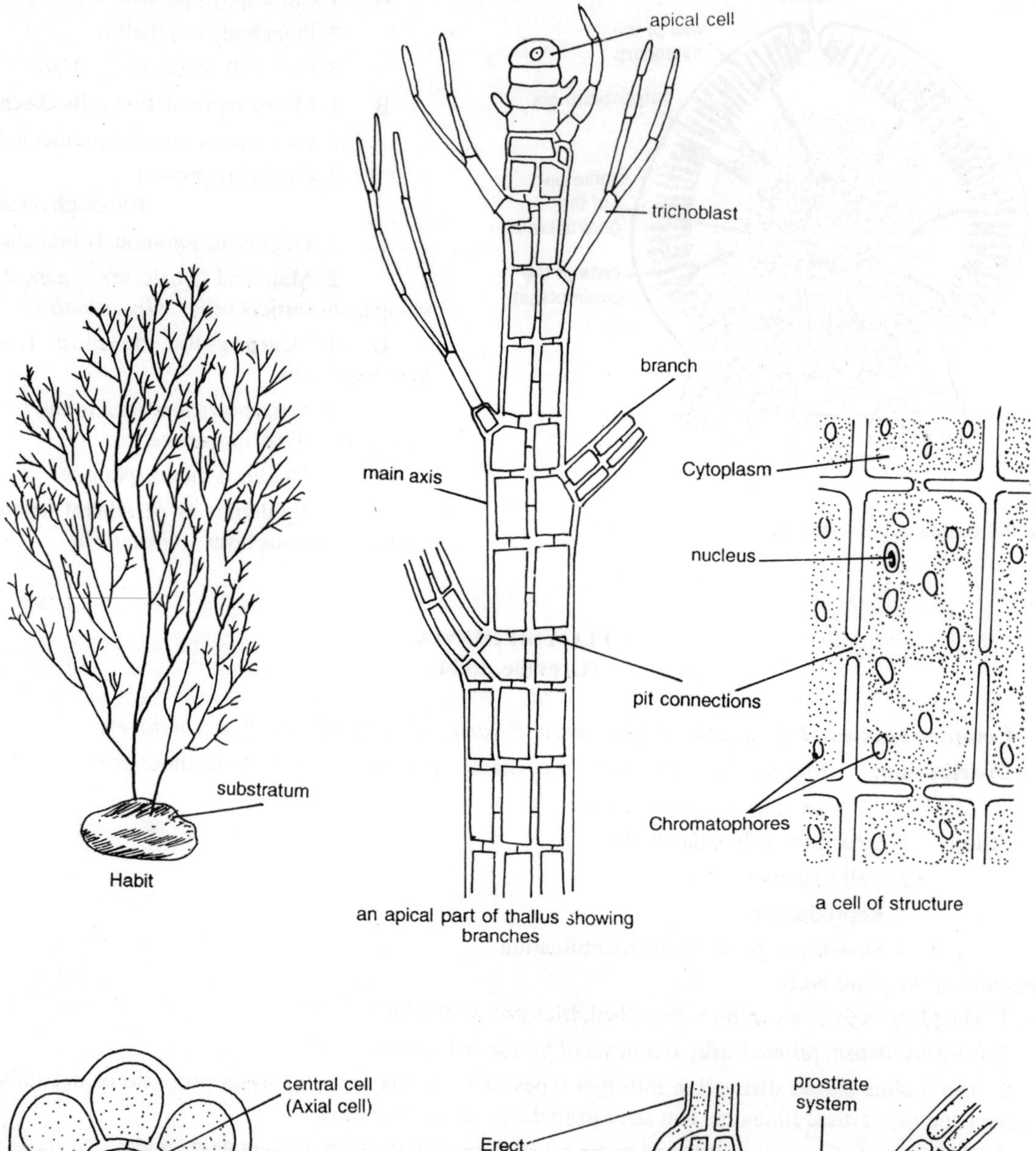

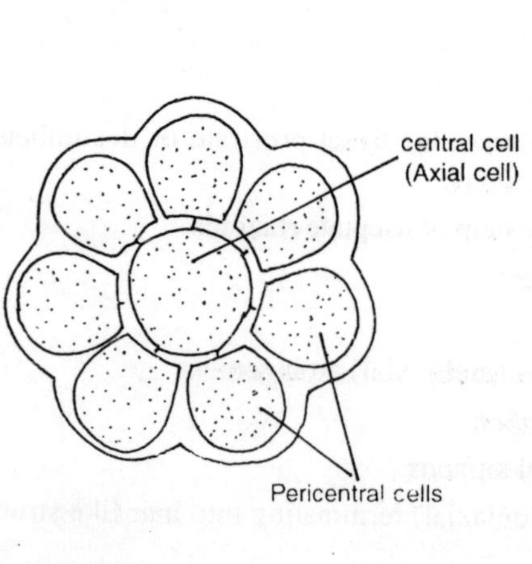

Fig. 9.6 Polysiphonia. T.S of axis

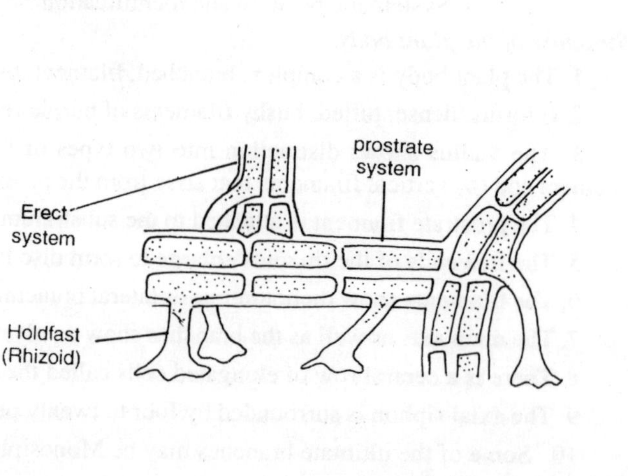

A part of thallus with rhizoids

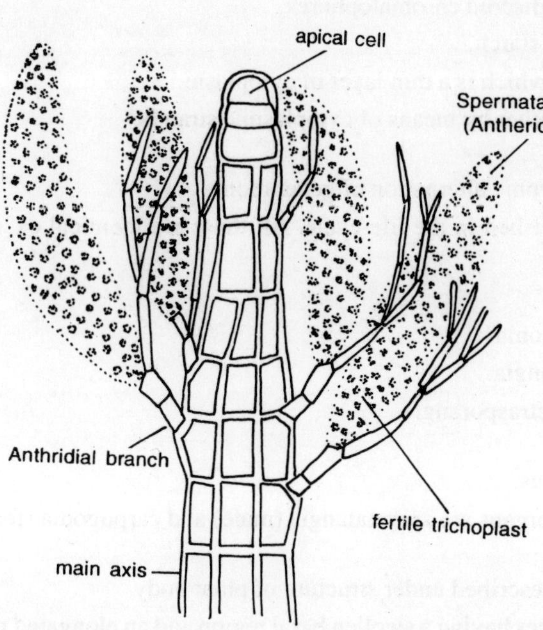

thallus bearing spermatangia

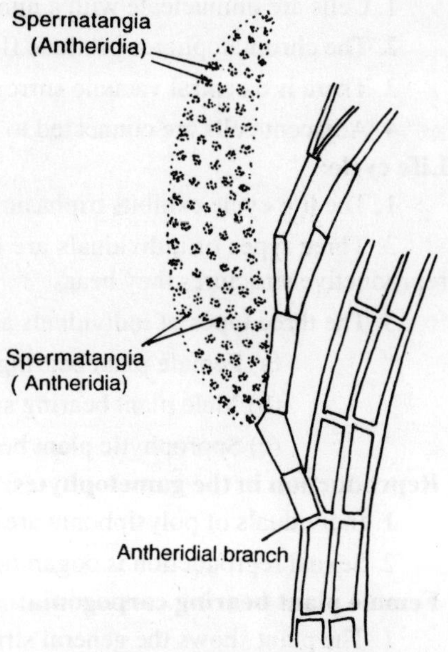

A fertile branch bearing spermatangium

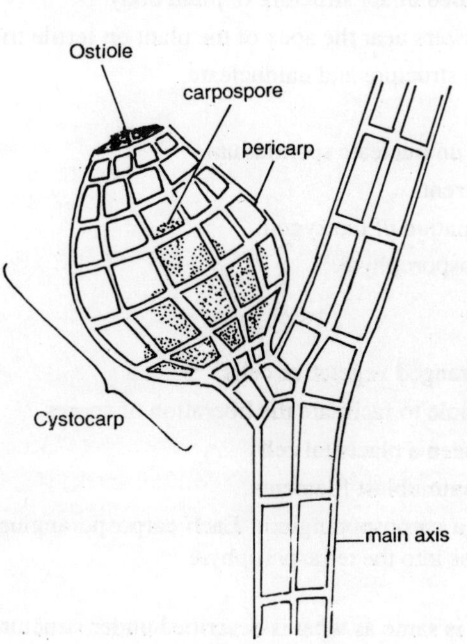

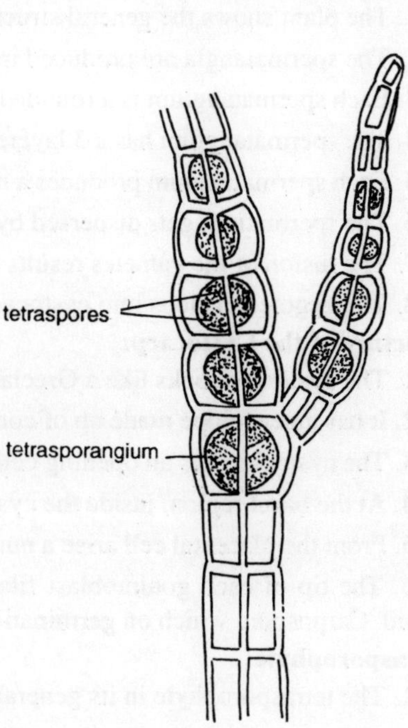

Fig. 9.7 Polysiphonia. A filament with cystocarp

A tetrasporophyte bearing tetrasporangia

Cell structure:

1. Cells are uninucleate with a number of discoid chromatophores.
2. The chromatophores produce floridean starch.
3. There is a central vacuole surrounding which is a thin layer of cytoplasm.
4. Adjacent cells are connected to one another by means of cytoplasmic strands.

Life cycle:

1. The life cycle exhibits triphasic, isomorphic alternation of generations.
2. Three types of individuals are distinguished in the life cycle. All these are identical except for the reproductive structures they bear.
3. The three types of individuals are:
   (a) Female plant bearing carpogonia.
   (b) Male plant bearing spermatangia.
   (c) Sporophytic plant bearing tetrasporangia

Reproduction in the gametophytes:

1. Individuals of polysiphonia are dioecious.
2. Sexual reproduction is oogamous, sex organs are spermatangia (male) and carpogonia (female).

Female plant bearing carpogonia:

1. The plant shows the general structure described under structure of plant body.
2. The carpogonia are flask shaped structures having a swollen basal region and an elongated neck called the trichogyne. There is a single egg at the base of carpogonium.
3. Carpogonia are produced either on carpogonial branches or on trichoblasts.

Male plant bearing Spermatangia:

1. The plant shows the general structure described under structure of plant body.
2. The spermatangia are produced in dense clusters near the apex of the plant on fertile trichoblasts.
3. Each spermatangium is a rounded or oblong structure and uninucleate.
4. The spermatangium has a 3 layered wall.
5. Each spermatangium produces a non motile uninucleate spermatium.
6. The spermatium gets dispersed by water current.
7. The fusion of the gametes results in the formation of the zygote.
8. The zygote develops into cystocarp or carposporophyte.

Structure of the Cystocarp:

1. The cystocarp looks like a Grecian urn.
2. It has an envelope made up of compactly arranged vegetative cells.
3. The cystocarp has an opening celled the ostiole to facilitate the liberation of spores.
4. At the basal region, inside the cystocrap is seen a placental cell.
5. From the placental cell arise a number of gonimoblast filaments.
6. The tip of each gonimoblast filament has a carposporangium. Each carposporangium produces a diploid. Carpospore which on germination develops into the tetrasporophyte.

Tetrasporophyte:

1. The tetrasporophyte in its general structure is same as what is described under structure of the plant body.
2. Tetrasporophyte reproduces asexually by means of tetrasporangia.
3. The tetrasporangia are spherical structures borne on one celled stalks.

4. The terasporangia arise laterally from the pericentral cells of the plant.

5. The tetrasporangium divides meiotically to produce four haploid tetraspores.

6. The tetraspores are arranged tetrahedrally. On germination they develop into gametophytes either male or female.

Diagrams to be drawn:

1. Habit sketch
2. A part of the plant enlarged to show branching.
3. T. S. of the thallus.
4. A few cells enlarged to show cytoplasmic connections
5. Male plant with spermatangium.
6. Female plant with carpogonium
7. Cystocarp
8. Tetrasporophyte

Systematic position and Identification:

A. 1. Chlorophyll present
2. Plant body is a thallus
3. Cell wall is cellulosic... *Algae*

B. 1. Motile reproductive cells absent
2. Chromatophores red pigmented
3. Cystocarp present... *Rhodophyceae*

C. 1. Plant body polysiphonous
2. Life cycle is triphasic ... *Ceramiales.*

D. 1. Cells uninucleate with discoid chromatophores
2. Trichoblast bearing spermatangia shows dichotomous branching
3. Carpogonia on unbranched trichoblasts ....*Rhodomelaceae*

E. 1. Plant body has a central siphon surrounded by many pericentral siphons.
2. Tetrasporangia borne singly on the thallus... *Polysiphonia*

PART TWO
FUNGI

CHAPTER 10

INTRODUCTION AND GENERAL CHARACTERS

Fungi represent one of the most important groups of plant kingdom. The study of fungi is known as Mycology. Etymologically mycology means study of mushrooms (Gr. *mykes* = mushroom; *logos* = study). Though the name mycology was originally meant to include only the mushrooms, nowadays it has a broad meaning. It encompasses all the fungi. The reason for mushrooms being the first to be studied among the fungi is that they are one of the largest among fungi and hence attracted the attention of naturalists even before the invention of microscope.

With Leunwenhoek inventing the microscope in the seventeenth century, the systematic sudy of fungi began. Pier Antonio Micheli- the Italian Botanist did a great lot to found the science of mycology and with justification he deserves to be called the founder of Mycology. In 1729 the published *Nova plantarum genera* in which he included his researches of Fungi.

What are Fungi?

It is really rather difficult to define the term fungi giving enough justice to its scope. Without entering into any hair splitting argument we can generally accept the definition given by Alexopoulos (1925). He defines fungi as "nucleated spore bearing acholorophyllours orgainsms which generally reproduce sexually and asexually whose filamentous, branched somatic structres are typically surrounded by cell walls containing cellulose or chitin or both."

Present day fungi include both common plants like mushrooms, puff balls, toadstools, morels rusts, smuts and host of other members.

Fungi and Mankind

Of what importance is the study of fungi to mankind? This question can be very easily answered by listing, both the good and bad done by fungi to man.

Fungi are friends as well as foes to mankind. Though the systematic study if fungi is only two and a half centuries old, the manifestations of this great group of plants on human beings, has been felt since time immemorial. Alexopolous in a lighter vein mentions "ever since the first toast was proposed over a shell full of wine, and the first loaf of bread was baked fungi are known to mankind." It is perhaps impossible to imagine how the world would have been without these fungi. Without fungi of course we would not have had such deadly plant diseases like rusts and smuts which take such a heavy toll of our wheat crop . Our potatoes would not have had the notorious late blight, areca palms would have been safe without kole roga, grapes would not have suffered with mildew disease, and we need not have bothered about food articles which now get the attack of molds; not much of care would be needed to maintain aspectic conditions in the laboratory. Our clothes, shoes would have been free from the green molds and the world would have been a real paradise to live in without these ugly creatures. But at the same time let us view the other side of the coin. Mankind would not have heard anything about alcoholic beverages; the famous champagne would be non existent, there would be no bakeries and perhaps there would be dead bodies of orgainisms all around for there would not be any saprophytic fungi to decompose them. There would not be that great potent drug Penicillin -the Cindrella of antibiotics and we would not have had such fine delicacies like the mushrooms and the morels , well these are a few of the favourite things we would have missed. And we really cannot afford to miss these. To provide both good and bad fungi are there and it is in his own interest that man is striving to have an intimate knowledge of these organisms, so that he can improve the races in which he is interested, and try to check the growth of others that are harmful to him. The above given explanation is just a representative of what the fungi are capable.

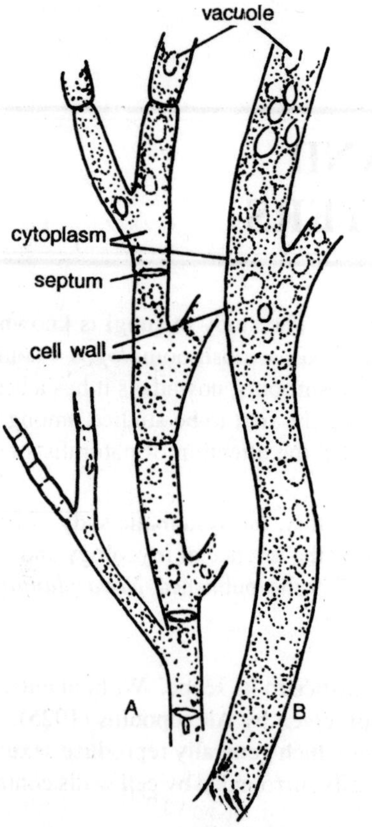

Fig. 10.1 Somatic structures of fungi. A-septate hyphae, B- aseptate and coenocytic hypha

General characters

Fungi comprise more than 3900 genera and 80,000 species.

Occurrence

Fungi have a great potentially for quick adaptation, hence they have colonised all possible types of habitats. Primitive fungi are aquatic in their habitat. Generally higher and higher fungi occupy amphibious and terrestrial habitats. Majority of the terrestrial forms are saprophytic in the soil or on animal and plant remains.

The fungi constitute a group of living orgainsms which do not have chlorophyll, roots, stems or leaves that are familiarly met with common plants. They are usually filamentous and multicellular (could be unicellular also). With few exceptions they exhibit little division of labour.

Nutrition and Growth

As fungi are achlorophyllous, they obtain nutrition by heterotropic mode.

This is of 2 types.

1. **Parasitic method**: [Gr.*Parasitos* = eating beside another] Parasitic fungi live on other organisms (plants, animals including human beings) and obtain their nutrition from a living host. In the process they may cause diseases to the host. Parasites may be ectoparasites, or endoparasites depending on their location in the body of the host. Ectoparasites live on the body of the host(e.g Erysiphales), while endoparasistes live inside the tissues of the host.

Parasites may also be classifed into obligate parasites and facultative parasites. Obligate parasits are those which can live only on a living protoplasm, while facultative parasites depending on the circumstances may live on dead organic matter.

2. *Saprophytic method*; (Gr. - *sapros*-rotten + *bios* - life). Fungi which obtain nutrition from dead and decaying organic remains are called saprophytes. Saprophytic fungi can be easily cultured in the laboratory. Saprophytes are also of 2 types (1) Obligate saprophytes and (2) Facultative saprophytes. Fungi which live on dead organic matter and are incapable of infecting living organisms are called obligate saprobes, those capable of causing disease on a living host or living on dead organic matter depending on the situation are called faculatative saprobes.

Mycorrhiza

Some fungi, enter into an association with the roots of higher plants and it is known as mycorrhiza (Gr- *Mykes*, mushroom; *rhiza*-roots). In many cases the association is mutually beneficial. Though much has been discovered (Kelley-1950) with regard to mycorrhiza the exact relationship is doubtful.

Fungi can synthesise all their nutritional requirements if they are initially supplied with an organic food such as carbohydrates. Laboratory-studies have indicated that fungi require all the essential and trace mineral elements for their growth. They can synthesise their vitamin requirements generally.

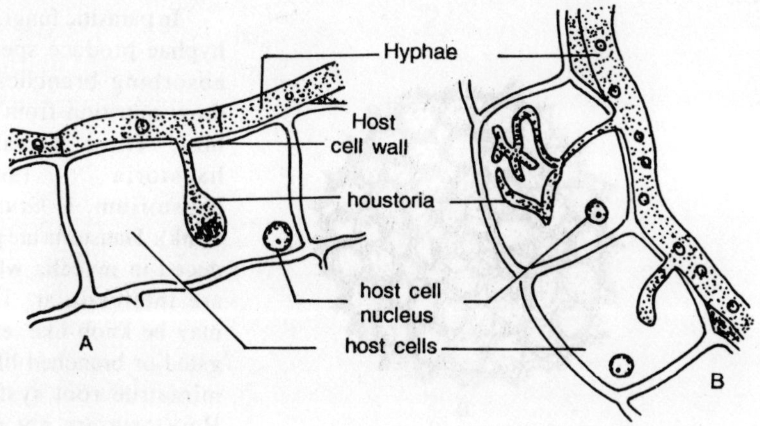

Fig. 10.2 Types of Haustoria in fungi. A-Saccate haustorium, B- Branched haustoria (In some cases, the hyphae are intercellular)

While some of the fungi are omnivorous and can grow on any moist organic matter, others are specific about nutrition and grow only on particular species.

Most fungi are capable of growing in the temeprature range between 0 and 35°C, the optimum temperature being 20-30°C. But they can withstand at least for few hours such low temperature as -195°C. As a contrast to bacteria fungi prefer an acidic medium with a pH of 6 being most suitable.

While light is not very much necessary for the growth it is needed at least at the time of spore formation.

Somatic Structure

The plant body of fungus generally consists of thread or tube like cylindrical filamentous structure called Hypha.(Pl.*hypae*, Gr-*Hypho*-web). A hypha consists a of thin wall and has inner lining of cytoplasm. A hypha may or may not have cross wall. The crosswalls are called Septa. (Sing -*Septum*. La-*Septum*-hedge or partitions). A hypha with septa is called *septate*, without it is called *aspetate*.

Cell wall Composition

The chemical composition varies in different fungi. While cellulose is the chief component in some, in most of the higher fungi the wall is mainly composed of chitin, Besides these, hemicellulose, lignin like substances have also been found in the fungal cell-wall. However, none of these occurs in isolation, the fungal cell wall being very complex. The composition of the wall may vary in young and old hyphae in different pH etc.

All fungi have a definite nucleus. During nuclear divcision definite chromosomes are organised. Some mycologists believe that nuclear division in fungi is not directly comparable to mitosis.

The aspetate hyphae usually have, number of nuclei distributed in them, the condition being termed coenocytic (Gr -*Koinos*-commonn+*Kytos,* a hollow vessel). In septate hyphae, the cell may be uninucleate, binucleate or multinucleate. Vacuoles, oil drops and other inclusions are generally present in the cytoplasm.

The fungal plant body usually consists of a mass of-hyphae called mycelium (plants-mycelia). In forming a plant body the mycelia may be loose or form thick stands called *Rhizomorphs* (Gr -*rhizo*-root *Morphat*-shape). In this mycelial strands are closely united to form root like structures exhibiting division of labour. Rhizomorphs are resistant to adverse environmental conditions.

In parasitic fungi, the hyphae produce special absorbing branches to draw nutrition from the host. These are called haustoria (sing-haustorium, L-*haustos* drink). Haustoria are produced in mycelia which are intercellular. They may be knob like, elongated or branched like a miniatrue root system. Haustoria are not produced in artificual culture.

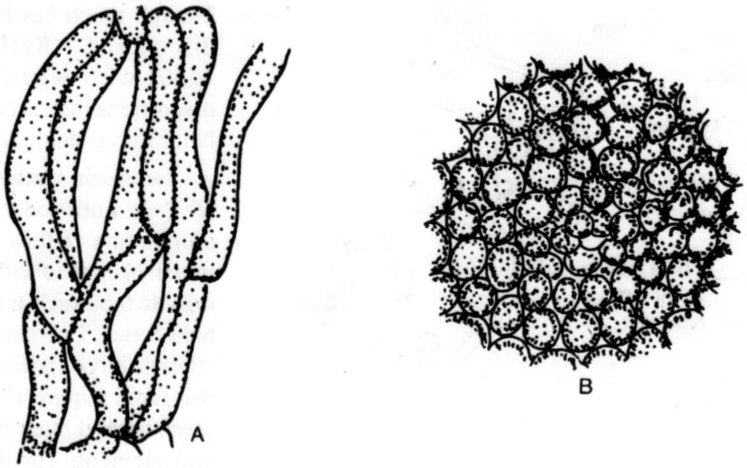

Fig. 10.3 Type of fungal tissues
A- plectenchyma, B- pseudoparenchyma (a cross sectional view)

Saprophytic fungi do not produce any haustoria as they absorb the food by direct diffusion causing disintegration of the organic matter.

Fungal hyphae have great longivity. As Alexopoulous states fungal colonies in nature, of more than 400 years old are known and it is possible that some mycelia are thousands of years old.

A mycelium starts its life as a short germ tube produced by the germination of a spore (Gr *spore*-seed). Spores are asexual propagating units of fungi.

In higher fungi, the mycelium has a tendency to organise into compact or loosely interoven tissues to form compact macroscopic strutcures. The general term plectecnchyma (or *Pleko* - weave + *enchyma*-infusion, i.e., a woven tissue) is used to designate all types of fungal tissues. There are 2 general types of *plectenchyma* - *prosenchyma* (Gr *Pros* towards+ *enchyma*-infusion ie., approaching a tissue) and *pseudoparenchyma.* In prosenchyma the hyphae are loose and they can be identified individually. They are arraged more or less parallel to to one another. Pseudoparenchyma (Gr. *Pseudo*-false +Parenchyma-a type of palnt tissue), consists of closely packed more or less isodiametric or oval cells resembling the parenchyma tissue of higher plants. In this type of tissue the hyphae lose their identity and are not distingusihable.

Prosenchyma and pseudo parenchyma tissues compose many somatic structure in higher fungi. Two such structures are *storma* (pl. Stromata, Gr *stroma* =mattress)and the *sclerotium.* (Pl-sclerotia, Gr *skeleros* =hards). A stromata, is a flat compact mattress like structure in which frucifications are formed. A sclerotium is a hard resting body which is resistant to adverse environmental conditions. It remains dormant and germinates during favourable conditions.

Reproduction: Fungi reproduce both asexually and sexually. In some fungi, during the formation of reprodctive structures (asexual or sexual) the entire thallus is used up. Such a case is common in lower fungi and they ae called *Holocarpic* (*Holos*-whole +*karpos*-fruit), In higher fungi however both vegetative and reproductive phases can occur together and these fungi are called *Eucarpic* (Gr Eu good, karpos fruit). Holocarpic nature is regarded as Primitive.

Asexual Reproduction: Asexual reproduction is of frequent occurrence in fungi than sexual and it is this which accounts for the rapid distribution of fungi. Asexual reproduction takes place by a variety of methods. They are (1) Fragmentation (2) Fission (3) Budding (4) Spore formation.

Fragmentation:

This is of 2 types : (a)Accidental and (b) purposive. In the first type when the mycelium accidentally breaks up into several fragments each one will develop into a new individual. In the second type, fragmentation of the mycelium, takes place as normal means of propagation. The hyphae breaks up into individual cells. Each such isolated cell is called Oidium. (Pl-oidia, Gr *oidion* = small egg) or Arthospore (Gr *Arthos*-joint spore- seed). Each arthosopre develops into a new individual. During unfavourable conditions arthospores surround themselves with thick wall. Such spores are called Chlamydospores (Gr *chalmyds*= mantle). They can withstand the inflavourable conditions.

Fission is the simple splitting of the cell into daughter cells. First, constriction of the nucleus of the cell takes place followed by the division of the nucleus and the formation of cell-wall. Fission as a means of reproduction is seen in unicelluair forms like yeasts and bacteria.

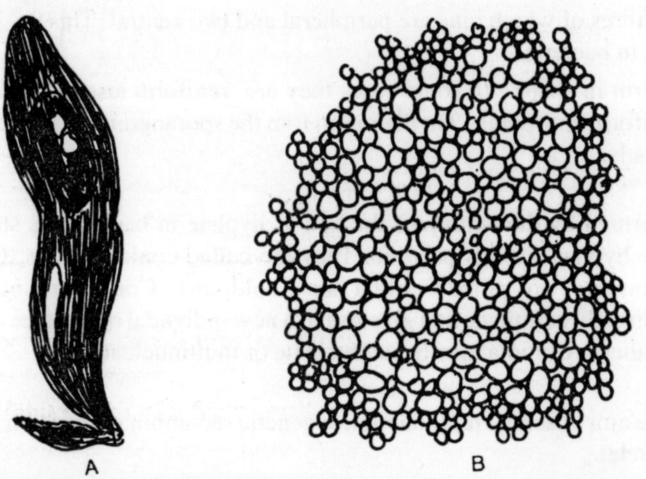

Fig. 10.4 Modifications of mycelium into a sclerotium. A- entire sclerotium, B- cross sectional view of a sclerotium

Budding: The production of a new cell by the outgrowth (bud) of a cell is called budding. As the bud is being formed, the nucleus in the parental cell divides with one of the daughter nuclei entering the bud. The bud eventually breaks from the parental cell and develops into a new individual.

Spore formation: By far the commonest mehtod of asexual reproduction is by spore formation. Spores vary in colour from hyaline through green yellow, orange, red, brown to black; in size from minute to large; in shape from globose through oval, long, needle shaped to helical; in number of cells from one to many; in the arrangement of cells and in the way in which spores themselves are borne (Alexopoulos 1952). As Alexopoulos states it is this infininte variety of spores that makes the study fascinating. While some fungi produce only one type others produce many types of spores.

Spore formation may usually be categorised into 2 types- *Internal* and *External*.

Internal spore formation:

The spores are formed internally in some parts of the hyphae. The parts of the hyphae which are set apart for spore production are called sporangia (sing-sporangium: Gr -*angion*-vessel). The spores themselves are called sporangio spores. A sporagium may be globse or oval or scarcely distinguishable from the rest of the hyphae. Generally it is a sac like structrure whose contents are transformed into many spores. At the time of sporangial formation the tip of a hypha usually swells into which cytoplasm and nuclei migrate. Later this is cut off from the remainder of the plant by a cross wall. Protoplasmic cleavage takes place resulting in the formation of spores. Sporangiospores may be motile. Usually lower fungi produce motile spores. Motile spores are called planospores or zoospores (Gr *zoo* animal). In higher fungi the spores are nonmotile and are called aplanospores. (Gr *a*= not + *plan* wander). The zoospopres have locomotory organs called flagella (singular-flagellum, *flagellum* = whip) which help in swimming. Zoospores may have one (uniflagellate) or two (biflagellate) flagella. The flagella found in fungi are of two types - *whiplash* and *tinsel*. The whiplash flagellum is unbranched and has a basal rigid portion much longer than the terminal flexible portion The tinsel flagellum is a branched feathery sturcture consisting of a rachis with lateral hair like projections.

Flagellar Apparatus:

This is a complex structure consisting of (a) flagellum (b) belpharoplast and (c) Rhizoplast. Blepharoplast (Gr. -*blepharis* = eyelash) which is the base of the flagellum is inside the cell. Rhizoplast is a thread like sturcure through which the flagellar base is connected to the nucleus. Electron micrographs have revealed that the flagellum consists of eleven fibres of which nine are peripheral and two central. This 9+ 2 arrangement is seen in all motile cells except in bacteria.

The zoospores are usually pyriform in shape. In some cases they are reniform also. The flagella are terminal in pyriform and lateral in reniform zoospores. On liberation fom the sporangium zoospores directly germinate to give rise to new individuals.

External spore Formation

In some cases, the spores are formed by the cutting of the tips of hyphae in basiptalous succession. There are specialised branches of the hyphae for this purpose. They are called conidiophores (Gr-*konis*= dust, *phore* = stalk). The spores themselves are called Conidia (sin Conidium). Conidia are non-motile. They are wind dispersed. Either they directly germinate and give rise to a new indivudal or produce zoospores which in turn develop into new individuals. Conidia may be uninucleate or multinucleate.

Sexual Reproduction:

As in other organisms the ultimate aim of sexual reproduction is genetic recombination which is accompanied by the fusion of compatible nuclei.

Fungi may be homothathic or heterothalic. In the former, both sex organs are formed on the same thallus and in the latter they are formed on different thalli.

The sex-cells are called Gametes. They are produced in sex-organs. Sex organs are called *gametangia*, if they produce gametes internally, or *gametophores* if they produce gametes externally. Gametophores however are found only in case of male (Gr, *gametos*=husband and *angion*-vessel). If gametangia produce only one type of gametes, they are called Isogametangia and the gametes themselves are called isogametes. If the gametes are of two different types such gametes are called anisogametes or heterogameetes and the gametamgia are called Heterogametangia. Heterogametangia are of two types, male and female. The male gametangium is called the antheridium and the female oogonium (Gr. *Antheros* =flowery;*oon* =egg+*gonos*= offspring). The female sex organ in Ascomycetes is ascogonium.

Having thus explained the nature of sexorgans, we may now learn the different types of sexual reproduction met within fungi. They are:

(1) **Planogametic Copulation**: This type is mostly seen in lower fungi. This involves the fusion of two motile gemetes (planogametes). The gametes may be isogametes or anisogametes. In this case, fusion is outside the gametangia. This is seen in some aquatic fungi belonging to lower phycomycetes.

(2) **Gametangial contact**: In a large majority of fungi, the male and the female gametes are undifferntiated uninucleate protoplasts. Here the sexorgans are antheridia and oogonia. In this method, the two sexorgans come in mutual contact. Usually to effect the contact, a conjugation tube or fertilization tube is formed from the antheridium which penetrates the oogonium. This forms a passage for the migration of male nuclei into the oogonium. After the nuclei migrate the antheridium disintegrates.

(3) **Gametangial copulation**: In this method, the entire contents of the contacting gametangia fuse. At the time of reproduction the two gametangia come nearer and become closely appressed. The walls between the two dissolve and the antheridical contents flow into the oogonium.

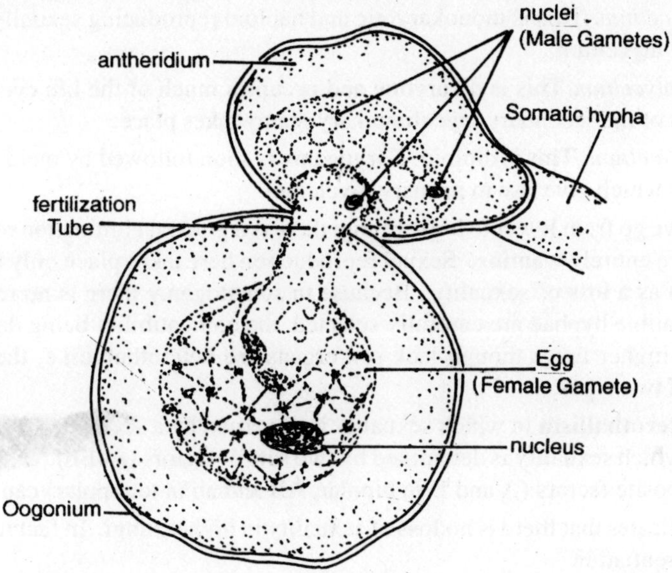

Fig. 10.5 Sex organs of fungi
Antheridium and oogonium showing gametangial contact

(4) **Spermatization**: In some higher fungi, special erect branches of the hyphae (spermatiophore) are produced which cut off uninucleate male reproductive cells called spermatia (Gr-spermatin-little seed). The spermatia are formed much the same way in which the conidia are formed. The spermatia are carried by wind or insects to the female reproductive structure- the receptive hypha the basal cell of which is fertile. The spermatial nucleus migrates into the receptive hypha.

Stages in Sexual Reproduction

Generally in lower fungi like phycomycetes, sexual reproduction is accomplished in three phases. In these forms haploid phase in predominant with a short diploid phase restricted to the zygote. The three phases are:

1. *Plasmogamy:* (Gr-*plasma*-molded object+ *gamos* - mariage union). In this, the two protoplasts (male and female) fuse bringing the nuclei closer.

2. *Karyogamy:* (Gr-*karyo*-nut, nucleus). In this phase, the two nuclei fuse.

3.*Meiosis*: (Gr-meiosis-reduction). This immediately follows karyogamy and restores the haploid complement of the chromosomes. In short, plasmogamy fuses the two protoplasts, Karyogamy brings about the fusion of two nuclei resulting in the formation of the diploid nucleus, meiosis restores the haploid condition.

In higher fungi (Ascomycetes and Basidimomycetes) however after plasmogamy, karyogamy is delayed for a very long time.

In the intervening phase the male and female nuclei just remain toghether without fusing. This condition is called Dikaryotic phase. Only at the end of the life cycle, kayogamy takes place which is soon followed by meiosis.

As a result of dikaryotic phase in the life history, three types of mycelia (with reference to nuclear complement) alternate. They are:

(1) *Primary mycelium:* This is monokaryotic and haploid reproducing sexually after plasmogamy. This produces secondary mycelium.

(2) *Secondary mycelium:*This is dikaryotic and occupies much of the life cycle. At the end of the life cycle, in some parts of the secondary mycelium karyogamy takes place.

(3) *Teritary mnycelium:* This is diploid. Karyogamy is soon followed by meiosis resulting in the formation of haploid cells which give rise to primary mycleium.

Sexuality: As we go from lower to higher fungi there is a gradual elimination of sex organs and in higher fungi, sex-organs are entirely wanting. Sexual reproduction here takes place only by somatogamy. But this should not be taken as a loss of sexuality. Because in somatogamy there is no random fusion of any two hyphae. The compatible hyphae are carefully selected, the compatibility being determined by genetic factors. Many of the higher fungi though lack sex-organs are heterothallic i.e, they are self incompatible. Heterothallism is of two types.

(1) **Bipolar heterothallism** in which sexualtiy is governed by a se[illegible] (2) **Tetrapolar heterothallism** in which sexuality is detemined by two sets of factors (AaBb). [illegible]wo cases only thalli which carry the opposite factors (A and B in bipolar, AB and ab in tetrapolar) can mate.

This clearly indicates that there is no loss of sexuality in higher fungi. In fact it is the most sophisticated type of sexual differentiation.

Classification: Fungi belong to the sub-division Thallophyta. Eichler (1886) divided Thallophyta into Algae and Fungi. Fungi included bacteria slime moulds and lichens.

Modern mycologists have placed fuingi under 3 different classes, namely:

(1)Schizomycetes (2) Myxomycetes (3) Eumycetes

Fitiz patrick (1930) removed Myoxmcyetes from fungi and divded Thallophyta into

1. Myoxthallophy (slime moulds)

2. Euthallophyta (include algae, fungi, bacteria and lichens). Engler clssified fungi as follows;

Eumycetes (true fungi),

Section A Lowerfungi

Class 1 : *Phycomycetes*

subclass i Oomycetes

subclass ii Zygomycetes

Section B Higher fungi

Class 2 : *Ascomycetes*

subclass i Hemiasci

subclass ii Euasci

Class 3 *Basidiomycetes*

subclass i Hemibassidii

subclass ii Eubasidii

subclas iii Heterobasidii

Section C *Deuteromycetes* or fungi imperfectii

Clement and Shear (1931) divided fungi into 5 divisions (1) Phycomycetes (20 Ascomycetes (3) Basidiomycetes (4) Promycetes (5) Deuteromuycetes.

Gwynne Vaughan and Barnes (1926) classified fungi into 5-classes, namely Phycomycetes, Archimycetes, Ascomycetes, Basidiomycetes and Deuteromycetes.

They divided phycomycetes into oomycetes and zygomycetes. Ascomycetes into plectomycetes, pyronomycetes and discomycetes. Basidiomycetes into protobasidiomycetes and Autobasidiomycetes.

Wolf and Wolf (1948) classifed fungi into Phycomycetes, Ascomycetes, Basidiomycetes, Myxomycetes and Deuteromycetes.

Gaumann and Dodge (1928) followed the classification of Gwynne Vaughan in the main divisions.

Tippu classified fungi into

1. Schizomycophyta Eg. Bacteria
2. Myxomycophyta Eg. Slime molds.
3. Eumycophyta-This includes four classes.
4. Phycomycetes. 2. Ascomycetes 3. Basidiomycetes 4. Deuteromycetes.

Alexopoulos (1956) places all fungi in the divisions Mycota. He however keeps away the order Acasiales and Labryintherlales form fungi as their affinitas are uncertain

The division mycota is divided into two sub-divisions (1) Myxomycotina and (2) Eumycotina

Myxomycotina has only one class-Myxomycetes

Eumycotina has the following classes

(a) Chytridiomycetes (b) Hypochytridiomycetes (c) Oomycets

(d) Plasmodiophoromycetes (e) Zygomycetes (f) Trichmycetes (g) Ascomycetes (h) Basidiomycetes and (i) Deuteromycetes.

CHAPTER 11

MYOXMYCETES

The members of the class myxomycetes are commonly called slime molds.They are true fungi in that they do not possess all the characters that we associate with fungi. In many of their structural attributes infact slime molds are in between plants and animals. Anton de Bary (1887) in fact considered them to be very close to animals and named them Mycetozoa (Gr *Mykes*-fungus, *zoon*-animal) Bessey (1950) another famous mycologist also followed de Bary.

The term myxomycetes, for the slime moulds was first coined by Macbride (1899). According to Macbride myxomycetes originated from a protozoan like ancestor.

Occurrence: Myxomycetes are found quite frequently in cool places. Decaying wood, humus rich soil, large fungi and decomposing debris of plant material are the favorite places to look for slime molds. A few of the myxomycetes are found on the lawn of grass. More than 450 speicies of slime moulds have been discovered so far.

The slime molds are universal in distribution, but a few of them are confined to tropics.

The vegetative stages of myxomycetes do not exhibit any plant charcaters at all. There is no cell wall. The cell wall makes its appearance only during the formation of spores.

The vegetative plant body of most of the myxomycetes is a naked, irregualr mass of protaplasm called plasmodium. The plasmodium is amoeboid in shape and multinucleate. The protoplasm can be divided into an outer ectoplasm and an inner endoplasm. Food vacuoles are present in endoplam. Nutrition is holozoic. The plasmodium moves with the help of pseudopodia.

The free living plasmodium becomes stationary during reproduction. Sporangia of various shapes and colours areproduced. The spores have a cell wall, the first indication of the plant nature of myxomycetes.

Classification: The class myxomycetes (equivalent to division myxomycota of some mycologists) is divided into three orders, namely plasmodiophorales, stemonitales and Acrasiales.

In this book *Plasmodiophora* and *Stemonites* are discussed.

PLASMODIOPHOIRA

CommonSpecies: *P.brassicae*

Distribtion: The fungus causes disease to cruciferous plants mainly to cabbage. The disease is known as Club root disease, finger and toe disease or clubbing disease.

Occurrence: The fungus is an obligate parasite occupying the underground portion of the cruciferous plants.

Study
1. Disease symptoms
2. Structure of the Pathogen
3. Reproduction
4. Systematic position and Identification.

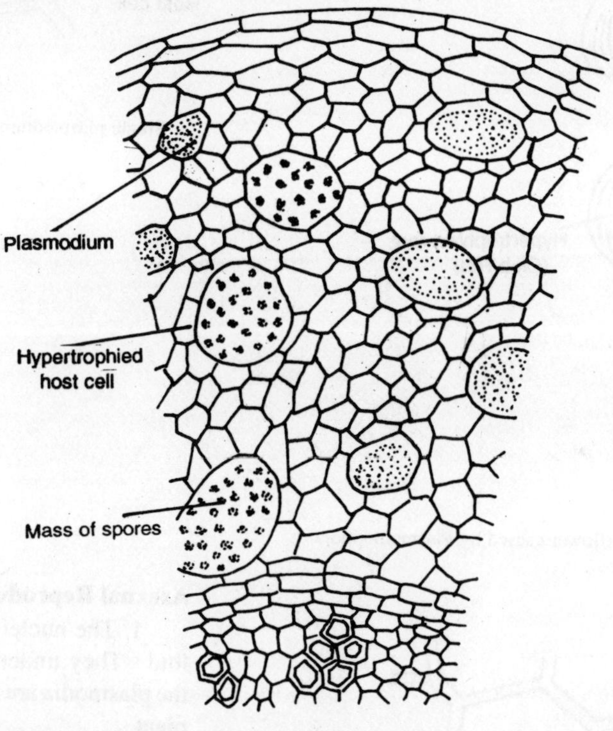

Fig. 11.1 T.S. of host root showing plasmodium and spores of Plasmodiophora

Disease symptoms

(1) The symptoms of infection are visible in the unnderground parts and mainly in the roots.

(2) Irregular growth or hypertorphy resultls in the root due to the infection by the fungus

(3) The hypertrophied roots get deformed appearing like club or knotted fingers and toes, hence the name of the disease.

Structure of Pathogen

1. The vegetative plant body of the fungus is a naked mass of protoplasm called Plasmodium

2. Multiple fission in the plasmodium results in the formation of a number of daughter plasmodia which develop into individual plants.

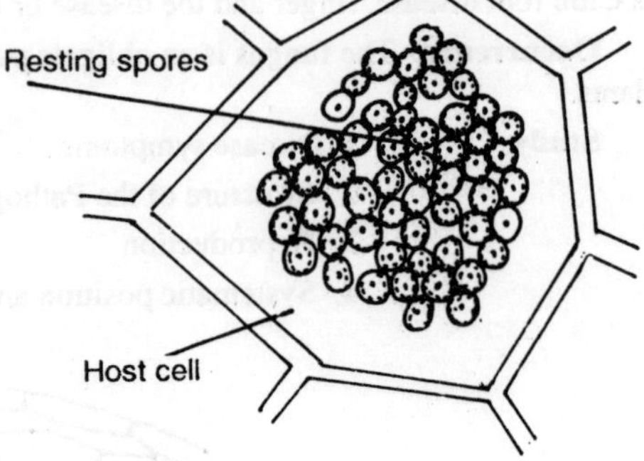

Single plasmodium

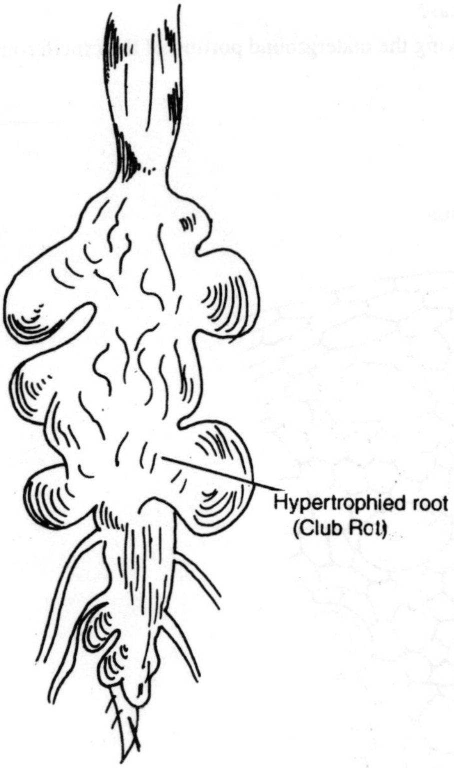

Fig. 11.2 Club root of cauliflower caused by Plasmodiophora

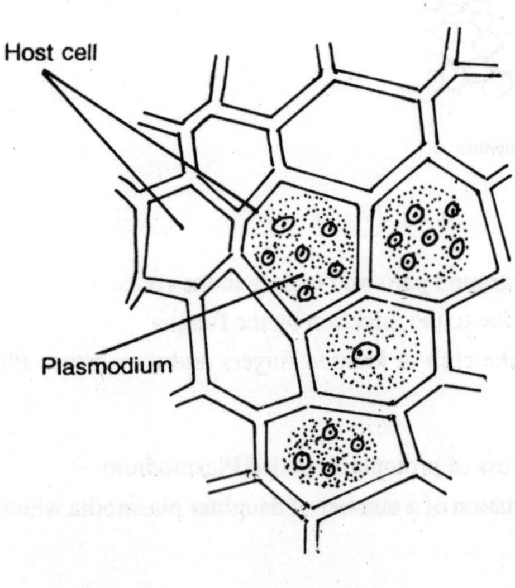

Fig. 11.3 Cells with spores masses

Asexual Reproduction

1. The nuclei in the plasmodium are diploid. They undergo reduction division when the plasmodia are in the cells of the root of host plant

2.After reduction, division, many individual, spherical spores are formed. The spores are non-motile and have a chitinous wall.

3. The death and decay of the root cells of the host sets of the spores free.

4. These spores (X) metamorphose themselves into biglagellate zoospores

5. The biflagellate zoospore infects the healthy plants, where it forms an irregularly shaped mxaymoeba.

6. On gaining entry into the host plant, they repeatedly divide to form a haploid plasmodium.

Sexual reproduction:

1. Each nucleus of the haploid plasmodium, gets isolated surrounds itself by a little cytoplasm and forms a gametangium.

2. The nucleus of each gametangium divides mitotically to form 8-10, biflagellate isogametes.

3. Two isogamates fuse and form a diploid zygote.

4. The diploid zygote divides mitotically and develops into diploid plasmodium.

Diagrams to be drawn

1. Habit sketch

2.Plasmodium

3. Host cell showing spores or zygotes.

Systematic Position and Identification:

A. 1. Chlorophyll absent
 2. Reserve food is glyucogen...*Fungi*

B. 1. Cell wall absent
 2. Protoplasm found in the form of plasmodium ...*Myxomycetes*

C. 1. No fructification
 2. Fungus is an endoparasite
 3. Vegetative development inside the host plant...*Plasmodiophorales*

D. 1. Gametes terminally flagellate
 2. Myxamoebae formed ... *Plasmodiophoraceae*

E. 1.Plant body is a plasmodium
 2. Both diploid and haploid plasmodia are found
 3. Infection causes club-shaped roots in the host ... *Plasmodiophora*

STEMONITES

Common species : *S. herbatica*

Distribution: The saprophytic fungus grows on dead and decaying wood, leaves, decomposed vegetable matter etc.

Study: 1. Structure of Plant body
 2. Reproduction
 3. Systematic position and Identification

Structure of plant body:

1. The vegetative plant body of *Stemonites* is a plasmodium.

2. It is a naked mass of protoplasm without any cell wall. The protoplasm has two ones an outer ectoplasm; and an inner endoplasm.

3. The plasmodium is multinucleate and also has many food vacuoles.

4. The plasmodium is amoeboid in shape and moves with the help of pseudopodia. Nutrition is holophytic

5. After a certain period of vegetative phase, the plasmodium becomes stationary on some substratum.

6. The entire contents of the plasmodium transform themselves into a sporangium

(It is is this sporangium growing on leaves or wood that is usually given in the laboratory as the specimen of *Stemonites*).

Reproduction:

1. Sexual reproduction is by spore formation-spores are produced in sporangia.
2. From the underlying plasmodium a sporanmgium develops

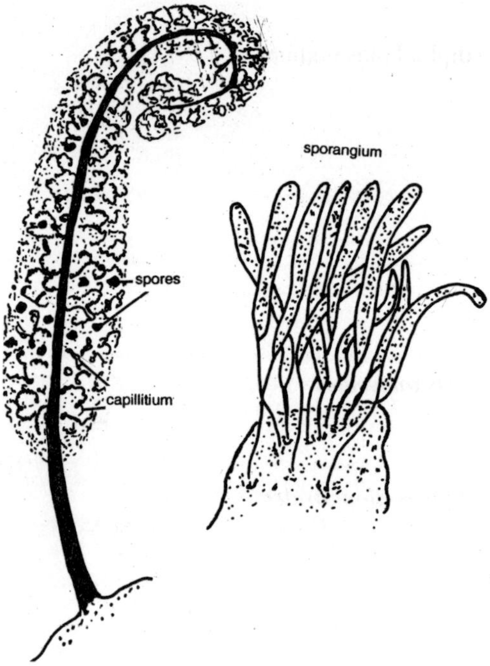

Fig. 11.4 Stemonites, habit

3. The sporangium has two parts, the stalk and the sporangium proper.
4. The stalk is called the stipe and it is an elongated structure
5. The sporangium proper has a central columella, which is the contninuation of the stipe.
6. The columella is surrounded by fine threads of cytoplasm which form a reticulum. The cytoplasmic threads are called the *capillitial threads.*
7. The surface (margin) layer of the capillitium develops a compact wall like layer called the Peridiium.
8. Within the meshes of the capillitial network are found many nuclei which undergo reduction division to form haploid spores.
9. The formation of the spores is the first stage in the life cycle indicating the plant characters of *stemonites*
10. The spores on liberation develop into amoeboid swarmars which are falgellate.
11. The swarm cells divide by means of fission to multipluy their number
12. The swarm cells fuse to form the zygote, which by repeated mitotic divisions develops into the Plasmodium

Diagrams to be drawn:

1. Habit skektch
2. Sporangium enlarged
3. A part of capillitum enlarged

Systematic position and Identification:

A. 1. Chlorophyll absent
2. Reserve food is glycogen....*Fungi*

B. 1. Cell wall absent
2. The vegetative pllant body is a naked mass of protoplasm called plasmodium ..*Myxomycetes*

C. 1. Plasmodium shows amoeboid movement
2. Fungus is a saprophyte ....*Stemonitales*

D. 1. The contents of the plasmodium get transformed into a sporangium
2. Plasmodium is always diploid....*Stemonitaceae*

E. 1. Sporangium consists of capillial network with a central columella
2. Spores are haploid and uninucleate....Stemonites.

CHAPTER 12

PHYCOMYCETES

The class Phycomycetes represents one of the lower classes of true fungi (Gr.Phycos-seaweed, myces-mushroom). They include about 254 genera and 1200 species. They are very common in occurrence. The bread mold (mucor), the water mold (saprolegnia), the white rust of mustard- all are Phycomycetes.

The plant body in Phycomycetyes is a conenocytic, asepate much branched mycelium, septa appearing either to demarcate the dead portions or at the time of formation of reprodcutive structures. Some of the lower forms are holocarpic, the rest Euçarpic.

Many primitive phycomycetes are predominantly aquatic in their distribution. Even higher forms (except Zygomycetes) show dependence on moisture. They may be parasites or saprophytes. Parasites grow on other fungi, algae, seed plants, ferns and animals also. Saprophuytic phycomycetyes occur in abundance in damp soil, dung and remains of plants and animals.

Reproduction*:* Phycomycetes reproduce by both asexual and sexual means. Asexual reproduction is brought about by the following ways (1) Fragmentation (2) Spore formation.

Phycomycetes produce both zoospores as well as condidia. *Saprolegnia, Phytopthora, Phythium* andthers produce zoospores. *Albugo* produces conidia. In *Mucorales*, asexual reproduction is by aplanospores. Chlamydospores are also found in some members.

Sexual reproduction*:*

Phycomycetes reproduce sexually by the following methods:

1. Planogametic copulation
2. Gametangial copulation
3. Gametangial contact.

In aquatic phycomycetes, sexual reproduction is by Planogametic copulation and in higher members it is by other two methods.

Stages in sexual reproduction:

There are only 3 stages, namely :

(a) Plasmogamy

(b) Karyogamy and

(c) Meiosis

In some cases as in *Mucor* and others, failure of fusion will result in the formation of parthenospores or Azygospores.

Classification

The filamentous phycomycetes were divided according to their mode of reproduction into two orders, namely: (1) Oomycetes (2) Zygomycetes. In oomycetes reproduction is oogamous and in zygomycetes it is isogamous (Martin 1950). The non-filamentous phycomycetes having round or lobed mycelical thallus were placed in Archimycetes (Gaumann 1949). But inclusion of all oogamous under omycetes and all isogamous forms in zygomycetes seems rather artifical. Gaumann (1952) divided the class into 5 orders based on mainly on the features of the thallus, flagellar type and reproduction. These are - Cytridiales, Blastocladiales, Monoblephoriales, Oomycetes and Zygomycetes.

Alexopoulos (1956) divides phcomycetes into 7 orders. Chytridiales, Monobleparidales, Plasmodiophorales Saprolegniales, Peronosporales, Mucorales and Entompohthorales.

Sparrow (1943) classifies phycomycetes into 3 series based on flagellar charcters.

Series 1-Uniflagellatae (flagellum one)

Series 2-Biflagellatae (flagella two)

Series 3-Aplanatae (no flagella)

Mc clean and Cook (1956) have divided phycomycetes into five orders, namely Chtryidiales, Acrasales, Saprolegniales, Blastocladiales and Monoblepharidales.

The following examples are discussed here

Synchitrium, Achyla Saprolegnia, Pythium, Phytophtora, Albugo, Sclerospora, Peronospora, Rhizopus, Mucor and *Pilobolus.*

ACHLYA

Occurrence: Most of the species of the fungus are aquatic sparophytes found in ponds, pools and stagnant waters. They are also common on decaying vegetable matter, dead flies etc.

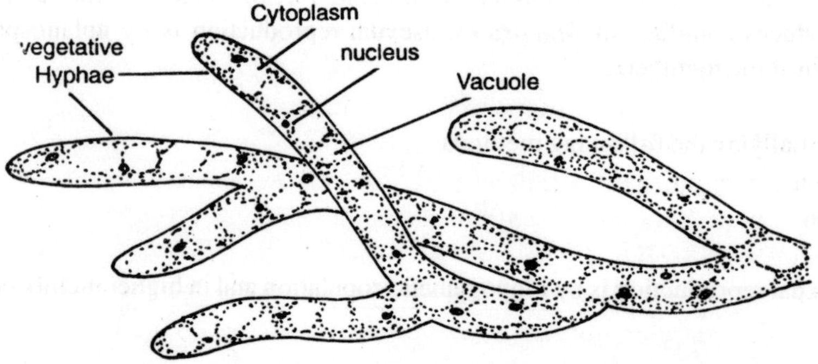

Fig. 12.1 Achylya. Vegetative mycelium

Study: 1. Culture in the laboratory.

2. Structure of the plant body.

3. Reproduction

4. Systematic position and identification

Method of culture in the laboratory:

The fungus can be cultured in the laboratory in the same way as *Saprolegnia*

Structure of plant body:

1. The plant body is mycelium consisting of numerous cottony white thread like structures.

2. The mycelium is aseptate, coenocytic and much branched.

3. The mycelium produces certain erect hyphae which grow vertically down into the substratum and help in the absorption of food material.

4. The hyphae are tubular and have a central continuous vacuole.

5. The cytoplasm is peripheral and has many nuclei. The cytoplasm also has oil globules and glycogen.

Reproduction :

1. Vegetative propagation is brought about by means of fragmentation. Each fragement develops into a new individual

2. Asexual reproduction takes place by zoospores formed in sporangia.

3. Sporangia are elongated cylindrical structures developed at the tips of hyphae.

4. The sporangia are separated from the rest of the hypha by a cross wall.

5. Each sporangium produces a large number of primary zoospores which are biflagellate and pyriform in shpae.

6. The primary zoospores undergo encystment and then give rise to kidney shaped secondary zoospores, which have laterally attached flagella.

7. The secondary zoospores give rise to teritary zoospores, which are similar in structure to the former.

8. The phenomenon of production more than two motile phases in the life cycle is called *Polyplanetism.*

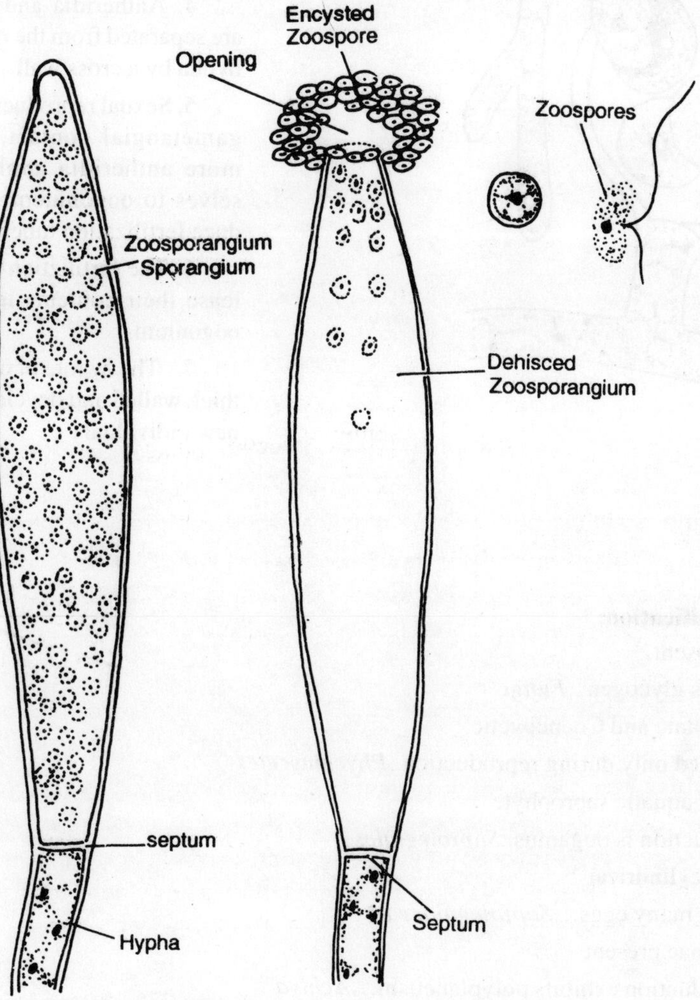

Fig. 12.2 Achylya. Single zoosporangium

Empty zoosporangium with encysted zoospores at the apex

9. The zoospores germinate to form a mycelium.

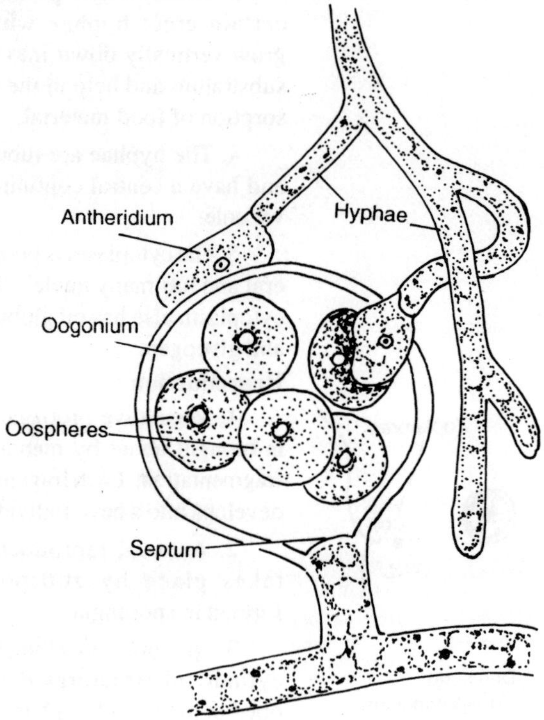

Fig. 12.3 Achylya. Sex organs

Sexual reproduction :

1. Sexual reproduction is oogamous. Species may be homothallic or heterothallic. The sex-organs are antheridia and oogonia.

2. Antheridia develop at the tips of hypha. They are multinucleate.

3. Oogonia are globose and multinucleate. They have many functional eggs.

4. Antheridia and oogonia are separated from the rest of the hypha by a cross wall.

5. Sexual reproduction is by gametangial contact. One or more antheridia apply themselves to oogoniuum and produce fertilization tubes.

6. The fertiliztion tubes release their contents inside the oogonium.

7. The zygote of oospore is thick walled and develops into a new individual .

Systematic position and Identification:

A. 1. Chlorophyll absent
   2. Reserve food is glycogen ..*Fungi*

B. 1. Mycelium aseptate and Coencoyctic
   2. Septa are formed only during reproduction..*Phycomycetes*

C. 1. The plant is an aquatic saprophyte
   2. Sexual reproduction is oogamus..*Saprolegiales*

D. 1. Zoosporangia cylindrical
   2. Oogonium has many eggs...*Saprolegniaceae*

E. 1. Rhizoidal hyphae present
   2. Asexual reproduction exhibits polyplanetism...*Achlya*

SAPROLEGNIA

Common species: *S.parasitica, S.terax* etc.

Occurrence: Saprolengia is an aquatic saprophyte. Majority of the species grow on dead animals and inside a water. *S. parasitica* is a parsitic species.

Study: 1. Culture method

2. Vegetative structure

3. Reproduction

4. Systematic position and Identification

Methods of culturing saprolelgnia in the laboratory:

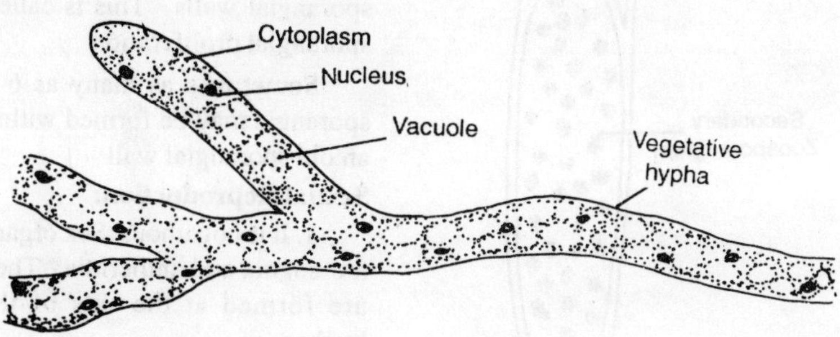

Fig. 12.4 Saprolegnia. Thallus structure

1. Select a clean petri dish.

2. Place one or two dead flies, small pieces of meat and some egg albumen

3. Add a little pond water

4. Close the petridish

5. Within a few days, the body of the housefly will be full of thread like mycelia of *Saprolegnia.*

Vegetative structrue:

1. The plant body is a mycelium. It is a soft whitish thread like structure.

2. The mycelium is coenocytic and profusely branched forming a white mold like growth on the substratum.

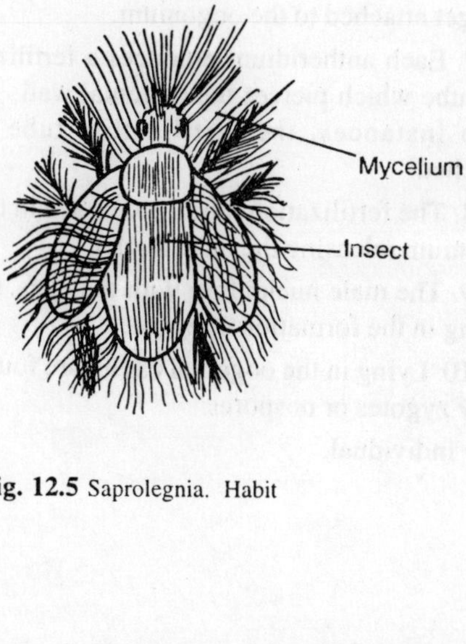

Fig. 12.5 Saprolegnia. Habit

3. Hyphae are aseptate and septa are formed only when the plants enter the reproductive phase.

4. The cytoplasm has many nuclei, oil globules and glycogen.

Reproduction:

1. *Saprolegnia* reproduces asexually as well as sexually.

2. Asexual reproduction is brought about by means of zoosporangia.

3. Zoosporangia are elongated club-shaped structures that develop at the tips of hyphae.

4. Each zoosporangium produces large number of pyriform biflagellate zoospores called primary zoospores.

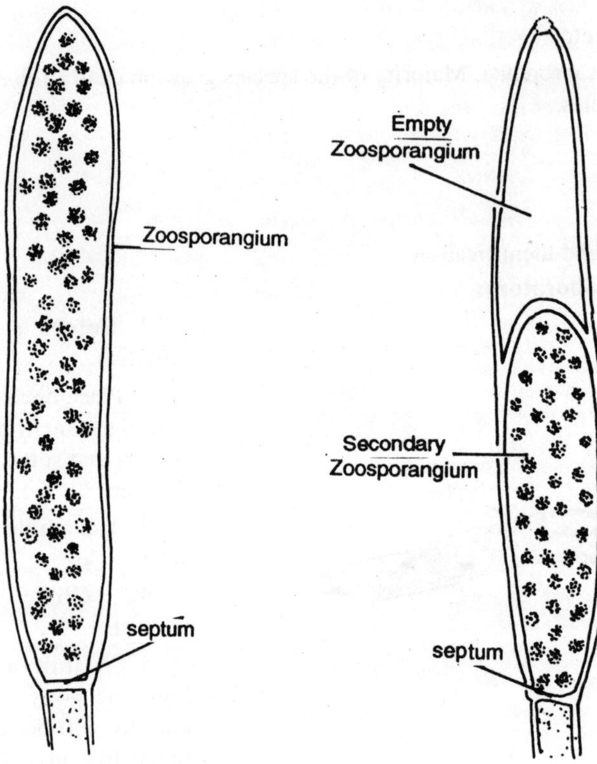

Fig. 12.6 Saprolegnia. Two zoosporangia

5. On liberation from the sporangium, the primary zoospores undergo encystment and then form secondary zoospores which are reniform in shape with lateralluy attached flagella. They develop into new individuals.

7. After the liberation of contents from a sporangium the cross wall beneath forms another sporangium, within the old sporangial walls. This is called sporangial proliferation.

Sometimes as many as 6-7 sporangia may be formed within an old sporangial wall

Sexual Reproduction:

1. It is oogamous. Sex organs are oognia and antheridia. They are formed at the tips of the hyphae.

2. Species of *Saprolegnia* are homothallic

3. The oogonium is a globose structure. At maturity it has 6-8 eggs or oospheres. All of them are functional.

4. The antheridium is a club shaped structure and has many male nuclei or sperms

5. Sexual reproduction is effected by means of gametangial contact.

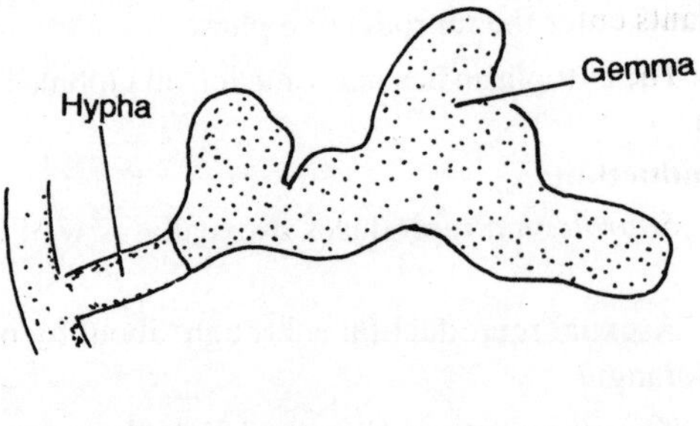

Fig. 12.7 Saprolegnia. Hypha with gemma

6. The antheridium becomes closely applied to the oogonium. One or more antheridia may get attached to the oogonium.

7. Each antheridium produces a fertilization tube which pierces the oogonial wall. In some instances, the fertilization tube is branched.

8. The fertilization tube bursts open in the oogonium releasing the male nuclei.

9. The male nuclei fuse with the eggs, resulting in the formation of zygotes.

10. Lying in the oogonial cavity are found many zygotes or oospores.

11. The zygote produces a germ tube and develops into a new individual.

Diagrams to be drawn:

1. Habit sketch
2. A part of the mycelium enlarged
3. Zoosporangium
4. Antheridia and Oogonia
5. Zygotes in the oogonial cavity

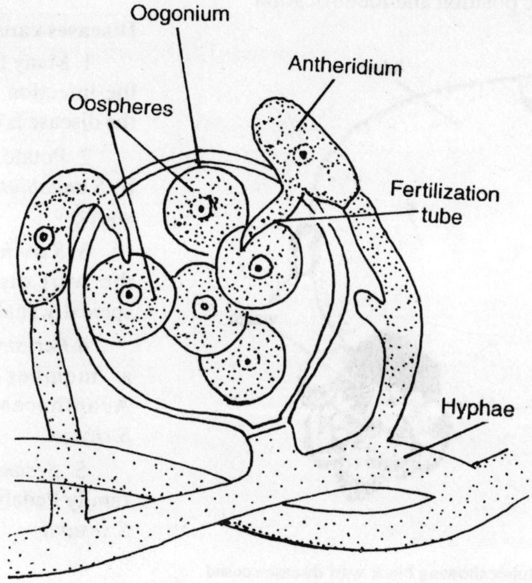

Fig. 12.8 Saprolegnia. Mycelium with sex organs

Systematic position and identification :

A. 1.Chlorophyll absent
2. Reserve food glycogen...*Fungi*

B. 1. Mycelium aseptate and Coenocytic
2. Septa are formed only during reproduction....*Phycomycetes*

C. 1. The plants is an aquatic saprophyte
2. Sexual reproduction is oogamous ...*Saprolegniales*

D. 1. Zoosproangia cylindrical
2. Oogonium has many eggs...*Saprolegniaceae*

E. 1. Sporangial proflieration is seen
2. Diplanetism is exhibited....*Saprolegnia.*

SYNCHITRIUM

Common species: *S. endobioticm, S.trichosanthoides, S.rhytzi, S. sesami* etc.

Occurrence: The fungus is parasitic and a pathogen causing several plant diseases.

Many of the economically important plants are infected by the fungus

Study: (1) Diseases caused by the fungus.

(2) Disease symptoms

(3) Structure of the fungus

(4) Systematic position and Identification

Diseases caused by fungus:

1. Many flowering plants get the infection by the fungus and the disease is called wart disease.

2. Potato tubers are infected by *S.endobioticum* causing potato warts.

3. *S.trichosanthoides* causes the wart disease of cucurbits speically snake gourds.

4. *Perystrophae bicalyculata*, a member of the family Acanthaceae is infected by *S.rhyizii.*

5. *Sesamum indicum* of the family Pedaliaceae is infected by *S.sesami*

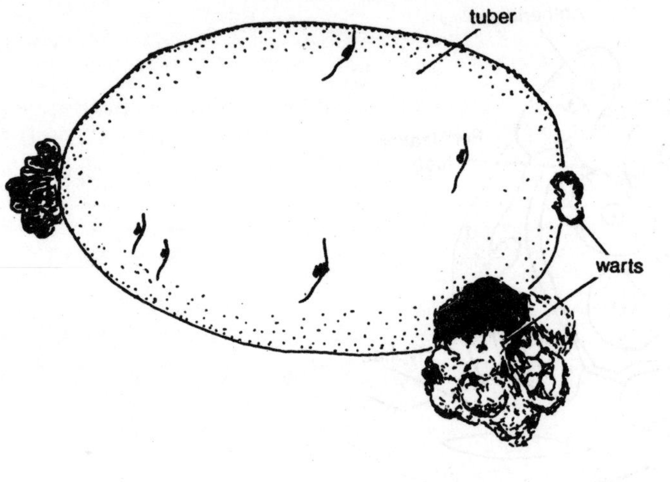

Fig. 12.9 Synchytrium. Infected potato tuber showing black wart disease caused by Synchytrium endobioticum

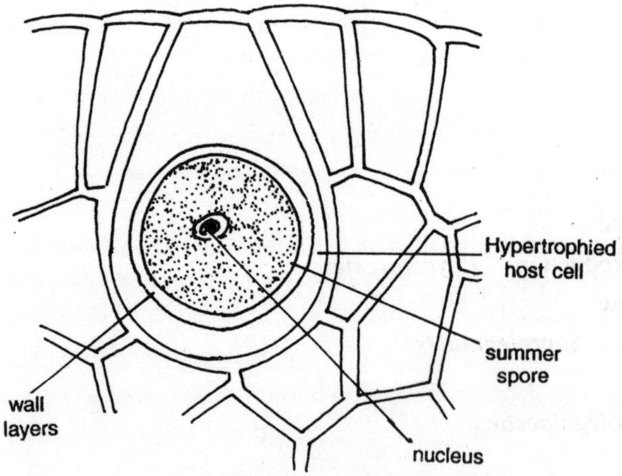

Fig. 12.10
Synchytrium.Summer spore within infected host cell

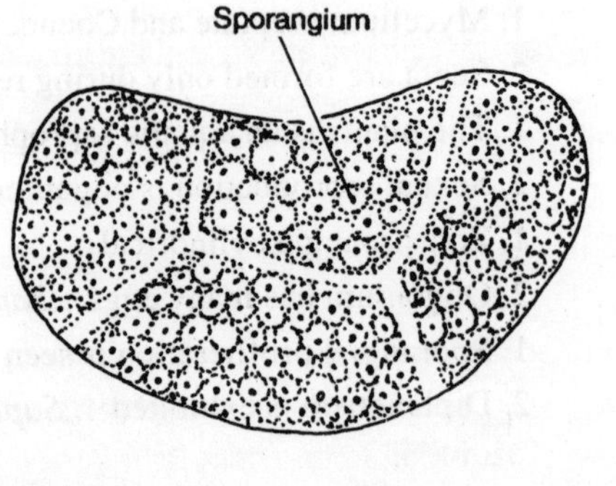

A sorus

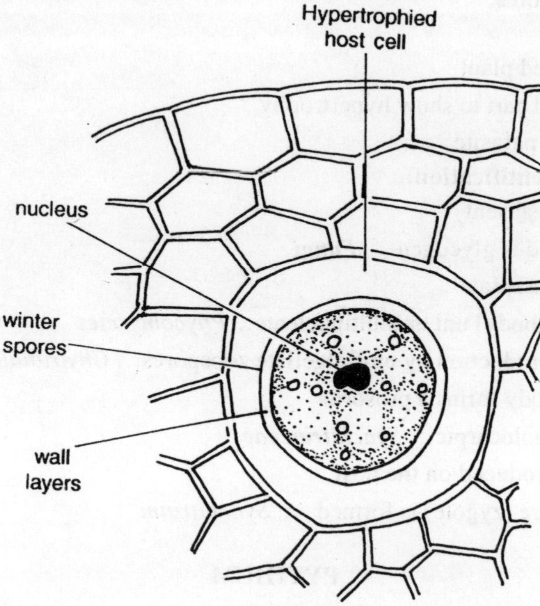

Fig. 12.11 Synchytrium. Winter sporangium within infected host cell

Disease symptoms

1. The characteristic symptoms of the disease are the development of small warts on the infected parts
2. In potato the tubers show the characeterstic warts
3. Many time hypertrophy and decay of host cells will also result

Structure of the fungus:

(1) The plant body of *Synchitrium* is unicellular.

(2) The cell remains within the host cell, i.e. intercellular., This condition is also said to be endobiotic.

(3) The thallus is globose and has a thick or sometimes thin chitinous wall surrounding it.

(4) The thick walled cell is also called Prosorus.

Asexual Reproduction:

1. The plant is holocarpic ie., the whole of the plant body gets converted into a reproductive structure.
2. Asexual reproduction is brought about the help of zoospores.
3. The zoospores can bring about a fresh infection.

Sexual Reproduction.

1. Sexual reproduction is resorted to only during unfavourable conditions. Otherwise, the usual method of multiplication is by zoospores.
2. The prosorus (the plant body) producers many gametes which are isogamous
3. The gametes fuse to form a diploid zygote which will be motile for sometime.
4. After a short free swimming period the zygote penetrates the host cell and undergoes a period of rest there. This is called the Resting spore.

5. At the return of favourable conditions, the zygote undergoes reduction division and produces a number of uninucleate haploid zoospores.

Diagrams to be drawn:

1. Habit sketch of infected plant
2. Section of the infected part to show hypertrophy
3. Host cell showing the parasite

Systematic Position and Identification:

A 1. Chlorophyll absent
2. Reserve food is glycogen .. ..*Fungi*

B 1. Plant body aseptate
2. Cells (plant body) uni or multinucleate...*Phycomycetes*

C 1. Asexual reproduction by uniflalgellate zoospores... *Chytridiales*

D 1. The plant body forms a prosorus
2. The cell is holocarpic....*Synchitriaceae*

E 1. Warts are produced on the host
2. Resting spore (zygote) is formed .....*Synchitrium*

PYTHIUM

Common Species : *P. debaryanum, P.graminicolum, P.myriotylum*

Occurrence: Many speices are amphibious. They mostly occur in moist humus soil and infect the hypocotyle region of seedling of plants like ginger, tobacco, mustard, wheat etc. After entering the host, the fungus acts as a parasite and causes disease like damping off, soft rot, foot root etc.

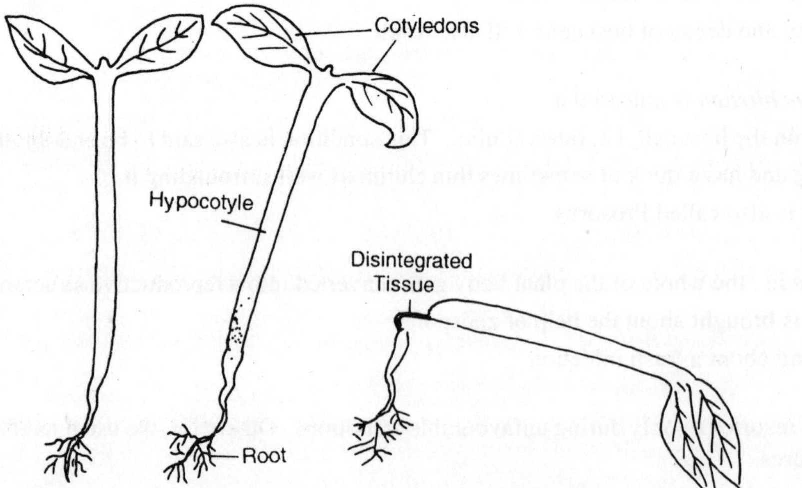

Fig. 12.12 Pythium. Host seedlings showing damping off disease caused by Pythium

Study: 1. Vegetative structure of Pathogen
2. Reproduction
3. Disease symptoms
4. Systematic position and Identification

Diseases caused by the genus:

1. *P.debaryanum*-Damping off tobacco seedlings.
2. *P.aphandiermatum*-Stem rot of papaya
3. *P.graminicola*-foot rot of what seedlings
4. *P. indicum*-Fruit rot of Lady's finger

Vegetative structures:

1. The plant body of the fungus is a mycelium

2. The mycelium is aseptate, coenocytic and much branched.

3. Hyphae have a vacuolated cytoplasm in which a large nuber of nuclei are embedded.

4. The mycelia superficially resemble the filaments of *Vaucheria*

5. Reserve food is in the form of glycogen

6. No haustoria are produced.

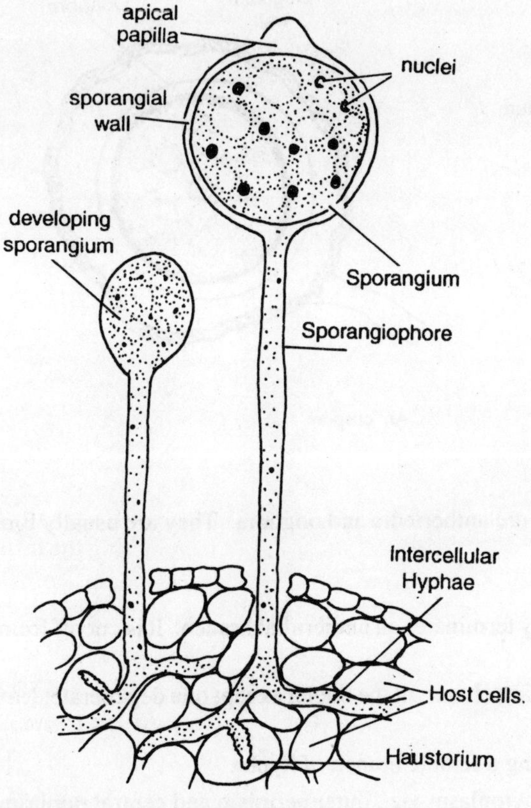

Fig. 12.13 Pythium. T.S. of host seedling showing zoosporangia

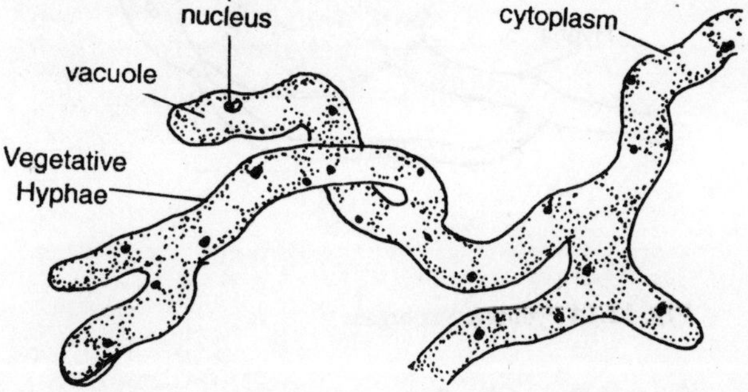

A portion of mycelium

Asexual Reproduction:

1. Asexual reproduction takes place by the formation of sporangia. Sporangia may be terminal or intercalary.

2. Sporangia are small, globular or sac like and are of two types viz., globose or spheroidal and elongate or filamentous.

3. Globose sporangia are globe like and terminal in position

4. Filamentous sporangia are indistingusihable from the hyphae.

5. The sporangia whether globose of filametnous are indistinguishable from the hyphae by a cross wall.

6. The contents of the sporangia form numerous zoospores which are reniform and bifflagellate.

7. Each sporangium has a beak like structure which enlarges to from a vesicle. It is into this vesicle that the zoospores are released first. Later the vesicle ruptures setting the zoospores free.

8. On germination each zoospore develops into a new individual

9. In some speices of pyuthium, the sporangia produce onon-motile conidia which germinate to give rise to new individuals.

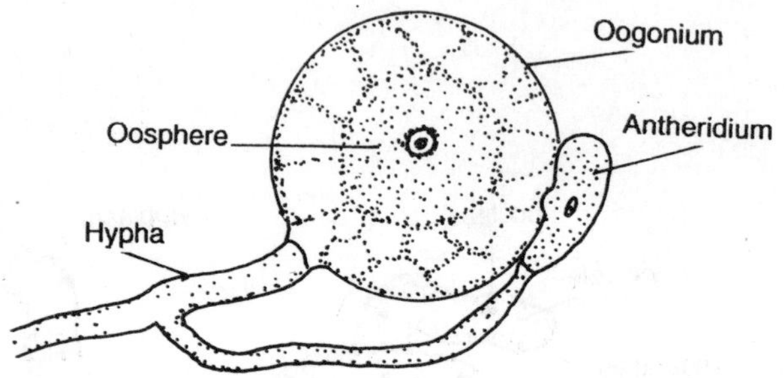

Fig. 12.14 Pythium. Sex organs

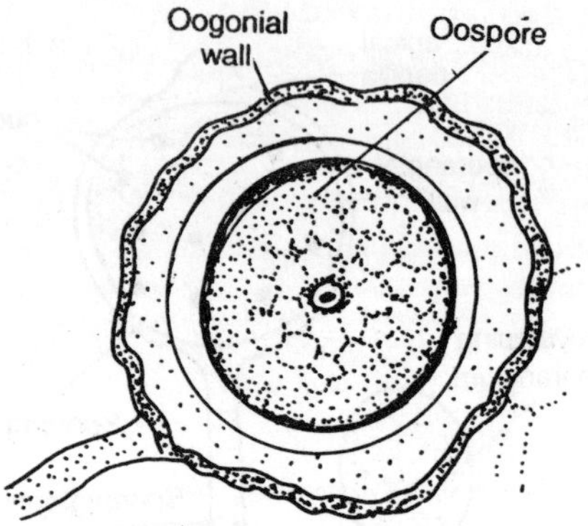

An oospore

Sexual Reproduction:

1. Sexual reproduction is oogamous. Sex-organs are antheriedia and oogonia. They are usually formed towards the end of the growing season.

2. Pythium is homothallic

3. Antheridium is small and club shaped; it usually terminates an antheridial branch. It is cut off from the rest of the hypha by a cross wall.

4. The antheridium is multinucleate in the beginning. Later all the nuclei except one degenerate, leaving only one functional male nucleus.

5. The oogonium is a globose structure terminating a female branch of hypha

6. The inside of the oogonium has two zones of cytoplasm viz., outer periplsm and central ooplasm.

7. The ooplasm has a single egg.

8. Sexual reproduction is effected by means of gametangial contact.

9. A fertilization tube develops from the antheridium. It penetrates the oogonium relasing the contents inside.

10. The fusion of the male and female gamete results in the formation of a zygote or oospore.

11. The oospore after undergoing a resting period develops into a new individual

Disease symptoms

The symptoms of the disease are noticeable only in the seedling stage after they have just come out of the soil. The seedlings become pale green. The stem at its base, i.e., just above the ground level exhibits decay of the tissues. The tissues turn brown. As aresult of the decay at the base of the seedling collapses.

In the fruit rot and root rot diseases, the tissues rot, and get decomposed resulting in the death of the plant.

Diagrams to be drawn:

(1) Habit of damping off of seedlings

(2) Zoosporangium

(3) Oogonium and antheridium

Systematic Position and Identification:

A. 1. Chlorophyll absent

2. Reserve food is glycogen ...*Fungi*

B. 1. Mycelium aseptate and coenocytic

2. Septa are formed only during reproduction

3. Oospore is thick walled...*Phycomycetes*

C. 1. Sexual reproduction not by Planogametic copulation

2. Oogonium has a single oosphere ...*Peronosporales*

D. 1. Sporangiophores not different from somatic hyphae...*Pythiaceae*

E. 1.Pathogen causes damping off of the seedlings.

2. Host stem has brown decay at the base

3. Zoospores are released into a vesicle

4. Facultative parasite...*Pythium*

PHYTOPHITHORA

Common species: *P.infestans, P.arecae, P.palmivora, P.colacasiae*

Occurrence: The fungus is a facultative parasite, occurs in soil as a saprophyte also gaining entry into the host plant at the opportune time.

The genus has more than 75 species and nearly all of them may be cultured artificially on the medium.

Study: 1. Disease caused by the fungus

2. Disease symptoms

3. Vegetative structure

4. Reproduction

5. Systematic position and Identification

Diseases caused by the fungus and symptoms:

Various species of *Phytophthor* a cause disease to economically important crops resulting in heavy damges. The following are some of the diseases

1. *Late blight of Potato*: This is caused by *P.infestans* and is of great significance as the crop is destroyed heavily. The disease appears periodically in the plains and hilly regions of Northern India.

The diseases symptoms are first noticeable on the leaves. The infected regions show small and pale coloured spots which eventually turn brownish or purplish discolouration become softened and the tuber decays.

As the disease symptoms appears late in the growing season. It is called late blight.

2. *Kole roga of Areca palms*: The disease is caused by *P.arecae* and is very common in Shimoga and other districts of Karnataka, where areca palms are cultivated extensively. The disease usually spreads rapidly during rainy season.

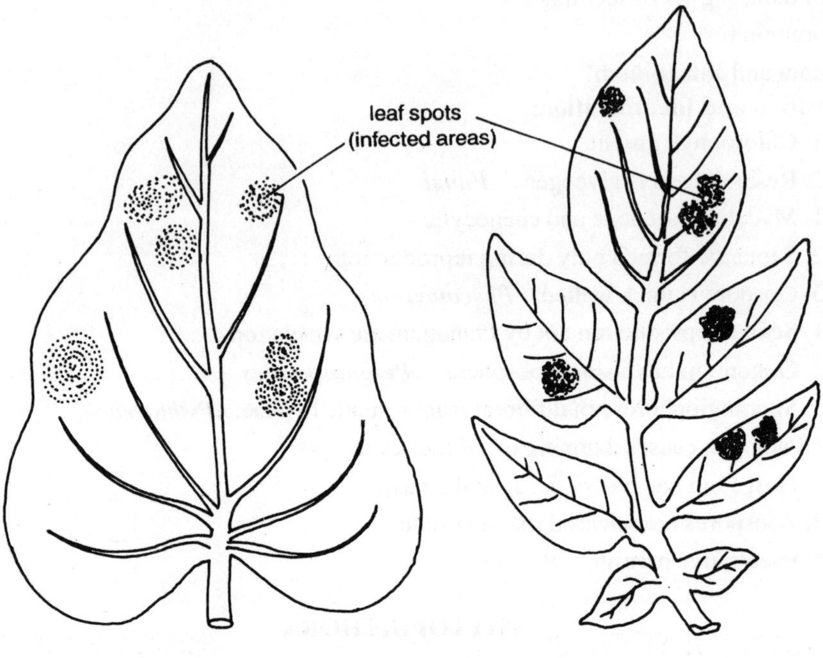

Fig. 12.15 Phytophthora. Colocasia leaf showing symptoms of late blight caused by Phytophthora — Potato leaf showing symptoms late blight caused by Phytophthora

The symptoms of the disease are first manifested on the nuts which show a water soaked area towards the base of which it gradually spreads. As a result of this , the nuts lose their green colour and young nuts drop down causing heavy loss. The disease also causes withering of leaves and drying up of the tops of trees.

3. *Bud rot of palms*: This is caused by *P.palmivora*. Trees like coconut (*Cocosnucifera*) and palmyra palm (*Borassus flabelliformis*) are infected by the fungus.

4. *Colacasia blight*: The disease is caused by *P.colocasiae*. Leaves and corm of colocasia plant are destroyed. Broad patches appearing on the leaves are the characteristic symptoms of the disease.

5. *Seedling blight*: The disease caused by *P.parasitica* destroys the seedling of castor (*Ricinus communis*) and many other plants.

6. *Bud rot of Cocoa*: *P.faberi* causes the decay of the cocoa plant in the young condition.

7. *Fruit drop*: This disease is brought about by *P.phaseoli*, which causes young fruits to drop down.

8. *Bud rot of tomato* : *P.megasperma*, *P.cryptogea* etc cause bud rot of tomato (*Lycopersicum esculentus*) and Betel leaf (*Piper betel*) etc.

Specimens given in the laboratory

1. Infected leaves or tubers of potato.

Observe the patches on the leaves, cut a section of the leaf to see the reproductive parts. Scrape an infected patch of the tuber and observe under the microscope.

2. Infected nuts of Areca

(Scrape the patches and observe the mycelia under the microscope)

3. Blight of colocasia leaves

(cut a section of the leaf passing through infected patch)

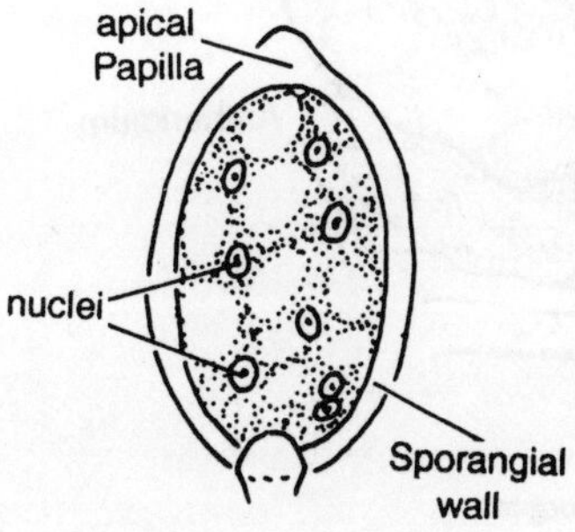

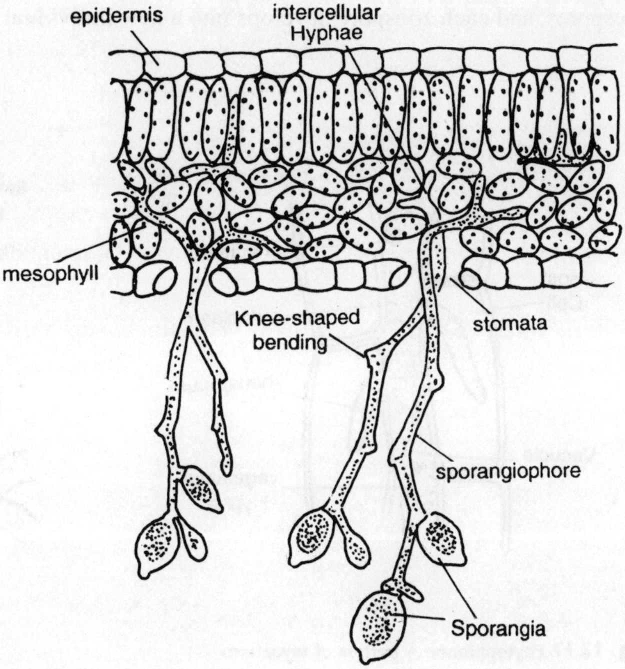

Fig. 12.16 Phytophthora. Single sporangium

T.S. of host leaf showing sporangiophores and sporangium of Phytophthora

Vegetative structure of the fungus :

1. The plant body is an aseptate, much branched, hyaline coenocytic mycelium.
2. The hyphae may be intercellular, but sometimes intracellular also.
3. Septa are formed only during reproduction.
4. The mycelium sends out haustoria into the host cells to absorb nutrition. Haustoria may or may not be branched.
5. Oil globules and glycogen present in cytoplasm as reserve food materials.

Asexual reproduction

1. Asexual reproduction is brought about by sporangia.
2. The sporangia are formed on sporangiophores which may be branched or unbranched.
3. In the case of infected potato leaves, the sporangiophores come out through the stomata of lower epidermis.
4. The sporangia are originally terminal in position on a sporangiophore. Later however the sporangia become lateral due to the process of evection.
5. Sporangia are lemon shaped and have a papilla or beak.
6. The sporangia are multinucleate and deciduous ie., they are detached from the sporangiophores and

are dispersed by wind.

7. On the availability of a suitable host plant a sporangium germinates directly into a new plant. In such a case the sporangium is called a conidium.

8. When moist conditions are prevailing in the atmosphere, the sporangium produces a number of zoospores, and each zoospore develops into a new individual.

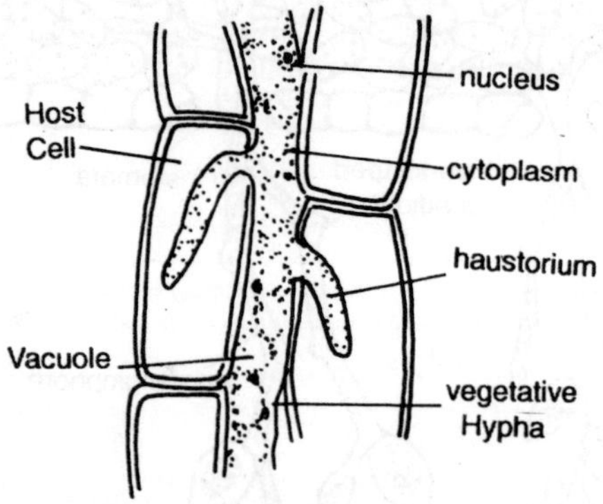

Fig. 12.17 Phytophthora. A portion of mycelium showing haustoria of Phytophthora

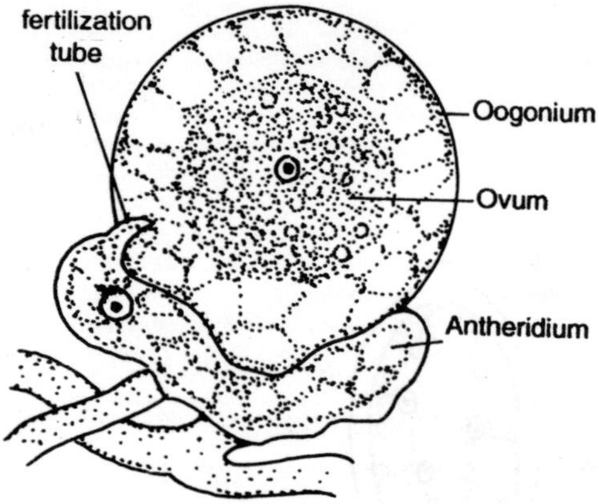

Sex organs

Sexual reproduction

1. Sexual reproduction is oogamous. Sex organs are antheridia and oogonia.

2. Species of phytopthora are both homothallic and heterothallic.

3. In homothallic species sex-organs are arranged as follows:

a. *Paragynous* : Antheridia and oogonia developed on different hyphae. They are placed side by side.

b. *Amiphigynous* : Antheridium decelops first. The oogonial hypha pierces through the antheridium. Antheridium is present like a collar at the base of the oogonium.

1. Antheridium is club shaped and multinucleate.

2. Oogonium is globose or oval and has a single egg.

3. Sexual reproduction is effected by means of gametangial contact. A slender fertilization tube develops.

4. Zygote or oospore is thick walled. It germinates to form a new mycelium.

Diagrams to be drawn :

1. Habit of infected potato, colocasia or arecanut.

2. Section through infected leaf (except for arecanut), to show mycelia and reproductive structures specially the sporangiphores.

3. Sexorgans to show paragynous and amphigynous arrangement.

Systematic position and identification

A. 1. Chlorophyll absent

2. Reserve food is glycogen ... *Fungi*

B. 1. Mycelium aseptate and coenocytic

2. Septa are formed only during reproduction

3. Oospore thick walled ... *Phycomycetes*

C. 1. Sexual reproduction not by planogametic copulation

2. Oogonium has a single oosphere ... *Peronosporales*

D. 1. Sporangiophores similar to somatic hyphae and have indefinite growth ... *Pythiaceae*

E. 1. Sporangiophores branched sympodially

2. Sporangia have an apical papilla and deciduous

3. Zoospores not released into a vesicle ... *Phytophthora*

ALBUGO

Common species : *A.candida, A.portulacae, A.evoluli, A.tragopgoonis* etc.

Occurrence : Albugo is an obligate parasite attacking mostly the cruciferous plants. The fungus causes a disease called White rust disease. Members of Amaranthaceae, Portulacaceae are also infected by this genus.

Study : 1. Disease caused by the fungus.

2. Disease symptoms.

3. Vegetative structure

4. Reproduction

5. Systematic position and identification

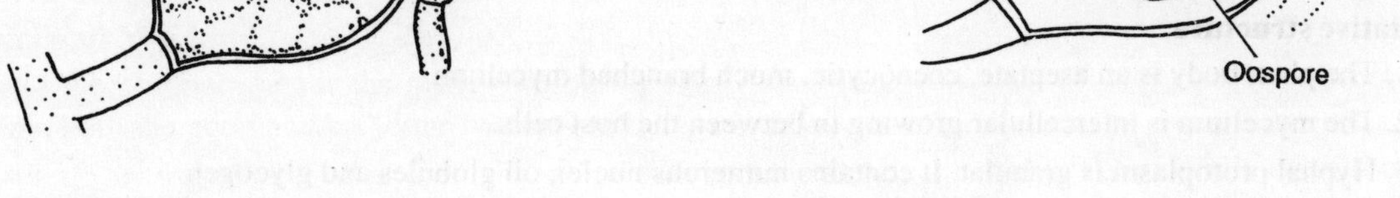

Fig. 12.18 Albugo. Sex organs An oospore

Diseases caused by the fungus.

Various species of *Albugo* cause diseases to many flowering plants. The following is a list of host and pathogenic species.

1. *A.candida* - white rust of cruciferous plants.
2. *A.evoluvli* - white rust of *Evoluvulus alsinoides*
3. *A.ipomoeae pandoranae* - white rust of *Ipomoea*
4. *A.bliti* - white rust of Amaranthaceae
5. *A.portulacae* - white rust of *Portulaca.*

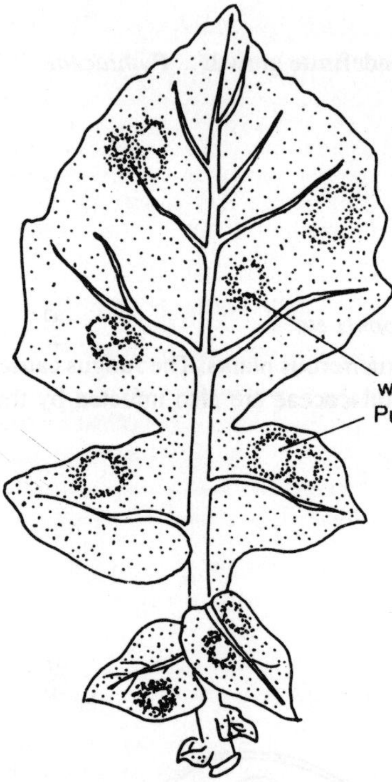

Fig. 12.19 Brassica leaf showing symptoms of white rust disease caused by Albugo

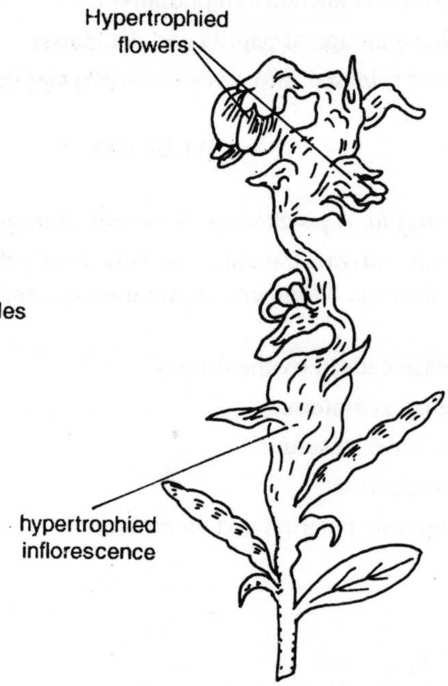

Hypertrophied inflorescence of Brassica infected by Albugo (Cystopus)

Disease symptoms : The young diseased plants show marked swelling blisters, pustules and twisting in their infected parts which may be leaves, stems, branches flowers or fruits except the roots. In extreme cases, the floral axis gets deformed, may show gall formation and even swell up to 12 - 16 times their nomral size. This is called phyllody. The leaves become succulent and distorted, while the floral organs show marked hypertrophy in as much they become fleshy and turn violet in colour. At lateral stages, the disease is easily recognisable by the white, smooth and mealy pathces that appear on the undersurface of the leaves. The pustules turn later powdery on account of the spores produced.

Vegetative structure

1. The plant body is an aseptate, coenocytic, much branched mycelium.
2. The mycelium is intercellular growing in between the host cells.
3. Hyphal protoplasm is granular. It contains numerous nuclei, oil globules and glycogen.
4. The mycelium produce short knob like haustoria into the host cells to absorb nutrition.

Asexual reproduction

1. Asexual reproduction is brought about by means of conidia.
2. A section of the infected leaf passing through a white pustule will show the asexual reproductiive structures.

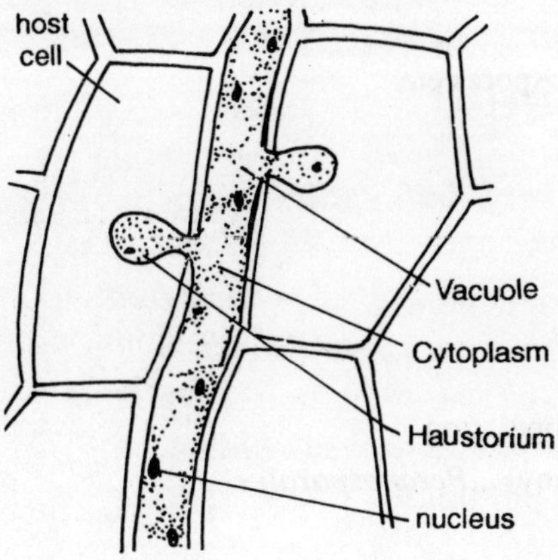

Fig. 12.20 Albugo. A portion of mycelium showing haustoria

3. The mycelium produces a number of erect, unbranched and club shaped hyphae. These are called conidiophores.

4. The conidiophores are multinucleate and have thickened side walls. They are not separated from the underlying mycelium by any cross septum.

5. The conidiophores at their tip produce a chain of conidia in basipetalous succession (older conidia at the top and younger ones below).

6. In between the conidia in the conidial chain is present a gelatinous disc or pad, which helps in the disjoining of the chain.

7. The conidia reproduction multinucleate and nonmotile.

8. As a result of the production of large number of conidia, a pressure is exerted on the epidermis which ruptures releasing them (conidia) outside.

9. The conidia are wind dispersed. On the availability of a suitable substratum, they germinate directly into a new individual.

10. If the atmosphere is having enough moisture, the conidia function like sporangia and produce many biflagellate zoospores. Each zoospore develops into a new individual.

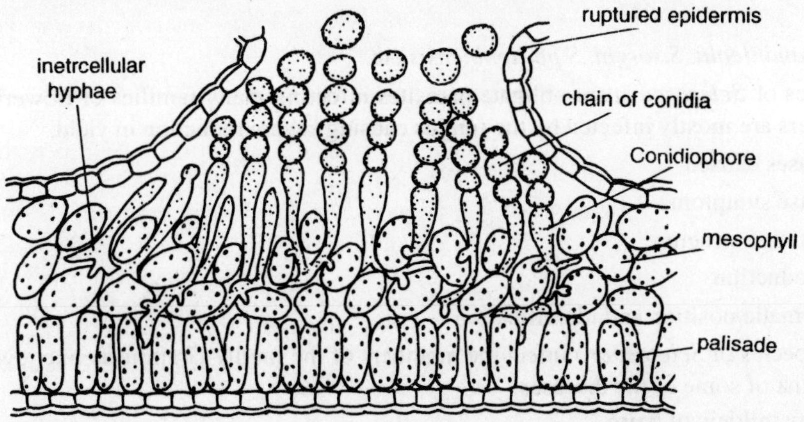

Fig. 12.21 T.S. of host leaf showing conidiophores and conidia of Albugo.

Sexual reproduction :

1. Sexual reproduction is oogomous. Sex-organs are antheridia and oogonia.

2. Sex-organs are generally seen in hypertrophied parts.

3. Antheridium is a small club shaped structure terminating a hypha. It has many male nuclei.

4. Oogonium is globose. A mature oogonium has one or many eggs. The oogonium with many eggs is called a Compound oospore.

5. Sexual reproduction is effected by gametangial contact. Fertilization tube developed from the antheridium helps in the transfer of male nuclei to the oogonium.

6. Zygote or oospore has a thick sculptured wall. The wall sculpturing is used as a taxonomic creiterion.

Specimens given in the laboratory

1. Infected leaves showing white blisters.
2. Hypertrophied stem, inflorescence etc.

Diagrams to be drawn

1. Habit sketch of infected plant parts.
2. Section of leaf showing conidiophores.
3. Section of a hypertrophied part to show sexorgans, oospores etc.

Systematic position and identification

A. 1. Chlorophyll absent
2. Reserve food is glycogen ... *Fungi*

B. 1. Mycelium aseptate and coenocytic
2. Septa are formed only during reproduction
3. Oospore thick walled ... *Phycomycetes*

C. 1. Sexual reproduction not by planogametic copulation
2. Oogonium has a single oosphere (rarely many) ...*Peronosporales*

D. 1. Obligate parasites
2. Conidiophores unbranched
3. Oosphore has a sculptured wall...*Albuginaceae*

E. 1. White blisters on the leaves
2. Hypertrophy of inflorescence
3. Haustoria button shaped
4. Conidia in basipetalous succession ... *Albugo*

SCLEROSPORA

Common species : *S. graminicola, S.sorghi, S.philippinensis* etc.

Occurrence : The species of *Sclerospora* are obligate parasites infecting many families of flowering plants. Graminaceous members are mostly infected by the fungus causing severe reduction in yield.

Study :
1. Diseases caused
2. Disease symptoms
3. Vegetative structure
4. Reproduction
5. Systematic position and identification

Diseases caused : The species of *Sclerospora* infect the members of the family Gramineae like jowar, bajra etc. The following is a list of some of the diseases:

1. Green ear disease or Downy mildew of bajra

The disease is caused by S.graminicola

The disease symptoms are of two types - green ear symptoms and downy mildew symptoms. The symptoms of the disease depend on the time of infection. If the infection takes place at the seedling stage, downy mildew symptoms are produced and if the infection is quite late ie., at the flowering stage, green ear symptoms are produced.

Downy mild symptoms :

1. Leaves are infected and proliferation of leaves takes place.
2. Leaves get etiolated as chlorosis sets in mainly on the upper surface.
3. Under surface of the leaf gets invested with soft whitish feathery mycelium. The pustules are very soft to touch like the down feathers hence the name downy mildew.

4. The plant appears stunted and bushy.

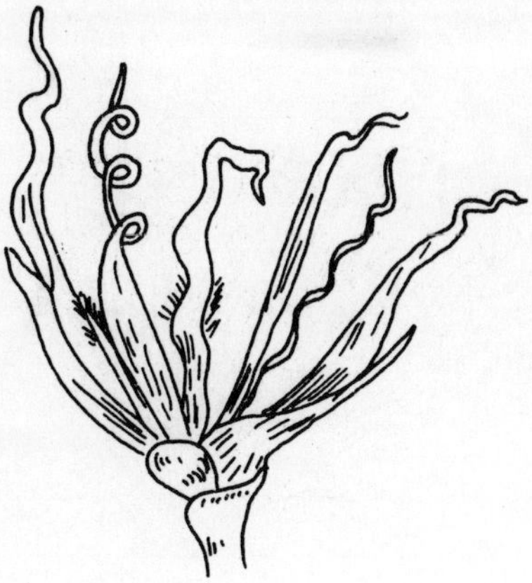

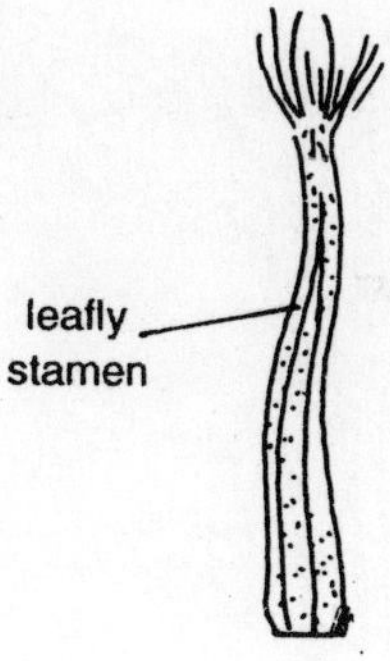

Fig. 12.22 Sclerospora.
Proliferated flower

Infected stamen

Green ear symptoms :

1. Leaves form a number of vegetative buds.

2. Inflorescence exhibits the green ear symptoms.

3. The entire inflorescence appears like a cluster of leafy appendages.

4. Stamens, awans, pistil etc become hypertrophied and distorted, complete sterility sets in and no grains are formed.

Downy mildew of jawar : This is caused by *S.sorghi* the disease is prevalent in the coastal regions of South India.

Disease symptoms are mainly visible as patches or pustules of downy growth of the leaves.

Downy mildew of maize : The disease is caused by *S.philippinensis*. The disease symptoms are almost similar to those of mildew of jowar.

Vegetative structure of fungus :

1. The plant body is a coenocytic aseptate, much branched mycelium.

2. The mycelium is intercellular and produces vesicular haustoria to absorb nutrition from the host cells.

3. Septa appear only at the time of development of reproductive structures.

Reproduction

1. Asexual reproduction takes place by the formation of spores or conidia on sporangiphores (conidiophores).

2. The sporangiosphores are erect, broad and branched hyphae bearing a number of sporangia.

3. Sporangiophores may be solitary or produced in groups of 3-4.

4. Sporangia are elliptical or oval and have a papillate outgrowth at the tip.

5. Sporangia vary in size; 13-34 metre long and 12-23 metre broad.

6. The sporangia develop during night and are shed during day time.

7. The sporangia are deciduous and germinate by producing reniform zoospores.

Sexual reproduction

1. Sexual reproduction takes place by means of oogamy. Sexorgans are antheridia and oogonia.

2. Fertilization results in the development of zygote. Zygotes are usually formed at the end of the growing season of host. They can be observed as brownish areas on the host plant.

3. Oospore (zygote) has an outer wall which is unevenly thick. On germination the oospore develops into a new individual.

Specimens given in the laboratory :

1. Leaves having downy pustules.

2. Green ear of Bajra.

(Cut the sections of leaves and tease open the green ear to study the malformations).

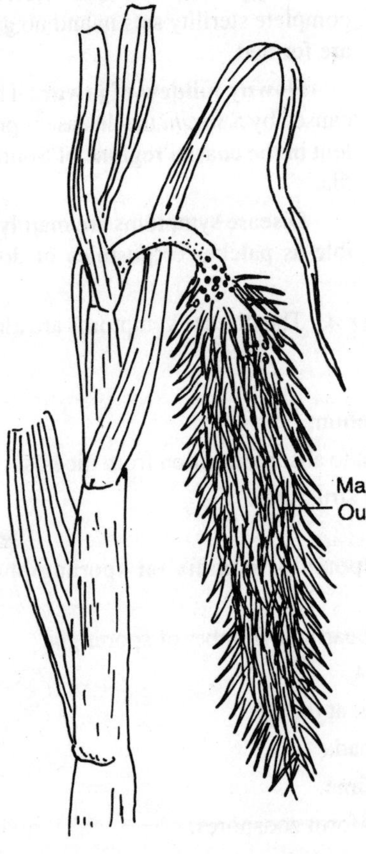

Fig. 12.23 Sclerospora. An infected ear of bajra showing green ear disease caused by Sclerospora

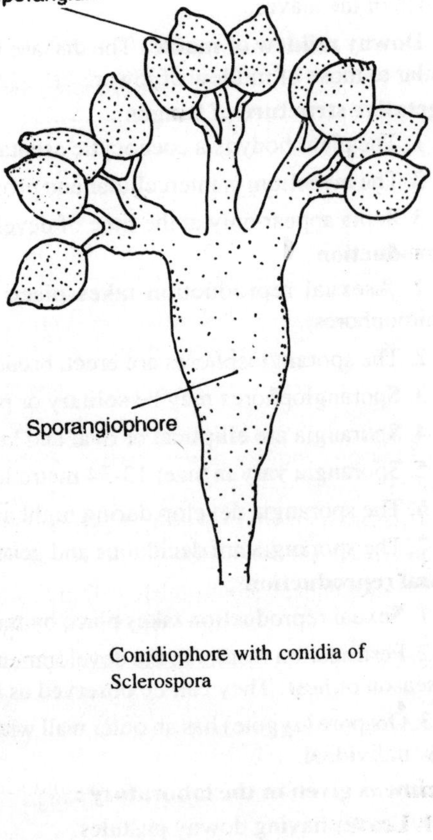

Conidiophore with conidia of Sclerospora

Diagrams to be drawn:

1. Habit sketch of downy mildew symptoms
2. Habit sketch of green ear symptoms
3. Deformed flower (enlarged)
4. Sporangiophores

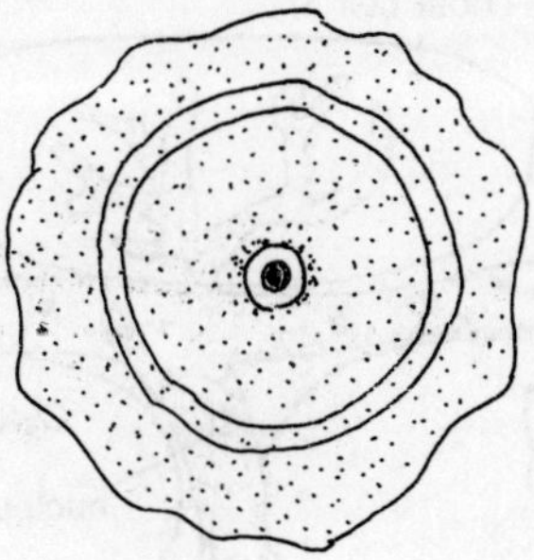

Fig. 12.24 Sclerospora. An oospore

Systematic position and identification

A. 1. Chlorophyll present
2. Reserve food is glycogen ... *Fungi*

B. 1. Mycelium aseptate and coenocytic
2. Septa are formed only during reproduction
3. Oospore thick walled ... *Phycomycetes*

C. 1. Sexual reproduction not by planogametic copulation
2. Sexual reproduction oogamous
3. Oogonium has a single oosphere ... *Peronosporales*

D. 1. Sporangiophores exhibit dichotomous branching
2. Sporangia borne singly on the sporangiophore ... *Pernosporaceae*

E. 1. Sporangiopores caducous and stout
2. Oospore wall united with that of the Oogonium ... *Sclerospora*

PERONOSPORA
(Ainsworth & Bisby 1954)

Common species : *P.pisi, P.parasitica, P.distructor* etc.

Occurrence : The genus *Peronospora* is an obligate parasite infecting many families of flowering plants. The disease caused is popularly called downy mildew because of the soft downy growth of the hyphae.

Study : 1. Diseases caused by the fungus
2. Disease symptoms
3. Vegetative structure
4. Reproduction
5. Systematic position and identification

Diseases caused by the fungus :

The genus peronospora has about 75sp most of which are pathogenic causing diseases to many economically important plants. Some diseases caused by them are given below :

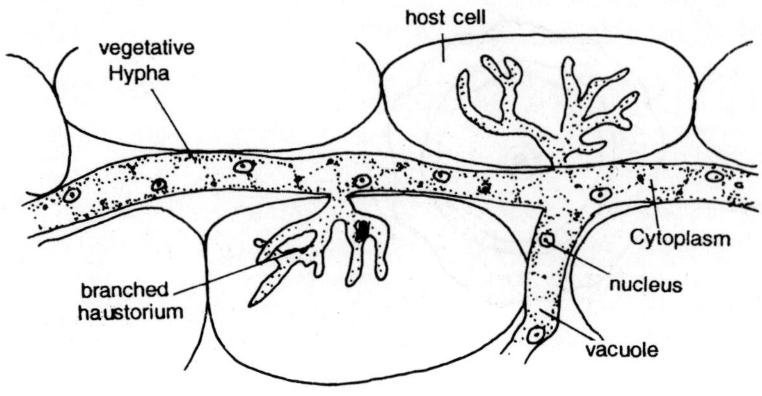

Fig. 12.25 Infected host cells showing haustoria of Peronospora

1. *Downy mildew of crucifers*. This disease is caused by *P.parasitica*. cruciferous plants like raddish (*Raphanus sativus*), Cabbage (*Brassica oleracea var capitata*) Cauliflower (*Brassica oleracea* var *botrytis*) mustard (*Brassica campestris*) etc.

2. *Downy mildew of peas. P.pisi* causes the downy mildew peas (*Pisum sativus*).

3. *Downy mildew of onion*. P.destructor causes the downy mildew of onion (*Allium cepa*).

4. *Downy mildew of cherianth*: This is caused by *P.cherianthi*. The host plant is *Cherianthus cherii*.

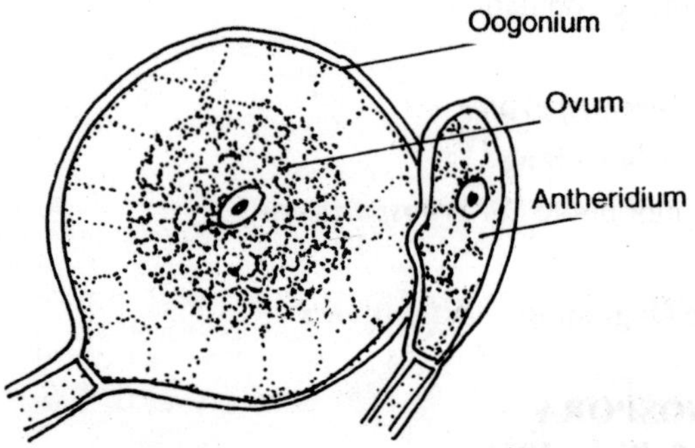

Fig. 12.26 Peronospora Sex organs

Diseases symptoms :

1. Downy mildew symptoms appear quite early in the development of the host plant.

2. When the plants are about 10-12 cms height, yellow brown patches of irregular shape appear on the upper surface of the leaves.

3. The lower surface of the leaves at these regions become yellowish invested with soft downy growth of the mycelia.

4. The down growth of the fungus consists of conidiophores.

5. The disease symptoms may appear also on the fruits resulting in the decay of the pericarp and reduction in the size of the seeds.

The appearance of brownish yellow patch on the upper surface of the leaf and downy growth of mycelia on the lower surface is charactristic of all the species of *Peronospora*.

Vegetative structure :

1. The plant body is a coenocytic, much branched, aseptate mycelium.

2. The mycelium is intercellular and produces knob like haustoria which absorb nutrition from the host cells.

3. Septa appear in the hyphae only during the formation of reproductive structures.

4. The granular cytoplasm is rich in reserve food materials like oil globules, glycogen etc.

5. The tubular hyphae have a central vacuole.

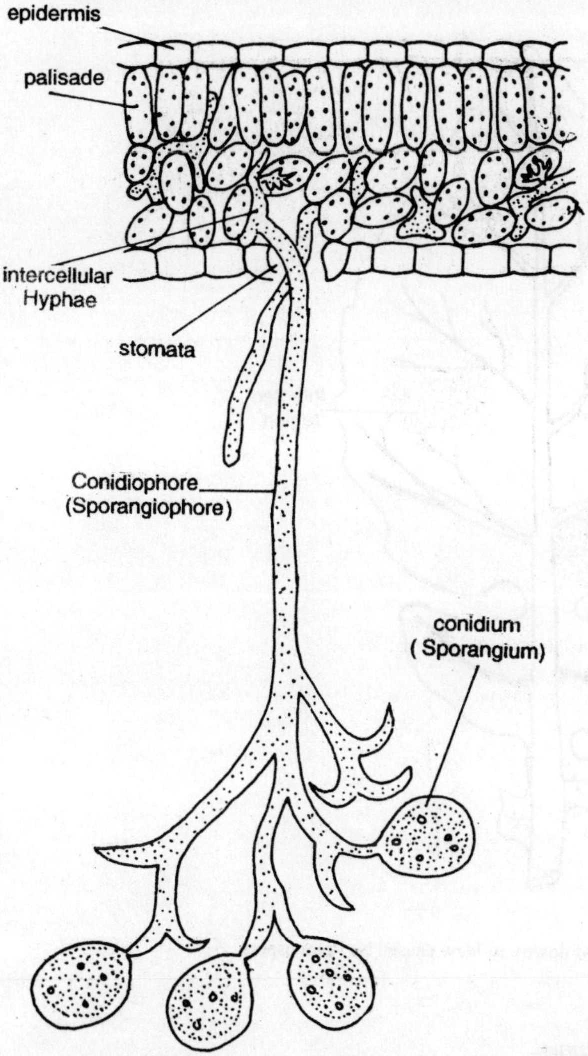

Fig. 12.27 T.S. of host leaf showing sporangiophore and sporangia of Perenospora

Reproduction :

1. Asexual reproduction is brought about with the help of conidia which are produced on conidiophores.

2. The conidiophores usually come out of the host leaf through the stomata present on the ventral surface.

3. The conidiophores give out branches at their apex; the branching is dichotomous.

4. The ultimate branches also called sterigmata are reflexed and bear conidia at their tips.

5. The conidia may be oval or elliptical in shape, greenish yellow or purple tinged in colour. The conidia are sensitive to dessication.

6. Conidia are wind dispersed and on germination develop into a new individual.

Sexual reproduction:

1. Sexual reproduction is oogamous. Sex-organs are antheridia and oogonia. The sex-organs are arranged paragynously (see-*Phytophthora*).

2. Antheridium is a small tubular structure. A mature antheridium has a single functional male nucleus.

3. Oogonium is a globose structure. It has a central ooplasm with an oosphere and a peripheral ooplasm.

4. Sexual reproduction is effected by means of gametangial contact. A slender fertilization tube develops from the antheridium. It pierces through the oogonial wall and releases its contents inside.

5. The zygote or oospore which is a result of fusion is light yellow in colour and has a sculptured wall called epispore. Oospores can be seen in the host tissues.

6. After a period of rest the oospore develops into a new individual by producing a germ tube.

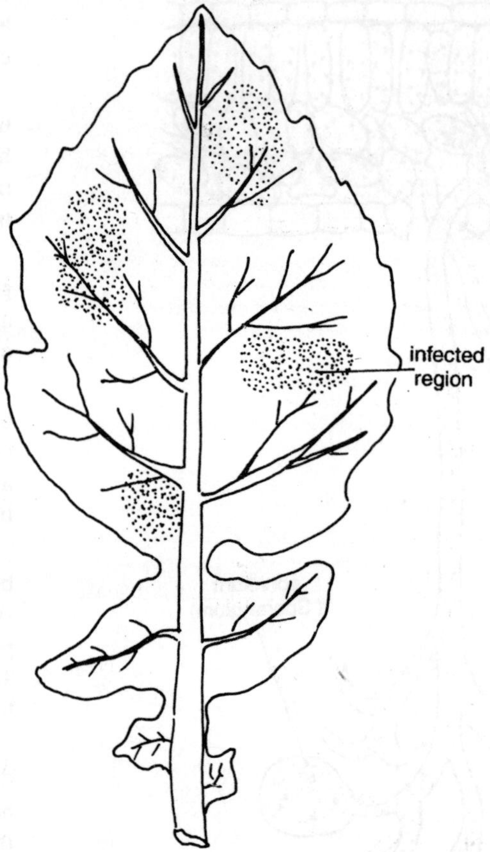

Fig. 12.28 Brassica leaf showing symptoms of downy mildew caused by Peronospora

Specimens given in the laboratory:

1. Infected leaves of pea or *Raphanus.*

Diagrams to be drawn:

1. Habit of infected leaves
2. Sporangiophores
3. Sexorgans and oospore

Systematic position and identification

A 1. Chlorophyll absent
2. Reserve food is glycogen ... *Fungi*

B 1. Mycelium aseptate and coenocytic
2. Septa are formed only during reproduction
3. Oospore thick walled ... *Phycomycetes*

C 1. Sexual reproduction oogamous

2. Oogonium has a single oosphere ... *Preonosporales*

D 1. Sporangiophores exhibit dichotomous branching

2. Sporangia borne singly on the sporangiophore ... *Peronosporaceae*

E 1. Ultimate branches of the sporangiophores are recurved

2. Conidia germinate by a germ tube

3. Oospore germination direct ... *Peronospora*

RHIZOPUS

Common species : *R.nigricans, R.stolonifer, R.arrhizus, R.artocarpi* etc.

Occurrence : Species of the genus are the most ubiquitous of the fungi. The members are saprophytes growing on any dead and decaying organic matter. *Rhizopus* grows very commonly on stale bread hence it is popularly called the bread mold. It also causes several plant diseases.

Study

1. Diseases caused and disease symptoms
2. Method of culturing
3. Vegetative structure
4. Reproduction
5. Systematic position and identification

Diseases caused : Some species of *Rhizopus* cause the fruit rot diseases in some flowering plants. The following are some of the diseases.

1. Apple rot .... *R.arrhizus*
2. Fruit rot of jack fruit .... *R.artocarpi*
3. Soft rot of strawberries .... *R.stolonifer*
4. Decay of bread .... *R.nigricans*

Disease symptoms of the fruit rot

1. Young fruits drop down prematurely
2. Infected peduncles get decolorised ultimately turning black
3. Mature fruits begin to rot and get decayed

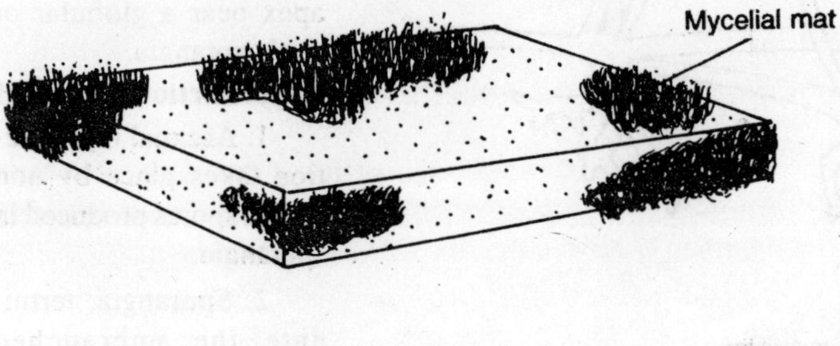

Fig. 12.29 Rhizopus. A slice of bread showing mould

Method of culturing in the laboratory

1. Take a few slices of bread. Add a little water to moisten them.

2. Keep the slices of bread in a petridish and cover it.

3. After few days a soft cottony growth of mycelium may be observed. This is the vegetative phase.

4. A little later small while pin head like structures appear from the mycelium. These are the young sporangia.

5. After few days the white pinheads turn black. This represents the mature sporangia liberating the dark colored spores.

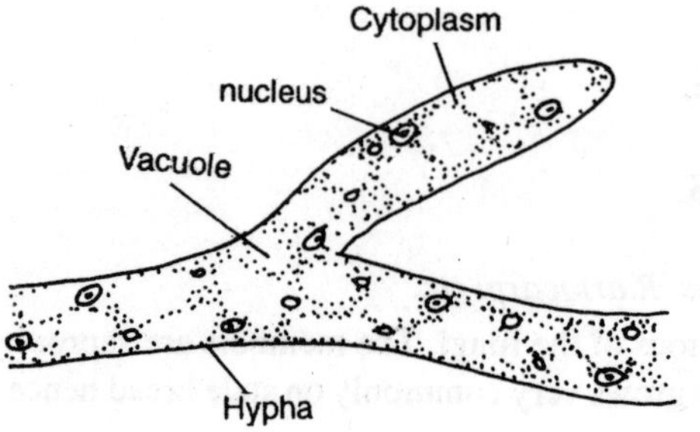

Fig. 12.30 Rhizopus. A portion of mycelium

6. It is generally difficult to observe the sexual stages.

Vegetative structure

1. The plant body of *Rhizopus* is a much branched, coenocytic, aseptate mycelium.

2. The hyphae are whitish and tubular. There is a centrally running continuous vacuole. Cell wall is chitinous.

3. The cytoplasm is peripheral. It has numerous nuclei and has oil and glycogen as reserve food material.

4. Cross walls appear only at the time of formation of reproductive structures.

5. The mycelium which is fairly old, has three types of hyphae. These are -

a. Stolon or distributive hyphae : These are horizontal tubular hyphae that grow above the substratum. There and there swollen portions are found on the stolon.

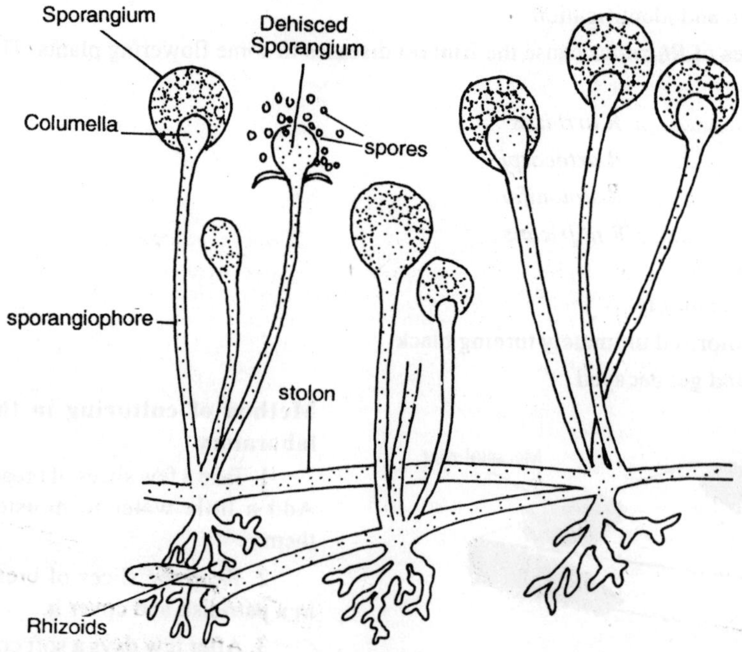

Fig. 12.31 Rhizopus. A portion of mycelium

b. Rhizoidal hyphae : From the swollen points of the stolon arise short branched hyphae called rhizoids. These grow vertically down into the substratum helping in anchorage and absorption.

c. Sporangiophores : These are erect, tubular, unbranched hyphae arising at the same points, where rhizoids arise. Sporangiophores at their apex bear a globular or oval sporangia.

Reproduction - Asexual

1. Asexual reproduction takes place by non motile spores produced in sporangia.

2. Sporangia, terminate the unbranched sporangiphores.

3. Occupying the central region of the sporangium is found a dome shaped columella, while protoplast is continuation with that of sporangiophore.

4. In between the columella and the wall of the sporangium a large number of spores are produced.

5. The spores are non motile and have a pigmented wall due to which they appear dark brown.

6. Spores are spherical, non motile and multinucleate. They are wind dispersed and on the availability of a suitable substratum develop into new individuals.

7. Occasionally thick walled chlamydospores may be formed as a means of asexual reproduction.

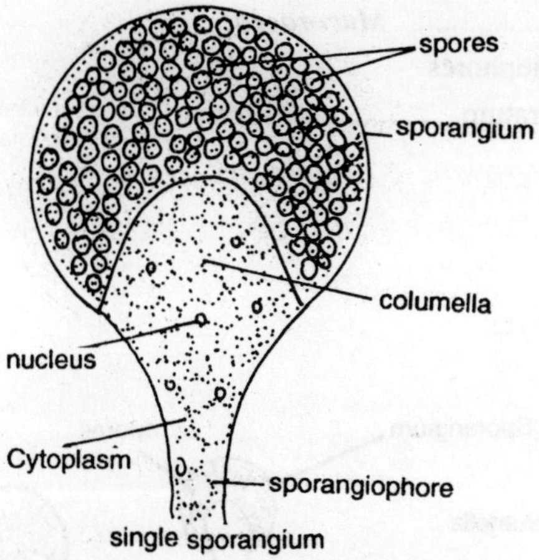

Fig. 12.32 Rhizopus. Single sporganium

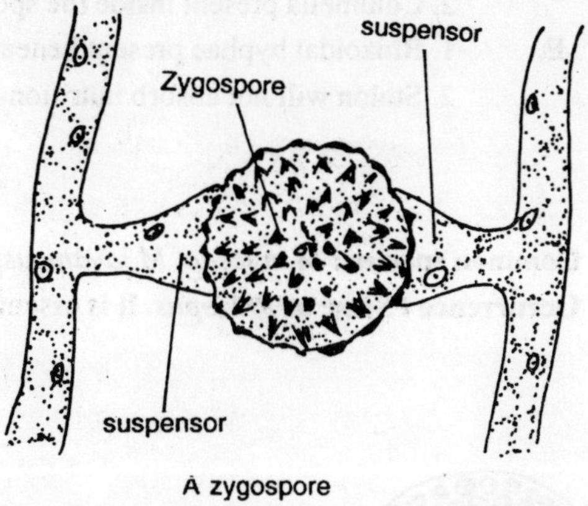

Rhizopus. A zygospore

Reproduction - Sexual

1. Sexual reproduction in *Rhizopus* takes place by means of gametangial copulation.

2. The gametangia are isogametangia, but the gametangia formed on the same mycelium will not fuse. This is due to the phenomenon of heterothallism.

3. The gametangia that fuse are morphologically similar but physiologically different.

4. Since the compatible gametangia cannot be distinguished into male or female, they are designated as + and - strains in order to show that they are different.

5. Hyphae belonging to + and - strains produce short outgrowths called progametangia.

6. The progametangia at maturity get divided into, a basal suspensor and a terminal gametangium.

7. The two gametangia of opposite strands fuse to form a zygospore.

8. The zygospore develops a thick ornamented wall.

9. After undergoing a period of rest, the zygospore germinates by producing an erect hypha, whose free end forms a sporangium.

10. The spores that are formed are of two kinds; 50% of them belong to + strain and the rest - strain. But morphologically the spores are same.

Diagrams to be drawn

1. Habit sketch
2. Plant body enlarged
3. Sporangium enlarged
4. Stages in sexual reproduction.

Systematic position and identification

A. 1. Chlorophyll absent
2. Reserve food is glycogen ....*Fungi*

B. 1. Plant body aseptate and coenocytic mycelium
2. Septa formed only during reproduction .... *Phycomycetes*

C. 1. Asexual reproduction by non motile spores
2. Nutrition saprophytic ....*Mucorales*

D. 1. Conidia present
2. Columella present inside the sporangium......................... *Mucoraceae*

E. 1. Rhizoidal hyphae present beneath the sporangiophores
2. Stolon will not absorb nutrition from the substratum ....*Rhizopus*

MUCOR

Common species : *M.mucedo, M.javanicus, M.priani* etc.

Occurrence : Same as *Rhizopus*. It is essentially a saprophyte.

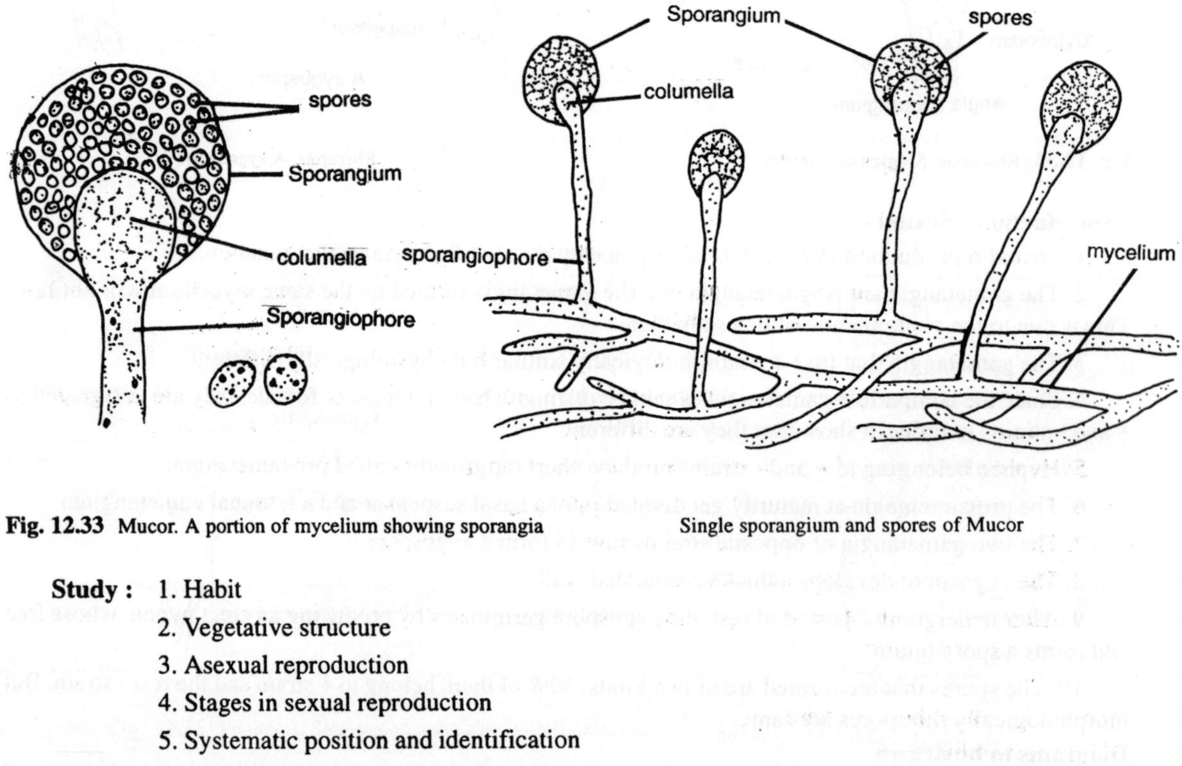

Fig. 12.33 Mucor. A portion of mycelium showing sporangia   Single sporangium and spores of Mucor

Study : 1. Habit
2. Vegetative structure
3. Asexual reproduction
4. Stages in sexual reproduction
5. Systematic position and identification

Habit : Same as *Rhizopus*

Vegetative structure : It is essentially similar to *Rhizopus* except for the following :

1. The plant body does not show differentiation into stolon and rhozoidal hyphae.

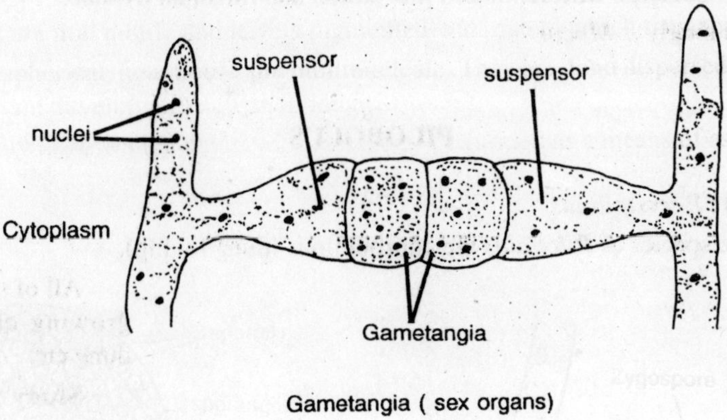

Fig. 12.34 Mucor.Gametangia (sex organs)

2. The branched mycelium produces erect unbranched sporangiophores which do not arise in groups.
3. Since rhizoidal hyphae are absent the general surface of the mycelium carries out absorption.

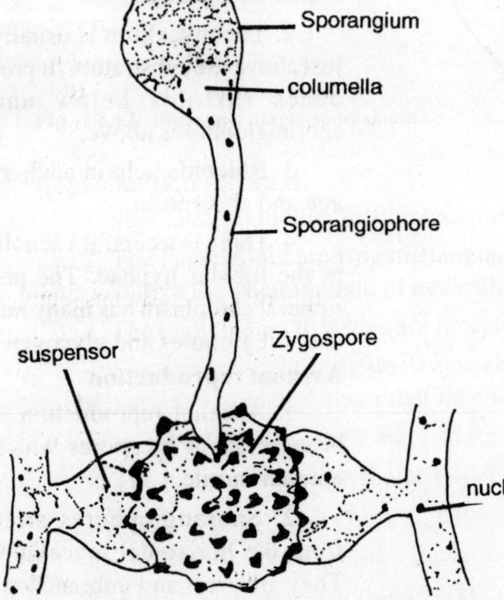

Fig. 12.35 Mucor. Germination of zygospore

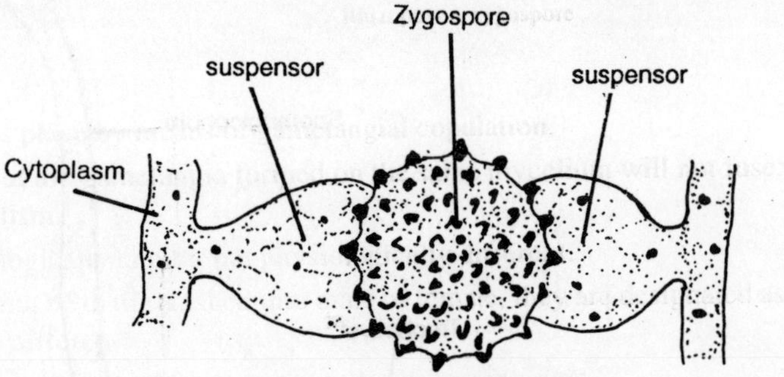

A zygospore

Asexual reproduction : Same as *Rhizopus*.

Sexual reproduction : Same as in *Rhizopus*.

Diagrams to be drawn : Same as in *Rhizopus*.

1. Plant body does not show differentiation into stolon and rhizoidal hyphae.
2. Sporangia arise singly ... *Mucor*

PILOBOLUS

Common species : *P. crystallinus*

Occurrence : The species of *Pilobolus* are coprophilous (dung loving). All of them are saprophytic growing on cow dung, horse dung etc.

Study : 1. Vegetative structure

2. Reproduction

3. Systematic position and identification

Vegetative structure

1. The plant body is an aseptate, coenocytic, much branched mycelium.

2. The mycelium is usually just above the substratum. It produces rhizoids below and sporangiophores above.

3. Rhizoids help in anchorage and absorption.

4. There is a central vacuole in the tubular hyphae. The peripheral cytoplasm has many nuclei, oil globules and glycogen.

Asexual reproduction

1. Asexual reproduction is brought about by spores which are non-motile.

2. Sporangiophores arise from the horizontal mycelium. They are erect and unbranched.

3. The sporangiophores have four parts. They are :

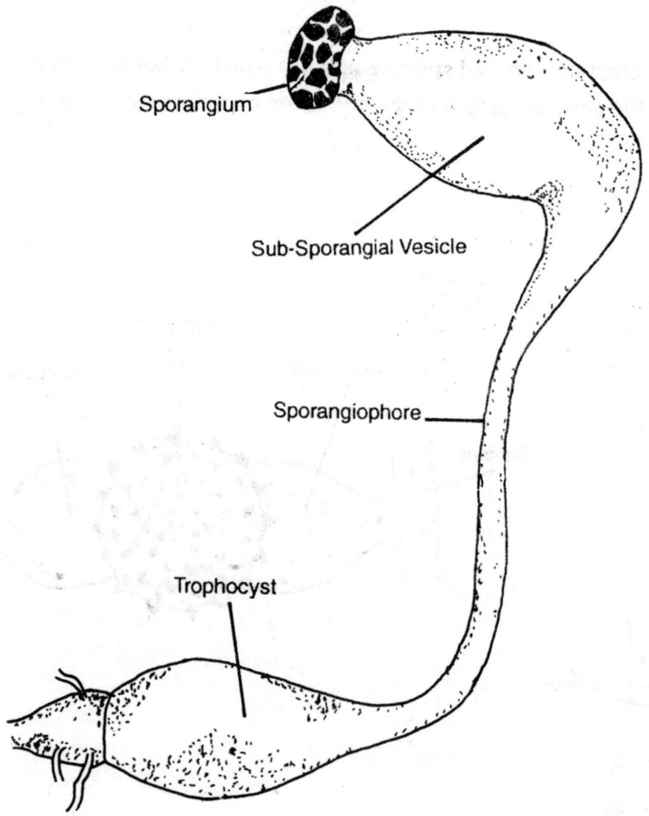

Fig. 12.36 Pilobolus. Single sporangiophore with sporangium

a. Trophocyst
b. Sporangiophore stalk
c. Sub-sporangial vesicle
d. Sporangium

1. The swollen base of the sporangiophore is called Trophocyst (Gr.*trophe* = food + *Kysis* = bladder).

2. The stalk of the sporangiophore is tubular and narrow.

3. The tip of the sporangiophore is swollen to form a vesicle called sub-sporangial vesicle.

4. The terminal part consists of the sporangium proper. The sporangia are somewhat longitudinally compressed and have a rigid dark coloured terminal wall.

5. Columella is present inside the sporangium.

6. The sporagnium is deciduous. It is discharged with explosive dehiscence mechanism.

7. The sub-sporangial vesicle ruptures in a vertical slit due to osmotic pressure. This causes a jet of sap to force out the sporangium.

8. The sporangia stick to grasses and other plants and when these are consumed by herbivorous animals, they (sporangia) are dissolved in the alimentary canal setting of the spores free. The spores come out through the fecal matter and grow into new individuals.

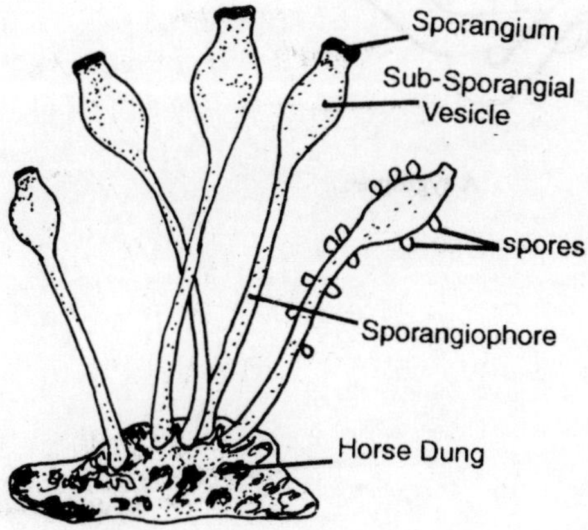

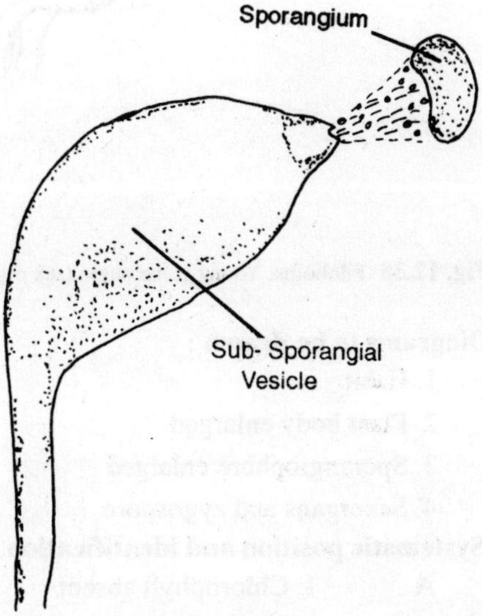

Fig. 12.37 Pilobolus. Discharge of spores from sporangium Fungus growing on dung

Sexual reproduction :

1. Sexual reproduction takes place by means of gametangial copulation.

2. The compatible gametangia are similar morphologically (Isogametangia).

3. The gametangia are borne on curved suspensors, when the two gametangia come nearer, they appear like a pair of tongs.

4. The gametangia fuse and form a zygospore.

5. The zygospore develops a thick and smooth wall. The wall is dark brown due to pigmentation.

Laboratory culture of Pilobolus :

1. Collect fresh dung of herbivorus animals, preferably of horse.

2. Keep it enclosed in a petridish. Add a little water, if the dung has become dry.

3. After few days, profuse growth of the mycelium may be observed.
4. A little later sporangiophores also will develop.

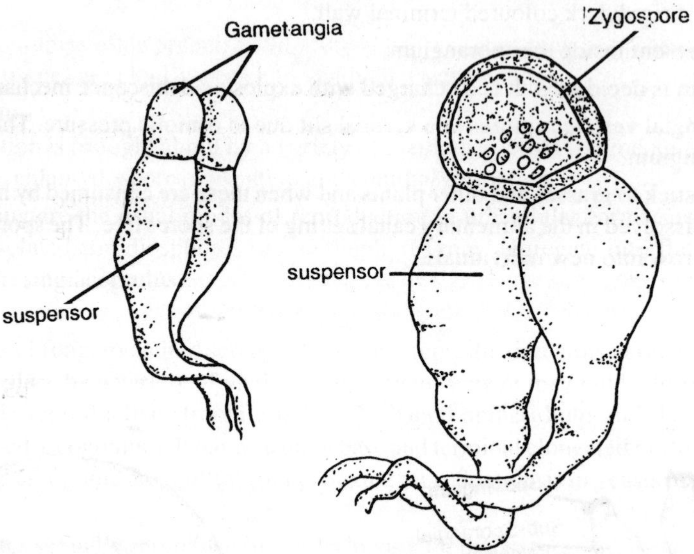

Fig. 12.38 Pilobolus. Young gametangia (sex organs) A zygospore

Diagrams to be drawn :

1. Habit
2. Plant body enlarged
3. Sporangiophore enlarged
4. Sexorgans and zygospore

Systematic position and identification

A 1. Chlorophyll absent
2. Reserve food is glycogen ... *Fungi*

B 1. Plant body, a septate, coenocytic much branched mycelium.
2. Sexual reproduction produces zygospores ... *Phycomycetes*

C 1. Members mostly saprophytes
2. Asexual reproduction by aplanospores ... *Mucorales*

D 1. Sporangia have small columella
2. Terminal wall of the sporangium is rigid and remarkable ... *Pilobolaceae*

E 1. Habitat is coprophilous
2. Subsporangial vesicle present
3. Sporangia are discharged by explosion mechanism ... *Pilobolus*

CHAPTER 13

ASCOMYCETES
(THE SAC FUNGI)

Ascomycetes together with basidiomycetes are often referred to as the higher fungi. They are considerably more advanced than the class phycomycetes in their complex structure and reproductive features, and are probably derived from lower fungi.

According to Aisnworth and Bisby (1950), there are about 1820 genera and 15,500 species of Ascomycetes. The Ascomycetes are called so, because they produce their spores in a special sac like structure called Ascus. In a great majority of Ascomycetes as a rule each ascus produces eight ascospores. But this may occasionally vary.

The yeasts, some of the green and black moulds, the ergot fungus, the powdery mildews, the cup fungi, the delicious morels, the toad stools and *Neurospora*- the Cinderella of geneticists are all Ascomycetes.

Occurrence. Ascomycetes are found in variety of habitats practically in all seasons of the year. The members may be parasitic or saprophytic. As parasites they are generally endoparasites; but some are ecto parasites (crysiphales). The saprophytic Ascomycetes usually live on the soil or on decaying logs of wood and leaves and produce large fructifications. A few forms are hypogean (gr-*hypo*-under; *geo*-earth) developing and remaining underground. Many members are coprophilous (Gr-*kopros;* dung + *phillen* to love) growing only on the dung of certain animals. Some of the Ascomycetes are also marine in distribution. Marine Ascomycetes may be saprobic on various types of organic matter sub-merged or floating or parasitic on marine algae and higher plants.

Ascomycetes and Human Welfare

Ascomycetes are important in human welfare in multifarious ways. Like other fungi, they also do both good and bad to man.

Chaetomium destroys the fabrics containing cellulose. In destroying crops, crop plants as well as timber and ornamental trees Ascomycetes are our worst fungous enemies causing such disease as apple scab, brown rot of stone fruits, powdery mildew, food rot of cereals, ear rot of corn etc. (Alexopoulos 1956). Ascomycetes have not spared human beings also. Aspergillosis (lung infection) and Ringworm disease are the common infections by Ascomycetes. *Claviceps purpurea* producing the ergot of rye is poisonous if the infected grains with sclerotia are consumed. On the other hand some Ascomycetes are highly beneficial. Who can forget the fermenting activity of yeast in bakeries and breweries, the drug penicillin obtained by *Penicillium*, the useful activities of *Aspergillus* in industry and fine delicacies such as morels and truffels.

Somatic structure

Except for members such as yeasts, the plant body of Ascomycetes is a septate much branched mycleium. Cell wall contains a large proportion of chitin. The cells of the hyphae may be uninucleate, binucleate or multinucleate. The cross septum dividing the cell is porous permitting the streaming of cytoplasm from one cell to another.

Ascomycetes mycelium is often organised into fungal tissues. They may be prosenchymatous or pseudo parehcnymatous. The tissues are chiefly associated with the fructifications.

Asexual reproduction

Asexual reproduction is brought about by a variety of methods such as fragmentation, fission, budding, arthospores formation, chlamydospores formation and conidial formation.

Fission and budding are the usual modes of reproduction in unicellular forms such as yeasts. Conidia may be produced in isolated conidiosphores or conidiophores may aggregate into flat mattess like hyphal masses called *Acervuli* (sing.acervulus).

Sexual reproduction

As in other classes of fungi even in Ascomycetes sexual reproduction aims at the fusion of two compatible nuclei. This takes place by a variety of methods. Sex organs may or may not be formed in Ascomycetes. When formed the male reproductive structure is a club-shaped antheridium and the female reproductive structure is a flask shaped ascogonium. It has a swollen base and terminal elongated portion called Trichogyne. Both the antheridia and ascogonia are multinucleate. Sexual reproduction as in yeasts may be by the fusion of isogametangia.

In general, Ascomycetes sexually reproduce by the following methods.

1. Gametangial copulation, 2. Gametangial contact, 3. Spermatization and 4. Somatogamy.

Stages in sexual reproduction

Unlike in Phycomycetes, there are four stages in Ascomycetes at which sexual reproduction is accomplished. They are plasmogamy dikaryotic stage, karyogamy and reduction division. As a result of this there is regular alternation between a monokaryotic mycleium, a dikaryotic mycelium and a short lived diploid mycellium. Sexually reproducing monokaryotic mycelium gives rise to secondary mycelium after plsmogamy. In this dikaryotic mycelium karyogamy takes place at a late stage producing a diploid generation which immediately unergoes reduction division to produce haploid cells (ascospores). These germinate to develop into the monokaryotic mycelium.

Life cycle pattern (the process of Ascus formation)

In all Ascomycetes, the life cycle pattern is uniform with marginal variations here and there.

At the time of reproduction, the two sex organs come nearer and the tip of the antheridium comes into contact with the tip of trichogyne. The separating walls dissolve and the antheridial contents migrate into the ascogonium. Plasmogamy takes place first. The male and female nuclei are irregularly distributed. Later they pair and the pairs of nuclei are gradually migrate to the periphery of the ascogonial base. At this stage, small protrusions develop from this ascogonial base which eventually elongate to form hyphae. These are called ascogenous hyphae as they produce ascus later. The pairs of nuclei (one male and female) migrate into the ascogenous hyphae; one pair into one ascogenous hypha. Herer the nuclei start dividing and produce many nuclei. But even the daughter nuclei remain in pairs and are not scattered. Always a male nucleus and a female nucleus will remain in pairs even among daughter nuclei. This type of division where there is mutual attraction between the nuclei of different genotype is called *Conjugate division*.

After the division, large number of nuclei (in pairs) are formed. The ascogenous hypha now gets divided into many cells. When cell walls are laid down, care is taken to see that a cell always encloses one male and one female nucleus.

Further development now proceeds usually in the terminal binucleate cell which produces a hook like structure known as *Crozier* towards a side. Now the two nuclei divide with a specific spindle orientation. In one of the nuclei the spindle is parallel to the long axis of the ascogenous hypha, in the other it is oblique so that one of the daughter nuclei enters the crozier. After the division four nuclei are produced, the cell itself gets divided into three cells : (a) antepenultimae cell having a single nucleus. (b) A penultimate cell having two nuclei (c) an ultimate cell (crozier) having a single nucleus. Penultimate cell is now called the ascus mother cell; as it is the one which produces the ascus. A protrusion develops from the ascus mother cell which elongates to form a sac like structure called Ascus. The diploid nucleus migrates into the ascus where it undergoes meiosis, followed by a mitosis to produce eight haploid nuclei. Each of the nuclei ultimately forms a spore- the Ascospore. After the spore formation the ascus breaks open releasing the ascospores which germinate and form the monokaryotic mycelium.

The above mentioned account gives a general picture of the process of ascus formation in Ascomycetes. While only the penultimate cells of the ascogenus hyphae form asci, other cells may also form them. Further, after the formation of ascus, the crozier and antepenultimate cell may fuse to form another terminal cell and the entire process is repeated.

In some forms, the antheridium is not formed or even when formed does not function. In such cases the ascogonial nuclei themselves pair and the process goes on as usual.

Structure and development of fructificaiton

While the asci are formed naked and the ascospores are libeated as soon as they are formed, in a great majority of Ascomycetes the asci are enclosed in a compact protective structure clled fruiting body or fructification. Usually when the asci and ascogenous hyphae are being formed, the vegetative hyphae grow round and form a protectie envelop or the sex organs are formed in the innerlining of the already formed fructification. The main function is to afford protection to these delicate reproductive structures. The general name given to a fructification in Ascomycetes is *Ascocarp*.

[Gr.*Askos* = sac + *karpos* = fruit]. Ascocarps vary in structure and compostion. On the basis of this they may be classified into following four categories [Alexopoulos, 1956].

1. **Cleistothecium**. (*Pl. cleistothecia* Gr.*kleistios* = closed + *theke* case). This is a completely closed structure, may be globose or ovoid. The wall is many layered and is called the peridium. Within this, the asci lie scattered. The ascospores will be liberated only on the decay of the peridium. In this type, there is no scope for the regular liberation of ascospores.

2. **Perithecium**. *Pl.perithcia*. (Gr.*Pere* - around). These are flask shaped structures possessing a small opening called ostiole to facilitate the liberation of ascospores. Lining the inner wall of the perithecium at both are numerous asci parallely arranged. Intermingled with the asci, sterile hair like processes are found called Paraphyses (Sing.paraphysis).

3. **Apothecium**. (Pl.apothecia. Gr.*Apothika* = store house). This is a open type of fructification and cup like with a broad opening sometimes saucer shaped also. It has a multilayered walled called the peridium. Lining the inner margin of the cup is the hymenial region (fertile) containing asci and paraphysis.

Of the three types of above named frutifications, apothecium is supposed to be the most advanced and cleistothecium most primitive.

4. **Ascotroma**. (Pl.ascostromata). In some, the asci are formed direcçly in a cavity in the stroma, so that stroma itself forms the wall of ascocarp.

In addition to the above, various other modifications of the fructifications occur.

Classification

The fructification is an important criterion in the classification of Ascomycetes. Their formation or non-formation and type of fructification when formed are all of great help in the classification.

Gwynnevaughan and Barnes (1937) divided Ascomycetes into three sub-classes namely.

Sub class : 1. Plectomycetes

(Ascocarps if present have no opening, asci are irregularly arranged)

Sub-class : 2. Pyrenomycetes

(Ascocarps flask saped when ripe; asci in parallel series)

Sub-calss 3. Discomycetes

(Ascocarp wide open when ripe, asci in parallel series)

But many modern mycologists classify Ascomycetes as follows:

Class - Ascomycetes

Sub-class 1. Hemi Ascomycetes

(Protoasomycetes) (no-fructification)

Sub-class 2. (Euascomycetes) (true Ascomycetes)

(fructification present)

Based on the nature of the fructification, the sub-class Euascomycetes is divided into three series namely

Plectomycetes (Cleistothecium)

Pyrenomycetes (Perithecium) and

Discomycetes (Apothecium)

SACCHAROMYCES (YEAST)

Common species : *S.cerveisiae, S.ludiwigii, Schizosaccharomyces octosporus*

Occurrence : The members are commonly called yeasts. They are widely distributed in nature. Yeasts are saprophytic and are commonly found in substrata rich in sugars like nectar of flowers, surface of fruits like grapes, apples etc. Milk and other food undergo fermentation due to yeasts.

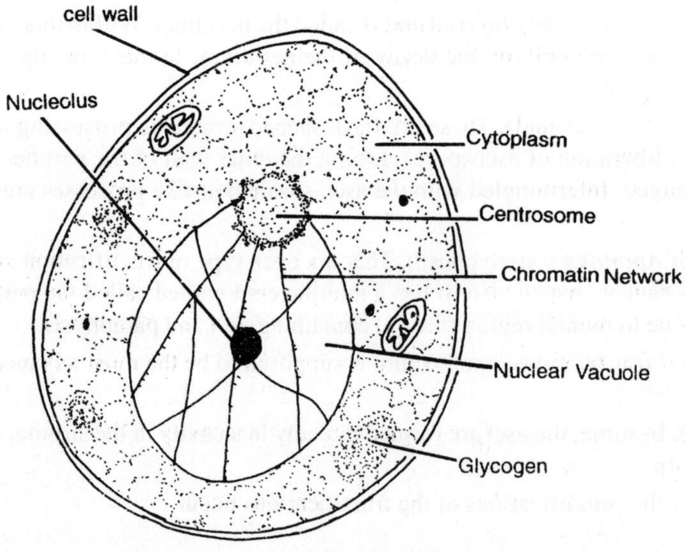

Fig. 13.1 Saccharomyces. A highly magnified cell of Saccharomyces

Study :

1. Method of culturing yeasts
2. Vegetative structure
3. Reproduction
4. Systematic position and identification

Culture of yeasts

1. Place small pieces of yeast cake in sugar solution
2. Yeast cells multiply and a dense cluster of cells is produced
3. After few days the yeast cells can be harvested

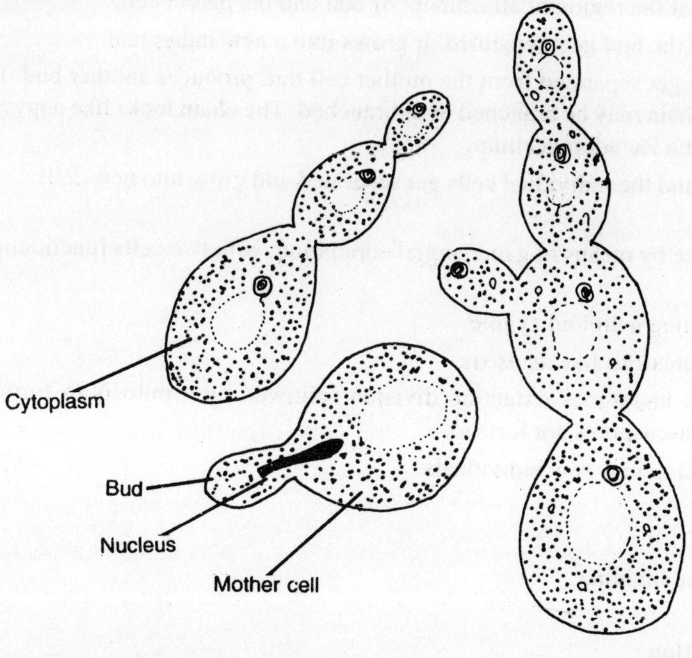

Fig. 13.2 Saccharomyces. Yeast cells showing budding

Vegetative structure

1. The thallus of yeast is made up of isolated cells (unicellular plant body)

2. The cells are usually oval or spherical ranging from 5-10μin diameter

3. Each cell has an outermost cell wall made up of chitin. Cellulose is absent.

4. The cytoplasm shows differentiation into outer ectoplasm and inner endoplasm

5. There is a large vacuole in the center

6. Placed immediately above the vacuole is the single nucleus (the nucleus was mistaken previously as the centrosome)

7. Cytoplasm is granular and contains reserve food materials like glycogen and oil

8. Mitochondira and starch grains are also found in the cytoplasm

Asexual reproduction

1. Yeast cells reproduce asexually by two methods viz., *fission* and *budding*.

2. Fission yeasts are called *Schizosaccharomyces* while budding yeasts are called *Saccharomyces.*

3. During fission, the single nucleus gets elongated.

4. A constriction appears in the center of the nucleus and it becomes dumb-bell shaped.

5. With the deepening of the constriction, the nucleus gets divided into two bits.

6. The two daughter nuclei migrate to opposite poles. A cross septum appears in the center of the cell dividing it into two.

7. The two daughter cells formed thus grow into adults cells.

8. Budding is another method of reproduction. During budding a small bulge arises from one side of the cell. The bulge increases in size and is now called a bud .

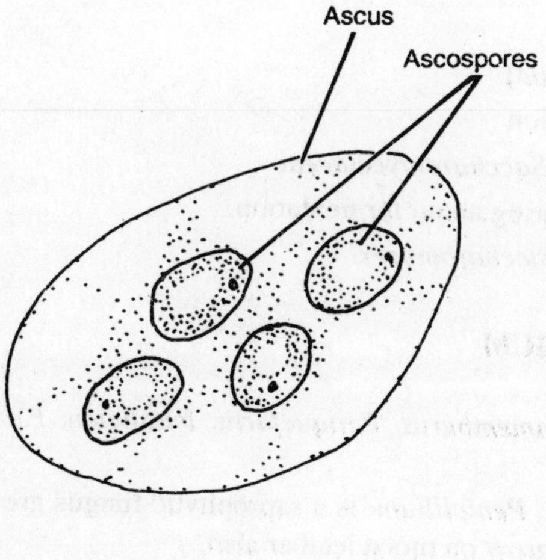

Fig. 13.3 Single ascus of Saccharomyces

9. Meanwhile, the nucleus of the mother cell divides into two daughter nuclei; one of these migrates into the bud.

10. A small constriction appears at the region of attachment of bud and the parent cell.

11. The constriction deepens and the bud gets separated. It grows into a new individual.

12. In some cases a bud does not get separated from the mother cell but, produces another bud. In this way a chain of buds is formed. This chain may be branched or unbranched. The chain looks like a mycelium hence it is called a Rhizomycelium or a Pseudomycelium.

13. Eventually the chain breaks and the individual cells get separated and grow into new cells.

Sexual reproduction

1. Sexual reproduction takes place by means of gametangial copulation, with two cells functioning like gametangia.

2. The two gametangia fuse forming a diploid zygote.

3. The zygote which is diploid functions like an ascus.

4. The diploid nucleus in ascus undergoes reduction division followed by a mitosis to form eight ascospores. In some cases only four ascospores are formed.

5. Ascospores on liberation develop into new individuals.

Diagrams to be drawn

1. Yeast cell magnified.

2. Yeast cells showing fission and budding.

3. Ascus with ascospores.

Systematic position and identification :

A. 1. Chlorophyll absent
2. Reserve food is glycogen ....*Fungi.*

B. 1. Mycelium is septate or plant body unicellular
2. Characteristic spores - ascospores are produced inside an ascus.
3. Ascospores usually eight, rarely otherwise......................*Ascomycetes.*

C. 1. No fruiting body.
2. Saprophytic nutrition
3. Asci not in clusters ....*Endomycetales*

D. 1. Asexual reproduction by budding or fission
2. Plant body unicellular ....*Saccharomycetaceae*

E. 1. Members produce enzyme zymase and bring about fermentation.
2. Plant body unicellular ....*Saccharomyces.*

PENICILLIUM

Common species : *P.notatum, P.chrysogenum, P.camembertii, P.roquefortii, P.italicum, P.digitatum* etc.

Occurrence : Commonly called green or blue mold, *Penicillium* is a saprophytic fungus growing on rotten fruits decayed vegetable, meat etc. Some of them grow on moist leather also.

Study

1. Diseases caused and symptoms
2. Method of culture in the laboratory
3. Vegetative structure
4. Reproduction
5. Systematic position and identification

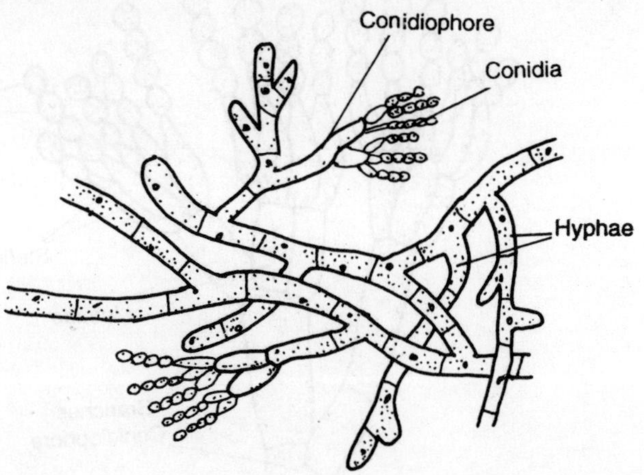

Fig. 13.4 Penicillium. A portion of mycelium bearing conidiophores

Diseases caused

1. *P.expansum*, causes decay of apples, pears and grapes in storage.

2. *P.italicium* and *P.digitatum* cause decay of citrus fruits like orange, sweet lime etc.

Disease symptoms are visible in the form of greenish or slightly bluish leisons on the fruit wall. Those are the places where reproductive structures are found.

Culture in the laboratory

1. Keep a stale orange in moist condition.

2. Sprinkle water to keep it wet.

3. Keep the orange enclosed.

4. After few days greenish blue coating appears on the rind.

5. Scrape the coating on to a microslide and observe under the microscope.

Vegetative structure

1. The plant body is a mycelium. It is slender composed of profusely branched septate hyphae.

2. The mycelium is divided into cells. Cells are either monokaryotic or dikaryotic.

3. The cell wall between the cells has an opening, through which cytoplasm streams from cell to cell.

Asexual reproduction

1. Vegetative propagation is brought about by means of fragmentation of hyphae. Each fragment develops into a new individual.

2. Asexual reproduction by conidia is the most common method of multiplication in *Penicillium.*

3. Vegetative mycelium produces conidiophores.

4. Conidiophores are erect and branched once or twice.

5. Each conidiophore or its branches bear at their tips a whorl of bottle shaped structures called Sterigmata or Phialides.

6. The branches of the conidiophores which bear phialides are called metulae.

7. The sterigmata at their tips produce a chain of conidia. Conidia are produced in basipetalous succession.

8. The entire structure consisting of conidiophores, metulae, sterigmata and conidia resembles a brush or broom, hence the name *Penicillium* (Gr.*Penicillus* = brush)

9. Conidia are unicellular. They may be uni or multinucleate.

10. Upon falling on a suitable substratum, conidia germinate into a new mycelium by producing a germ tube.

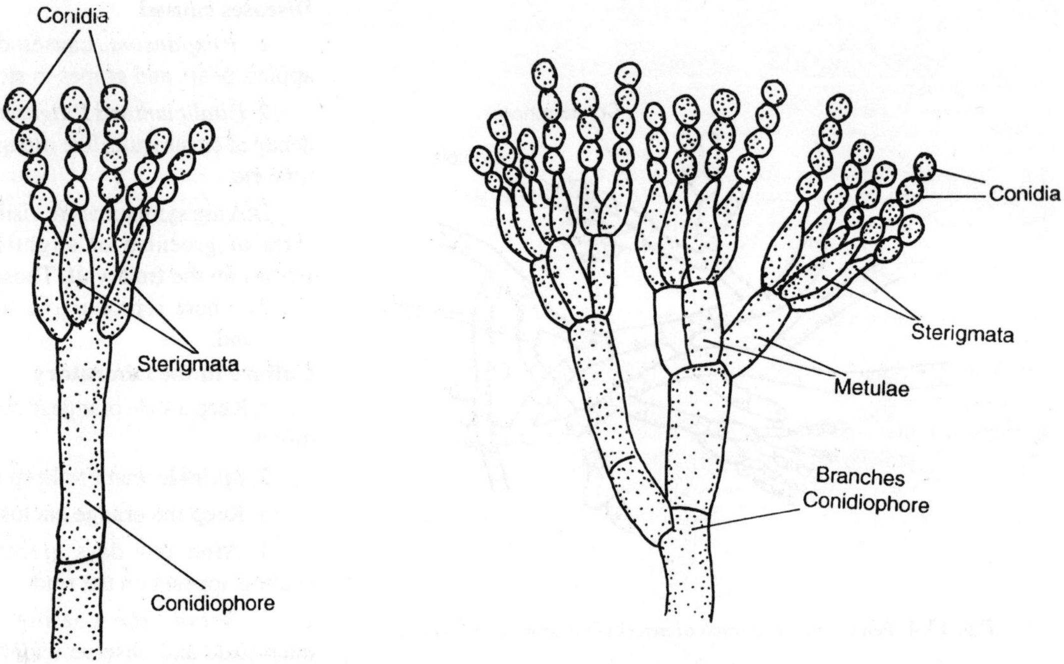

Fig. 13.5 Penicillium. Single conidiophore (monoverticillate) Single conidiophore (biverticillate)

Sexual reproduction

1. Sexual reproduction is by gametangial contact. Sex organs are antheridia and ascogonia.
2. Sexual reproduction is very rare.
3. The fruiting body produced by penicillium is a cleistothecium.
4. The clesitothecium is a completely closed structure. It has a mutlilayered wall called peridium.
5. Lying scattered in the cavity of cleistothecium are many asci. Between the asci, sterile hyphae called paraphyses are found.
6. Each ascus has eight ascospores. The ascospores come out of the cleistothecium by the rupturing of the wall of the latter. On germination they develop into a new mycelium.

Diagrams to be drawn

1. Habit of infected host
2. Mycelium
3. Condiophore
4. Cleistothecium

Fig. 13.6 Penicillium. Sex organs

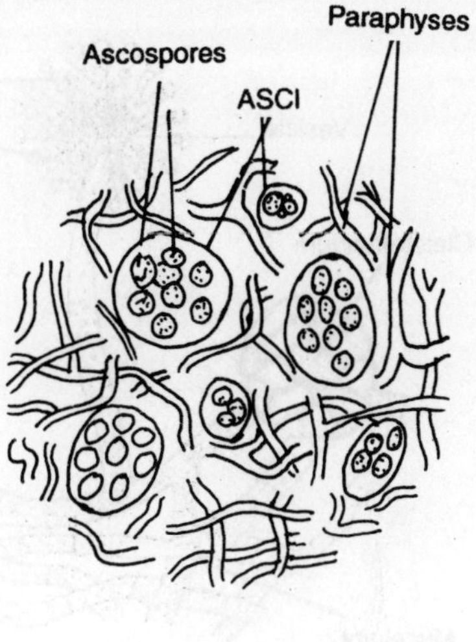

Cleistothecium in section

Systematic position and identification

A. 1. Chlorophyll absent
2. Reserve food is glycogen ....*Fungi.*

B. 1. Mycelium is septate
2. Ascospore produced in ascus ....*Ascomycetes*

C. 1. Conidia are chief mode of multiplication
2. Asci scattered inside the cleisothecium ....*Aspergillales*

D. 1. Fructifications not stalked
2. Peridium made up of pseudoparenchyma ....*Aspergillaceae*

E. 1. Conidiophores branched, but no vesicle
2. Asexual structures resemble a broom ....*Penicillium*

ASPERGILLUS

Common species : *A.niger, A.fumigatus, A.flavus* etc.

Occurrence : The species of *Aspergillus* are generally saprophytic growing on stale bread, decaying vegetable, meat etc. It also grows on jams, jellys, footwear, clothes etc.

A few species of *Aspergillus* are parasitic causing lung diseases.

Study

1. Diseases caused and symptoms
2. Laboratory culture
3. Vegetative structure
4. Reproduction
5. Systematic position and identification

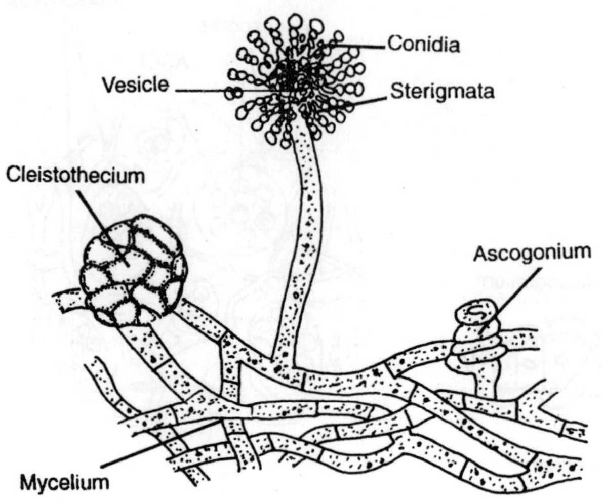

Fig. 13.7 A portion of mycelium showing conidiophore cleistothecium and sex organs of Aspergillus

Diseases caused and symptoms

1. Rot disease : This is caused by *A.niger*. The hosts that are usually attacked are pomegranate (*Punica granatum*), Dates (*Phoenix dactylifera*), Figs (*Ficus carica*) etc.

2. Aspergillosis : This is an allergic type of lung disease caused by the spores of *A.niger, A.fumigatus* and *A.flavus*, cattle and other animals and human beings also get the infection.

3. Otomycosis : A type of ear disease in human beings caused by several species of *Aspergillus.*

Disease symptoms

1. Saprophytic species produce symptoms in the form of greyish green or bluish green growth of the mycelium.

2. Fruits get rotten and decomposed due to the entry of fungus through wounds.

3. The symptoms of *Aspergillosis* resemble those of tuberculosis.

Laboratory culture

1. Place a piece of bread or citrus fruit under a bell jar.

2. Keep it moist

3. After few days greenish blue or yellowish mycelia grow profusely.

Vegetative structure

1. The plant body of the fungus is a much branched, septate closely interwoven mass of mycelium.

2. Hyphae are divided into cells. Cells are multinucleate suspended in a granular cytoplasm.

3. Oil globules are present in the form of reserve food.

4. While most of the myeclium is above the substratum, certain hyphae grow into the substratum, possibly to help in anchorage and absorption.

Asexual reproduction

1. Asexual multiplication may take place vegetatively by fragmentation of the hyphae.

2. The chief method of sexual increase however is by the formation of conidia.

3. From the vegetative mycelium are produced many unbranched erect hyphae called conidiophores.

4. Each conidiophore arises from a foot cell of the mycelium.

5. The tip of each conidiophore swells to form a dome shaped vesicle.

6. The vesicle from its surface gives rise to a number of sterigmata or phialides, which are bottle shaped structures. The sterigma are multinucleate.

7. Sometimes a second layer of sterigmata may arise from the first layer in which case, the first layer is called primary sterigmata and the second layer is called the secondary sterigmata.

8. The sterigmata of the second layer (first layer, if there are no secondary sterigmata) produce a chain of conidia. The conidia are produced in basipetalous succession.

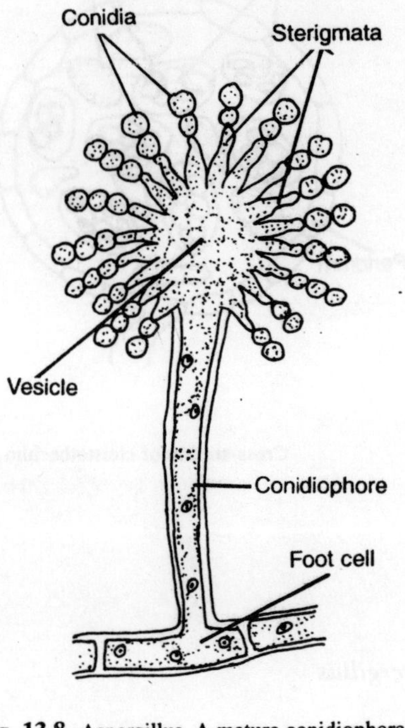

Fig. 13.8 Aspergillus. A mature conidiophore

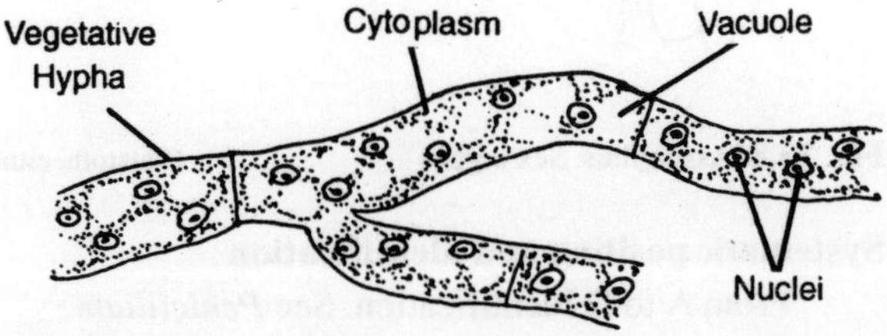

Hypha enlarged

9. The conidia are small, rounded unicellular structures. They may be uni or multinucleate. The conidial wall is pigmented in hues of black, green and blue.

10. The conidial wall is double layered. The outer epispore is spiny, while the inner endospore is smooth. On germination the conidium develops into a new individual.

Sexual reproduction

1. Sexual reproduction is by means of gametangial contact. Sex organs are antheridia and ascogonia.
2. Antheridium is small elongated and multinucleate.
3. Ascogonium is a flask shaped structure with a bulbous base and a narrow neck called trichogyne.
4. Fruiting body is a cleistothecium. Many asci lie scattered in the cavity of the fruiting body. Each ascus has eight ascospores.
5. On liberation from the cleistothecium, the ascospores germinate to give rise to new individuals.

Diagrams to be drawn

1. Habit of infected material
2. Mycelium with condiophores
3. A part of the condiophore enlarged
4. Cleistothecium.

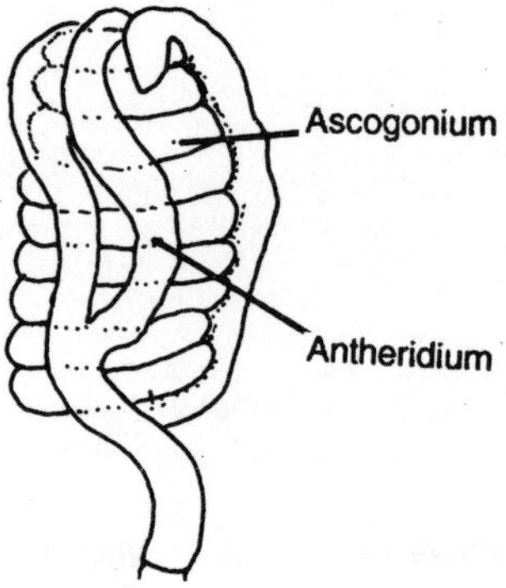

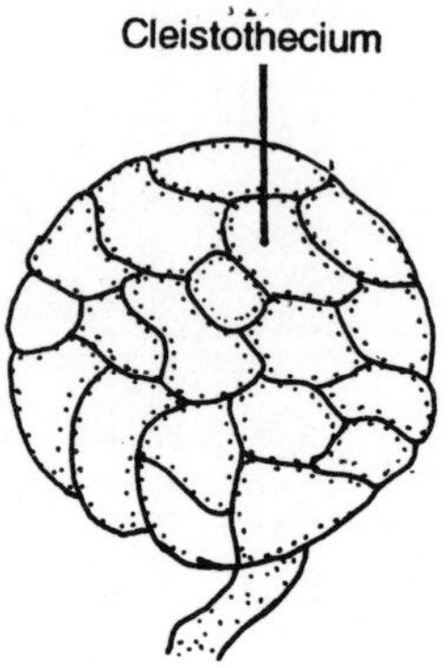

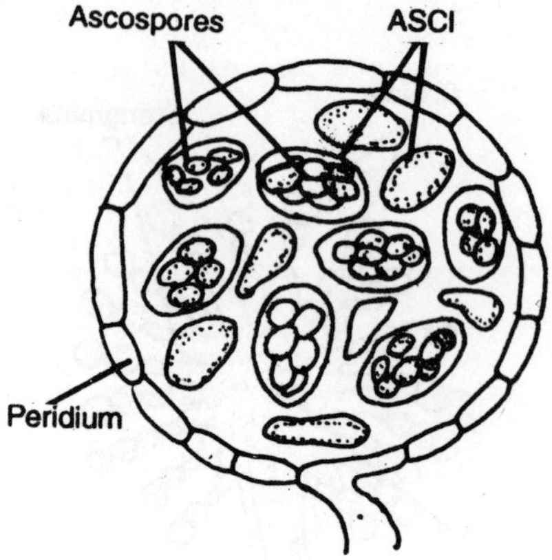

Fig. 13.8A Aspergillus. Sex organs Cleistothecium Cross-section of cleistothecium

Systematic position and identification

From A to D identification. See *Penicillium*.

E. 1. Conidiophores unbranched

2. Tip of the conidiophore swells to form a vesicle

3. Secondary sterigmata present sometimes.......... *Aspergillus*

ERYSIPHAE

Common species : *E.polygoni, E.graminis, E.cichoriacearum* etc.

Occurrence. The species of *Eryisphae* cause a disease which is commonly called powdery mildew disease. The members are ectoparasites sending only haustoria into the host. The superficial growth of the cottony white fungus gives the appearance of sprinkled powder hence the name powdery mildew.

Many families of flowering plants like Gramineae, Papilonaceae, Caricaceae etc are infected.

Study : 1. Diseases caused and disease symptoms.

2. Vegetative structure

3. Reproduction

4. Systematic position and identification

Diseases caused and disease symptoms

Various species of *Erysiphae* cause diseases to economically important plants. The following is a brief description of some of the diseases and their symptoms.

1. Powdery mildew of peas. The disease is caused by *E.polygoni*. It makes its appearance when the plants are fruiting during the month of Jan-Feb. The disease becomes rampant during dry weather.

Disease symptoms are visible first on the leaf. Later they spread to the other parts. Powdery patches appear on either surface of the leaf. Gradually the patches spread to tendrils also. The mycelium of the fungus is seen at the region of patches.

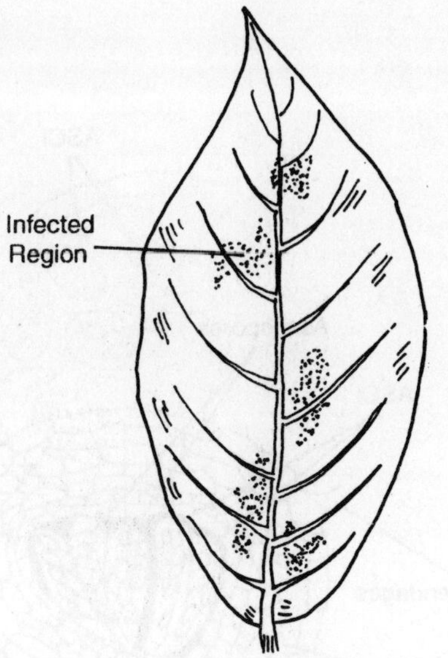

Fig. 13.9 Erysiphe. Host leaf infected with powdery mildew caused by Erysiphe

2. **Powdery mildew of barley**. The disease is brought aboty by E.graminis, Varhordei. Besides barley, the fungus also infects wheat, rye, oats and other graminaceous plants. Though the damage caused by the fungus is not very extensive, death of the leaves, considerably reduces the photosynthetic efficiency of the plant.

Discasc symptoms are first visible on the leaves. Both surfaces of the leaves, specially the upper surface exhibits powdery patches. Gradually the disease spreads to the stem and inflorescence also. Infected plants show increased rate of respiration and transpiration. Photosynthesis gets retarded.

3. **Powdery mildew of cucurbits**. Many cucurbits specially pumpkin, ribbed gourd are infected by the fungus *E.cichoracearum.* This pathogen also infects plants like tobacco, potato, lettuce, sun-flower, mango, castor etc.

Disease symptoms become prevalent when the atmosphere is moist. Small spots appear on leaves and stems, gradually become wider giving the characteristic powdery appearance. Dark fruiting bodies of the fungus are visible on the leaves at later stages. Plants get defoliated and fruits drop down prematurely due to infection.

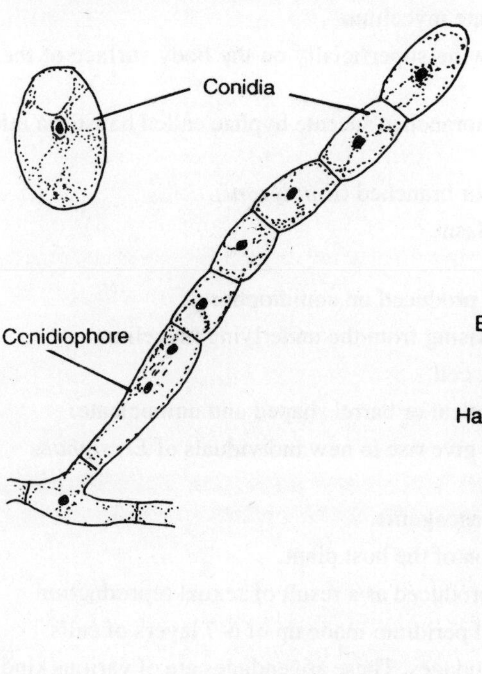

Fig. 13.10 Erysiphe. Single conidiophore with chain of conidia

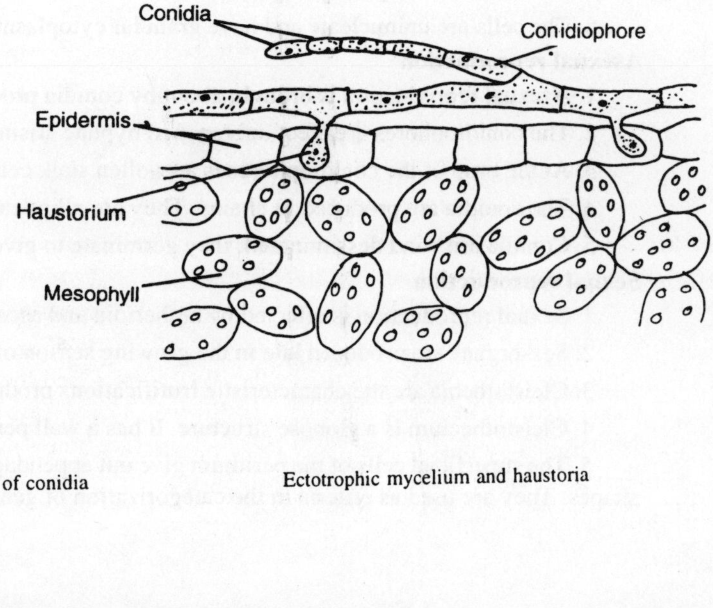

Ectotrophic mycelium and haustoria

Specimens given in the laboratory

1. Leaves of pea, pumpkin or papaya etc.
2. Scraping of the powdery patches will reveal the mycelium.
3. Free hand section of the leaf will show the haustoria.

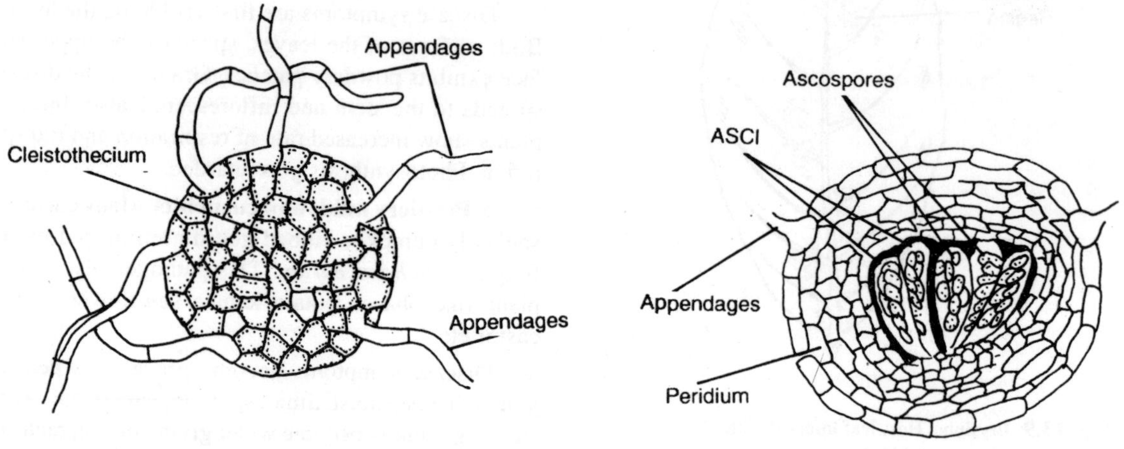

Fig. 13.11 Erysiphe. Single cleistothecium V.S. of cleistothecium

Vegetative structure

1. Plant body of the fungus is a much branched, septate mycelium.

2. The fungus is an ectoparasite; the mycelium growing superficially on the body surface of the host plant.

3. The superficial mycelium however sends erect, unbranched saccate hyphae called haustoria into the cells to absorb the nutrition from the host.

4. The haustroia may be saccate (*E.cichoracearum*) or branched (*E.polygoni*).

5. The cells are uninucleate and have granular cytoplasm.

Asexual reproduction

1. Asexual reproduction is brought about by conidia produced on conidiophores.
2. The conidiophores are erect unbranched hyphae arising from the underlying mycelium.
3. At the base of the conidiophores is a swollen stalk cell.
4. The conidia are produced in chains. They are elliptical or barrel shaped and uninucleate.
5. Conidia are wind dessiminated; they germinate to give rise to new individuals of *Erysiphae*.

Sexual reproduction

1. Sexual reproduction is effected by antheridia and ascogonia.
2. Sex-organs are produced late in the growing season of the host plant.
3. Cleistothecia are the characteristic frutifications produced as a result of sexual reproduction.
4. Cleistothecium is a globose structure. It has a wall peridium made up of 6-7 layers of cells.
5. The superficial cells of the peridium give out appendages. These appendages are of various kinds and shapes. They are used as criteria in the categorization of genera and species.

6. Lying in the cavity of the clistothecium are found a number of asci. Each ascus usually has eight ascospores.

7. The ascospores are unicellular, uninucleate, haploid and on coming out of the fruiting body (the peridium breaks open irregularly) germinate to form a new individual.

Diagrams to be drawn:

1. Habit of infected leaves.
2. Mycelium with conidia.
3. Cleistothecia with appendages.

Sysematic position and identification

A. 1. Chlorophyll absent
2. Reserve food is glycogen ... *Fungi*

B. 1. Mycelium is septate
2. Ascospores produced in Ascus ... *Ascomycetes*

C. I. Members are ectoparasites
2. Powdery patches appear on the host ... *Erysiphales*

D. 1. Mycelium tubular and hyaline
2. Condidia produced in large numbers ... *Erysiphaceae*

E. 1. Ascospores unicellular
2. Many asci in cleistothecia
3. Appendages on cleistothecia myceloid ... *Erysiphae*

PHYLLACTINIA

Common species : *P.dalbergiae*.

Occurrence : The fungus is a semi-endo parasite. It grows commonly on the leaves of *Dalbergia, Cassia, Morus acacia, Betula* etc.

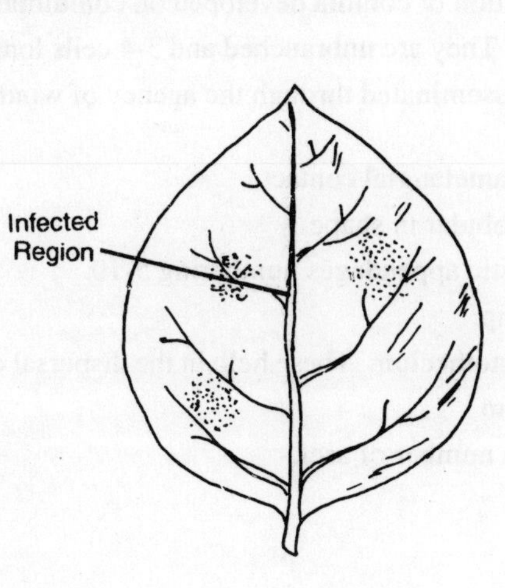

Fig. 13.12 Dalbergra leaf infected with powdery mildew caused by Phyllactinia.

Powdery patches appear on the leaves generally on the lower surface.

Study : 1. Disease symptoms
2. Vegetative structure
3. Reproduction
4. Systematic position and identification

Disease symptoms :

1. Powdery pathces of white hyphae seen on the leaves.

2. In the older stages of the leaf orange coloured cleistothecia are seen. Mature cleistothecia are dark coloured.

Vegetative structure

1. The plant body is a septate, much branded mycelium.

2. The hyphae of the mycelium are not completely superficial, they are semi endoparasites.
3. The hyphae produce haustoria, which penetrate the mesophyll cells of the host and absorb nutrition.
4. Hyphae have uninucleate cells with granular cytoplasm.

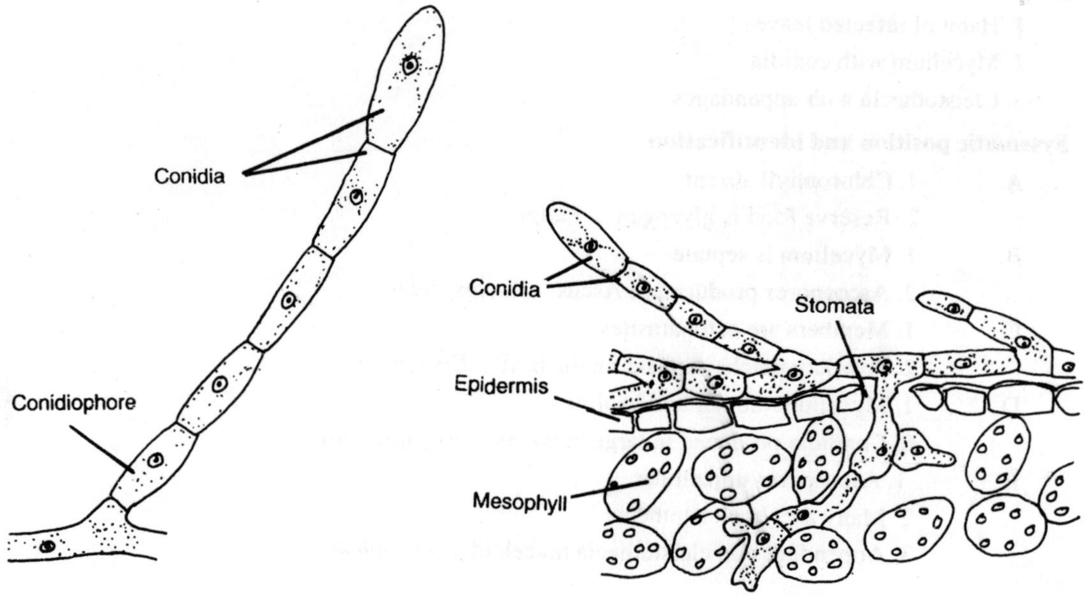

Fig. 13.13 Phyllactinia. Single conidiophore with chain of conidia Ectotrophic mycelium

Asexual reproduction

1. Asexual reproduction is brought about by the formation of conidia developed on conidiophores.
2. Conidiophores arise from the underlying mycelium. They are unbranched and 3-4 cells long.
3. Conidia are unicellular and uninucleate. They are disseminated through the agency of wind.

Sexual reproduction

1. Sexual reproduction is brought about by means of gametangial contact.
2. Fruiting bodies are called cleistothecia which are globular in shape.
3. Cleistothecia are dark coloured and have characteristic appendages numbering 5-10.
4. Each appendage has a bullbous base and a pointed tip.
5. Some sap oozing hyphae are also present on the cleistothecium. Thes^ help in the dispersal of spores.
6. Cleisothecium has a pseudoparenchymatous peridium.
7. Lying in the cavity of the cleistothecium are found a number of asci.
8. Each ascus has only two ascospores.

Diagrams to be drawn

1. Habit sketch of infected leaves.
2. Mycelium with conidiophores.
3. Cleistothecium showing appendages.

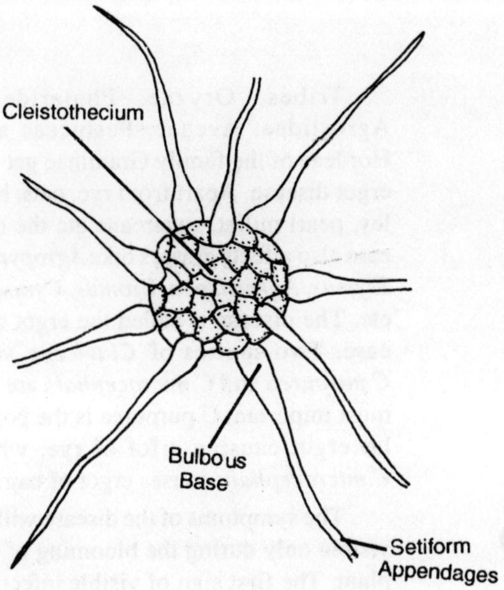

Fig. 13.14 Phyllactinia. Single cleistothecium

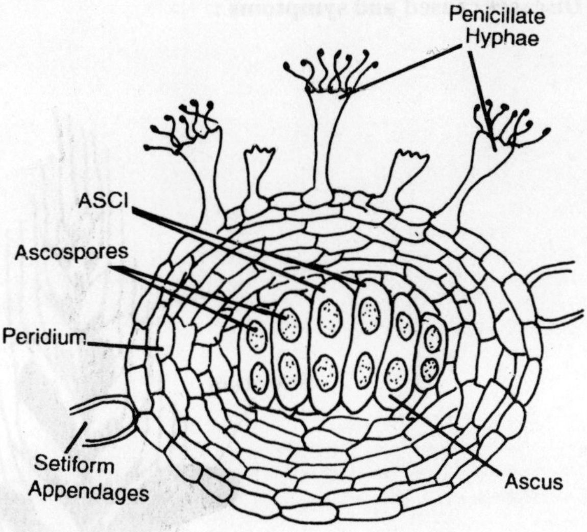

V.S. of cleistothecium

Systematic position and identification

A. 1. Chlorophyll absent
   2. Reserve food is glycogen ... *Fungi*

B. 1. Mycelium septate
   2. Ascospores produced in ascus... *Ascomycetes*

C. 1. Mycelium tubular and hyaline
   2. Powdery patches appear on the host ... *Erysiphales*

D. 1. Mycelium tubular and hyaline
   2. Conidia produced is large numbers
   3. Asci found in cleistotheia ... *Erysiphaceae*

E. 1. Plant body semiendoparasitic
   2. Fructification is a cleistothecium
   3. Appendages on the cleistothecium have a bulbous base and a pointed tip.
   4. Only two ascospores produced is the ascus ... *Phyllactinia.*

CLAVICEPS

Common species : *C.purpurea, C.microcephala* etc.

Occurrence : *Claviceps* is an obligate parasite infecting the members of Gramineae, especially rye, oats, barley, etc. The disease is popularly called the Ergot disease.

Study : 1. Diseases caused and symptoms.
2. Vegetative structure
3. Reproduction
4. Systematic position and identification

Diseases caused and symptoms :

Fig. 13.15 Claviceps. Single sclerotium

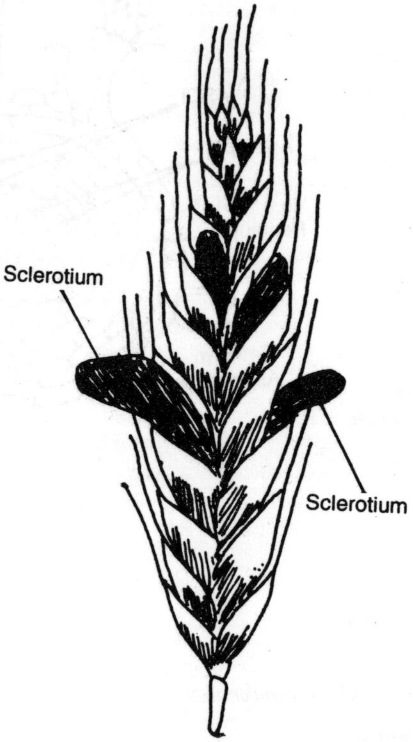

Infected inflorescence of rye showing sclerotia of Claviceps

Tribes, Oryzae, Phalarideae, Agrostidae, Avenae, Festuceae and Hordeae of the family Graminae get the ergot disease. Apart from rye, oats, barley, pearl millet, sugarcane etc the disease also affects grasses oike *Agropyron, Elymus, Agrostispoa, Bromus, Cynodon* etc. The disease is called the ergot disease. Two species of *Claviceps* viz., *C.purpurea* and *C.microcephala* are the most important C.purpurea is the popular ergot causing erfot of rye, while *C.microcephala* causes ergot of bajra.

The symptoms of the disease will be visible only during the blooming of the plant. The first sign of visible infection is the exudation of sweet sap from the infected ears. The characteristic symptom of the disease however is the appearance of dark brown coloured hard masses of fungal tissue called sclerotia. It is these sclerotia called ergots that have given the name to the disease.

Consumption of ergots mixed with the grains results in a disease called *Ergotism.* Ergotism affects cattle as well as human beings. In cattle loss of horn, hooves, tail etc takes place, while in human beings loss of hair and nails or even gangrene may set in.

Vegetative structure

1. The plant body of the fungus is noticeable in the ovarian tissues of the host plant.
2. The mycelium is septate, much branched and the cells are multinucleate.
3. The mycelium develops as a result of the germination of ascospores.

Asexual reproduction

1. The fungus reproduces asexually by the formation of conidia.
2. Conidiophores are produced inside the ovary of the host plant. They are aggregated into clusters called ascervulus.
3. Conidia are uninucleate and unicellular. On liberation from the ovary (usually by insects as there is nectar) they infect other healthy plants.

4. The asexual stage is called the *Sphacelia stage* or *honeydew stage*. The former name is a misnomer. Originally this stage was identified as a separate individual and named *Sphacelia*.

Because of the production of nectar the stage is called the honey dew stage.

Sclerotium stage.

1. This is the characteristic stage of the fungus. Also known as the ergot, the sclerotium is a hard brown mass of fungal tissue, slightly larger than the grain which it replaces. The ergots will be peeping out of the inflorescence. They are rich in alkaloids.

2. The sclerotium is pink or purple in colour sometimes.

3. The sclerotiums stage constitutes the resting stage of the fungus. Usually sclerotia reach the soil during the threshing of grains.

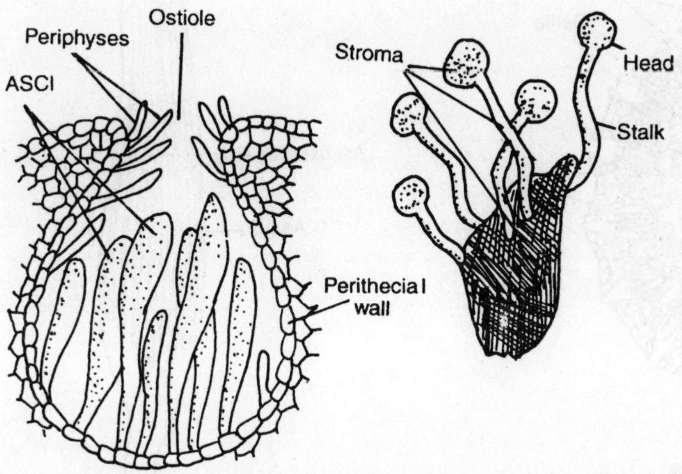

Fig. 13.16 Claviceps. V..S. of ascocarp   Germinating sclerotium of Claviceps

Ascostroma stage

1. After overwintering, the sclerotium germinates by putting forth small pin head like structures called ascostroma.

2. Each ascostroma is about an inch long. It has a stalk and a rounded head.

3. In a mature ascostroma, small cavities appear which later develop into perithecia.

4. Each perithecium is a flask shaped structure. It has a small opening called the ostiole.

5. Lining the base of the perithecium inside are a number of asci and sterile hairs (paraphysis).

6. Each ascus has eight ascospores.

7. Ascospores are long, thread like and acicular (needle like). On liberation from the perithecium, they are dispersed by wind. They germinate on the stigmatic surface of a suitable host plant.

Specimens given in the laboratory

1. Infected ears of host plant
2. Sclerotia
3. Sclerotia with ascostroma

Diagrams to be drawn

1. Habit of infected ear
2. Section showing the conidiophores, conidia etc
3. Sclerotium (ergot) sectional view
4. Sectional view of ascostroma.

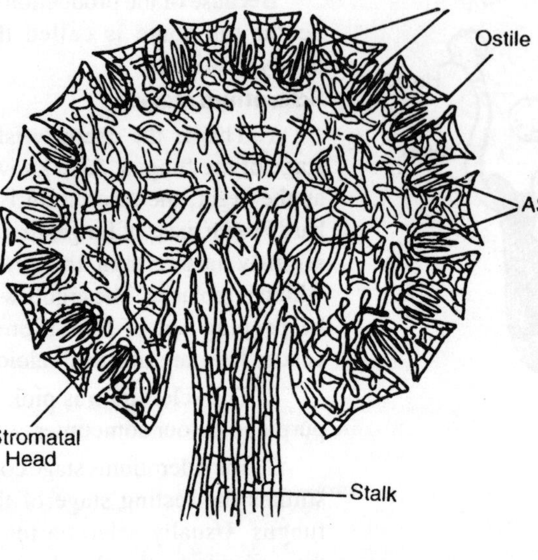

Fig. 13.17 Claviceps. V.S. of stromata

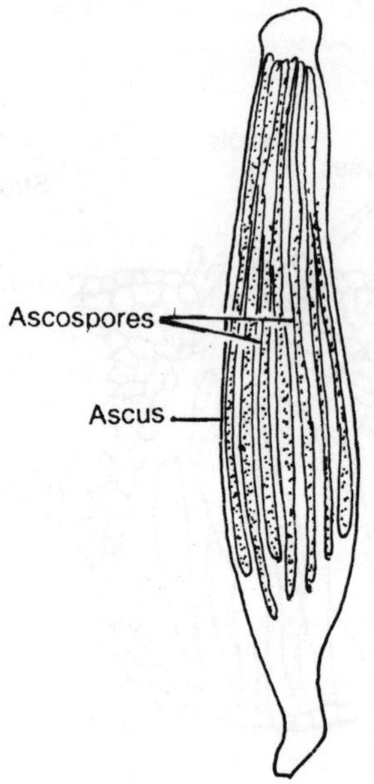

An ascus containing ascospores

Systematic position and identification

A. 1. Chlorophyll absent
   2. Reserve food is glycogen ... *Fungi*

B. 1. Mycelium septate
   2. Ascus and ascospores produced ... *Ascomycetes*

C. 1. Ascospores acicular
   2. Perithecia found in ascostroma ... *Clavicipitales*

D. 1. Ascus has a perforated lid to facilitate spore dispersal ... *Clavicipitaceae*

E. 1. Ergots dark brown in colour and hard
   2. Ascospores acicular and arranged parallely in the ascus ... *Claviceps.*

XYLARIA

Common species : *X.hypoxylon, X.mali, X.polymorpha* etc

Occurrence : Xylaria is cosmopolitan in distribution. It usually grows as saprophyte on decaying wood. *X.hypoxylon* causes root rot of apples while *X.mali* causes black rot of apple trees.

Study : 1. Structure of the plant body
2. Reproduction
3. Systematic position and identification

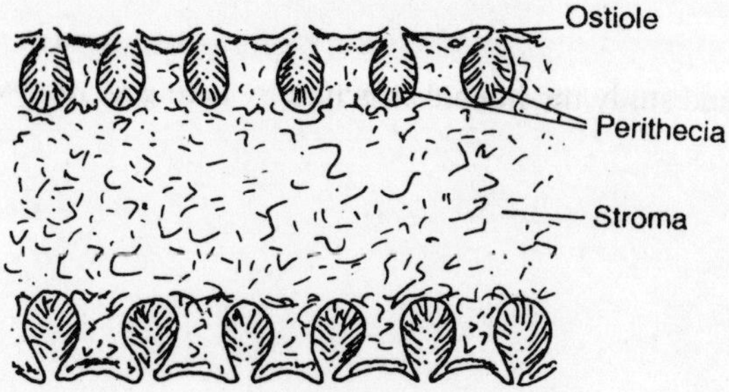

Fig. 13.18 Xylaria. Stromata in cross-section

Structure of the plant body

1. The mycelium is well branched and at intervals forms thick strands.

2. These strands are heterotropic and grow towards light.

The hyphae are septate and branched. Cells are uninucleate.

Reproduction :

1. The strands of mycelia develop into dark coloured stroma.

2. These are erect stout and cylindrical structures.

3. These may be branched or unbranched.

4. The stoma develop numerous perithecia which are arranged at right angles to the surface of the stroma.

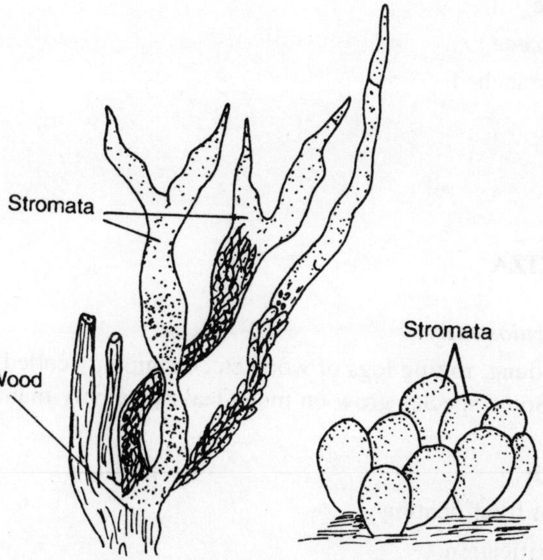

Fig. 13.19 Xylaria. Stromata growing on rotten wood

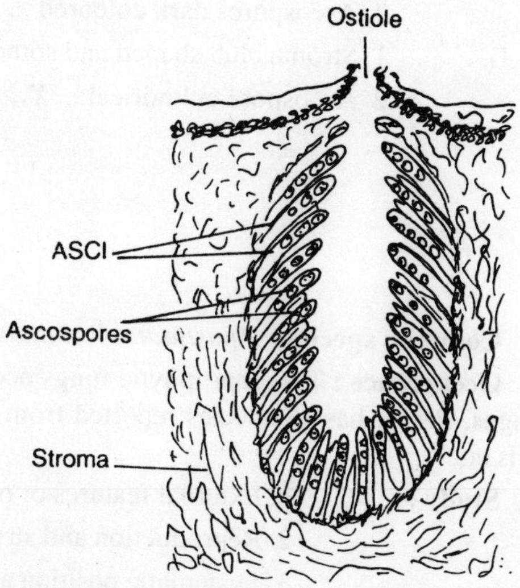

V.S. of perithecium

5. During early stages stroma is covered by conidiophores.

6. A transverse section of the stroma reveals a circular outline and a wall made up of closely packed hyphae.

7. Near the peripheral region are found a number of small flask shaped cavities called perithecia.

8. Each perithecium opens to the exterior, with the help of a small opening called the ostiole.

9. Lining the base of the perithecial cavity are found a number of asci. Intermingled with the asci are found the paraphyses.

10. Each ascus has eight cylindrical ascospores. On liberation from the ascus, they germinate to give rise to new individuals.

Specimens given in the laboratory

1. Stroma of *Xylaria* (observe the external details and study the intenal structure by taking a cross section).

Diagrams to be drawn

1. Habit of stroma
2. Ground plan of transverse section
3. A perithecium enlarged

Systematic position and identification

A. 1. Chlorophyll present
2. Reserve food is glycogen ... *Fungi*

B. 1. Mycelium septate
2. Ascus and ascospores are formed ... *Ascomycetes*

C. 1. Perithecia dark coloured and pyriform.
2. Sunken perithecia ... *Sphareiales*

D. 1. Dark coloured stroma of fungal tissue
2. Ascospores dark coloured ... *Xylariaceae*

E. 1. Stroma club shaped and sometimes branched
2. Ascospore cylindrical ... *Xylaria.*

PEZIZA

Common species : *P.pustutata, P.catinus, P.vesiculosa* etc.

Occurrence : This saprophytic fungs occurs on dung, rotting logs of wood etc. Commonly called cup fungus, *Peziza* has 13 species reported from India. Some species grow on moist leaves, heavily manured soils etc.

Study : 1. External features of plant body.
2. Reproduction and structure of fructification
3. Systematic position and identification.

External features :

1. The plant body is a mycelium. It is extensively branched
2. The hyphae entangle among themselves to form a network
3. The mycelium is septate and each cell is multinucleate.

Asexual reproduction

1. This is brought about by conidia and thick walled chlamydospores
2. Conidia are produced on conidiophores and they are budded off exogenously.
3. Chlamydospores are produced from the cells of the mycelium

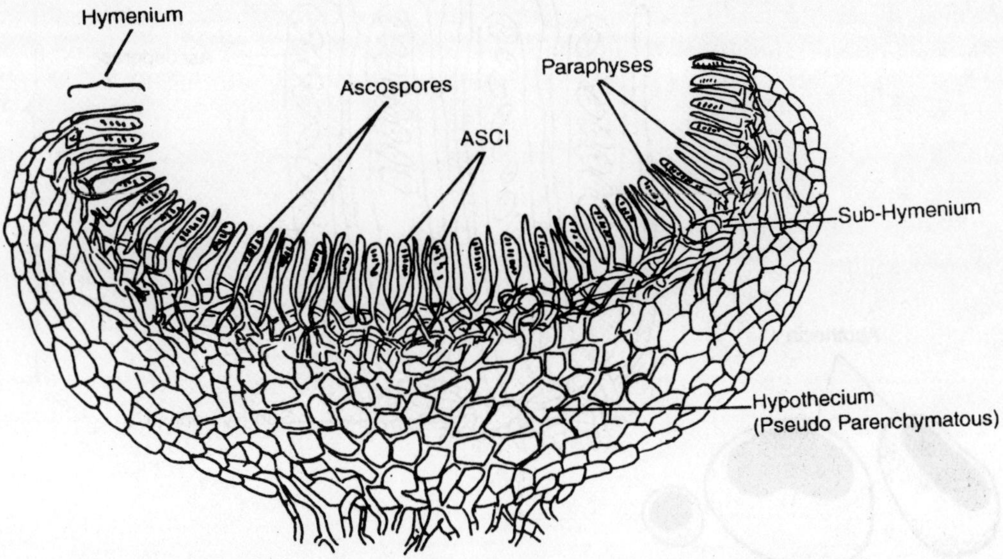

Fig. 13.20 Peziza. V.S. of apothecium

Sexual reproduction and structure of fructification

1. Sexual reproduction takes place by means of somatogamy
2. The asci are arranged in ascocarp called Apothecia
3. Apthoecium is a small cup shaped structure and it is dull white in colour.
4. A vertical long section of the apothecium will reveal the following structure.
5. The saucer or disc like fruiting body has a short stalk. Occasionally it may be sessile also.
6. The wall of the apothecium is made up of closely packed fungal hyphae. The wall is called *Peridium*
7. Internal to the peridium is the medulla made up of loosely interwoven hyphae
8. The cavity of the cup is made up of hymenial layer
9. Below the hymenium is the hypothecium or the subhymenium
10. The hymenium is the fertile layer. It conists of parallely arranged asci and paraphyses (sterile hairs)
11. Each ascus has eight ascospores arranged uniseriately in an oblique manner
12. Ascospores on germination will develop into a new mycelium.

Specimens given in the laboratory

1. Apothecia
2. Long section of apothecium (slide)

Diagrams to be drawn

1. Habit of apothecium
2. Long section of apothecium
3. A few asci enlarged

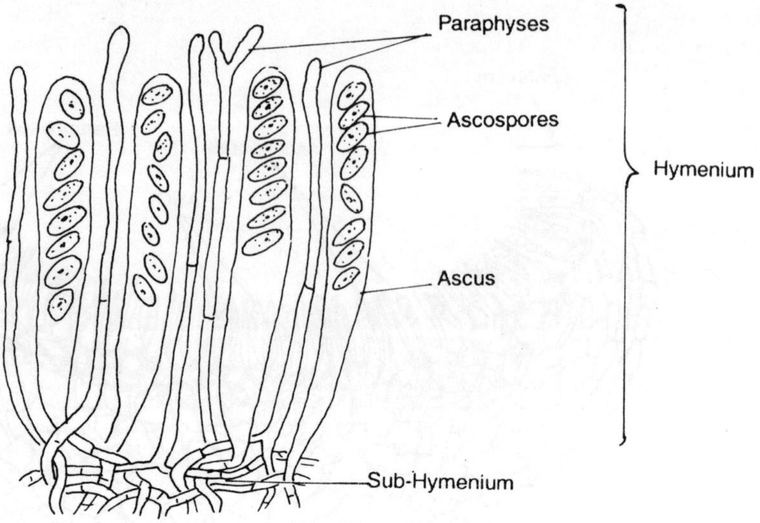

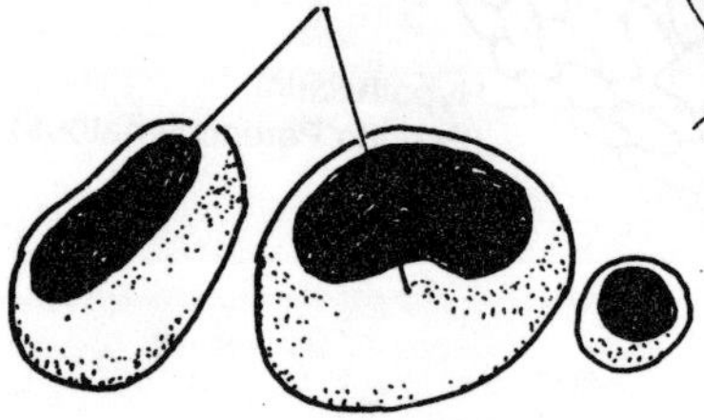

Fig. 13.21 Peziza. Cups (ascocarps)

A portion of hymenium enlarged

Systematic position and identification

A. 1. Chlorophyll absent
2. Reserve food is glycogen ... *Fungi*

B. 1. Mycelium septae
2. Asci and ascospores produced ... *Ascomycetes*

C. 1. Fructifications not found in stroma
2. Apothecia fleshy ... *Pezizales*

D. 1. Fructification not differentiated into stalk and terminal pileus.
2. Fructification disc or saucer shape ... *Pezizaceae.*

E. 1. Apothecia stalked or subsessile
2. Hyphae subterranean
3. Apotheica not large (usually 1-5 cms in diameter... *Peziza.*

PYRONEMA

Common species : *P.confluens.*

Occurrence : A saprophytic fungus usually growing on humus soil rich in mineral matter. Occasionally it also grows on dead and decaying logs.

Study : 1. Structure of the plant body
2. Reproduction
3. Systematic position and identification

Structure of the plant body

1. The plant body of the fungus is a mycelium.

2. The mycelium grows on the substratum. It is septate and much branched. The cells are multinucleate with 6-10 nuclei in each cell.

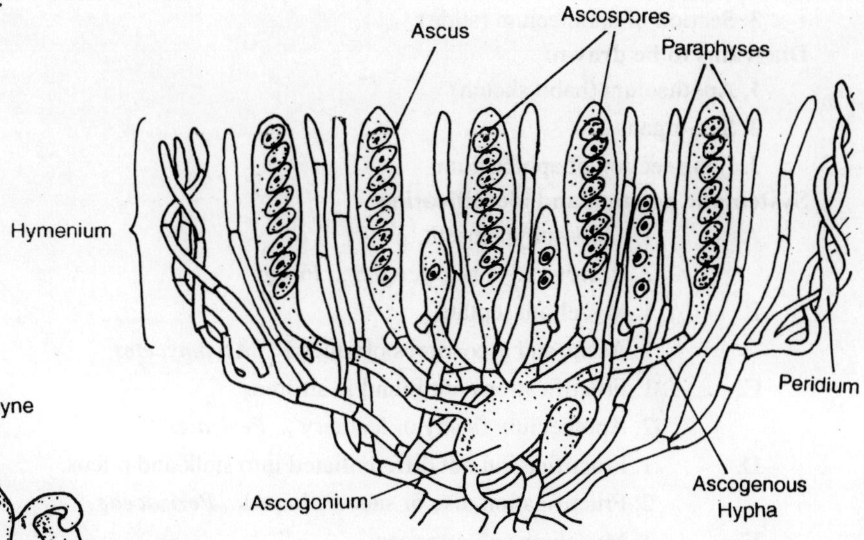

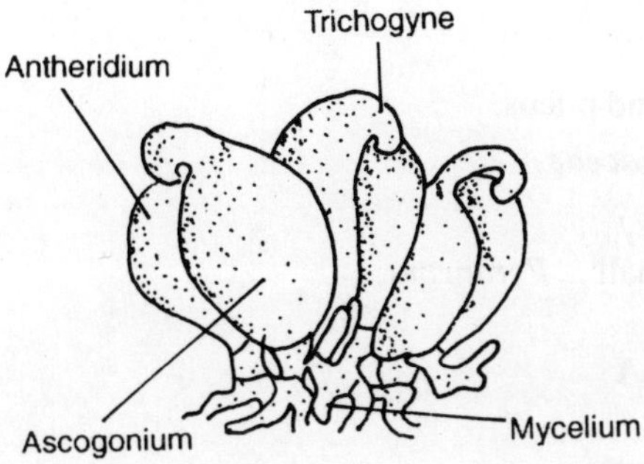

Fig. 13.22 Pyronema. A portion of mycelium with sex organs of Pyronema.

L.S. of apothecium

Reproduction

1. Asexual reproduction takes place by means of fragmentation

2. Sexual reproduction is of the oogamous type. It is effected by means of gametangial contact.

3. The sex-organs are antheridia (male) and ascogonia (female). Species of *Pyronema* are homothallic.

4. Antheridia are club-shaped structures developing on short branches of the hyphae. They have many male nuclei.

5. Ascogonium is a bit elliptical or sub globose in shape. It has a short curved appendage at its apex known as trichogyne.

6. Ascogonium has a number of functional nuclei at its base.

7. Sexual reproduction leads to the formation of the fruiting body called the apothecium.

8. The apothecia are small disc shaped structures surrounded by a wall called peridium.

9. Lining the apothecial cup internally is the hymenial layer.

10. The hymenial layer consists of parallely arranged asci intermingled with sterile hairs called paraphyses.

11. Each ascus has eight ascospores arranged in a uniseriate manner. Ascospores are uninucleate.

Specimens given in the laboratory

1. Apothecia
2. Sex-organs (slide)
3. Section of apothecium (slide)

Diagrams to be drawn:

1. Apothecium (habit sketch)
2. Sex-organs
3. Long section of apothecium

Systematic position and identification

A. 1. Chlorophyll absent
   2. Reserve food is glycogen ... *Fungi*

B. 1. Mycelium septate
   2. Ascus and ascospores produced ... *Ascomycetes*

C. 1. Fructifications not found in stroma
   2. Apothecium fleshy or leathery ... *Pezizales*

D. 1. Fructification not differentiated into stalk and pileus.
   2. Fructification disc or saucer shaped... *Pezizaceae*

E. 1. Mycelium subterranean
   2. Sex-organs prominent, apothecium very small ... *Pyronema.*

MORCHELLA

Common species : *M.esculenta, M.conica, M.deliciosa* etc.

Occurrence : Popularly known as Morel, *Morchella*, has a delicious and highly priced fructification (*M.esculenta*). The species of *Morchella* are found growing in decidious forests on decaying wood or on soil with plenty of leaf litter. Attempts to cultivate the edible species have been met with only partial success.

Study : 1. Structure of the plant body
2. Structure of fructification
3. Ascus and ascospores
4. Systematic position and identification

Structure of the plant body

1. The plant body of the fungus is a subterranean mycelium growing a few inches beneath the soil surface
2. The mycelium is much branched, with septate hyphae; cells are multinucleate.

Structure of the fructification

1. The underground mycelium produces a compact, macroscopic pseudoparenchymatous ascocarp (fructification)
2. The ascocarp is aerial ie., it comes above the ground and about 2-5 inches in height.
3. The ascocarp has hues of grey or dark brown which depends on species and stage of growth.
4. The ascocarp may be distinguished into two regions, viz the stipe and pileus, which are of equal length.
5. The stipe is the stalk, made up of pseudo parenchymatous tissue
6. In the centre of the stipe is found a hollow cavity

7. The pileus is a conical structure and represents the fertile portion of the asocarp.

8. The surface of pileus is smooth at young stages, while in a mature pileus it consists of many cavities (depressions) and ridges which look like a broad meshed network.

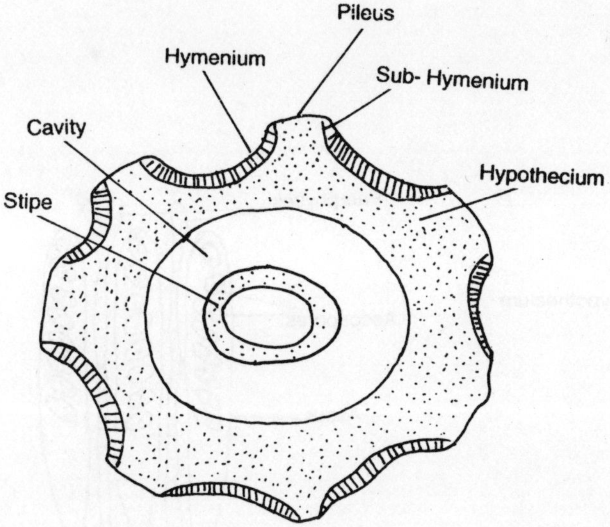

Fig. 13.23 Morchella. Cross-section of pileus

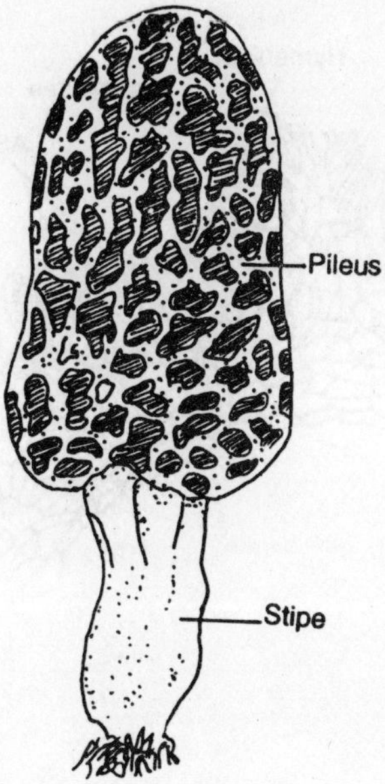

An ascocarp

9. The cavities represent the fertile regions (asci) while the ridges are sterile.

10. A section of the pileus passing through the ridges and depressions shows the arrangement of asci.

11. Extending into the cavities from the margin are found a number of asci and paraphyses. This region is the hymenial region.

12. At the base of the hymenium is the subhymenium, where the hyphae are closely interwoven.

13. The arrangement of the asci is at right angles to the ridges.

14. Asci are arranged parallely in a single layer and have eight uniseriately arranged ascospores.

Specimens given in the laboratory

1. Ascocarp
2. Section of the ascocarp (slide)

Diagrams to be drawn

1. Habit of ascocarp
2. Sectional view of pileus
3. A few cavities enlarged
4. Ascus with ascospores and paraphyses

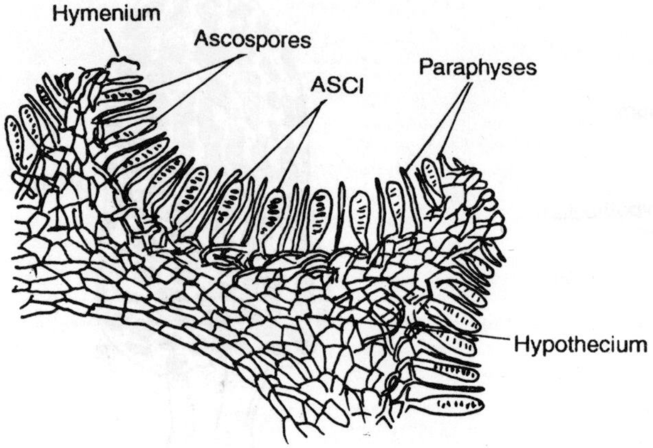

Fig. 13.24 Morchella. A cross-section of a portion of pileus enlarged

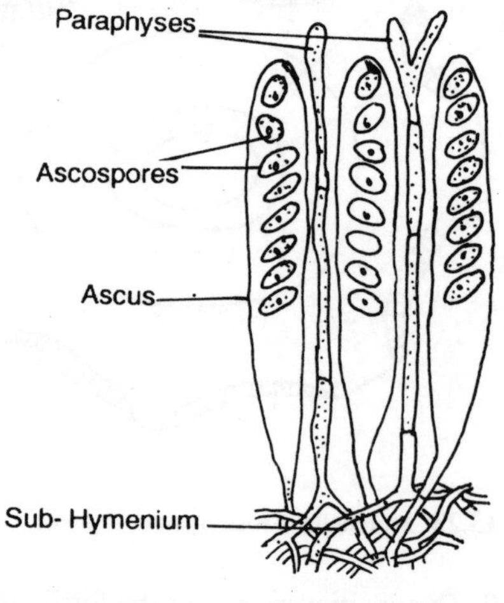

A portion of hymenium enlarged

Systematic position and identification

A. 1. Chlorophyll absent
2. Reserve food is glycogen ... *Fungi*

B. 1. Mycleia septate
2. Asci and ascospores produced ... *Ascomycetes*

C. 1. Fructifications not found in stroma.
2. Apothecium fleshy or leathery ...*Pezizales*

D. 1. Ascocarp distinguishable into stipe and pileus
2. Apothecia not saucer shaped ... *Helvellaceae*

E. 1. Pileus has cavities and ridges
2. Stipe is sterile and hollow ... *Morchella*

CHAPTER 14

BASIDIOMYCETES

Basdiomycetes resemble one of the largest classes of fungi with nearly 550 genera and 20,000 species. They differ from all other fungi in that they produce spores called Basidiospores on the outside of the spore producing club shaped hypha called basidium. Basidiospores are uninucleate and are very much comparable to ascospores.

Occurrence and economic importance

Basdiomycetes are both parasites and saprophytes. Some of them are coprophilous also. Parasitic members of Basdiomycetes are most dangerous from the point of view of food crops. The rusts and smuts are so notorious that they have been a great menace to world's agriculture. The saprophytic Basdiomycetes cause great damage to timber and rail road sleepers in that their mycelia gradually eat away into the tissues.

However some of the higher Basdiomycetes are fine delicacies. Notable among these is the filed mushroom, *Agaricus campestris bisporous*, which is cultivated large quantities to meet the heavy demand of mushrooms livers the world over.

Somatic structure

The plant body is a septate, much branched mycelium, that grows into the substratum in search of nutrition. The mycelium is usually white, bright, yellow or orange coloured. In some forms, mycleia fuse to form thick strands called rhizomorphs.

The mycelium of Basdiomycetes passes through three stages mainly primary mycelium, secondary mycelium and tetriary mycelium before the life cycle is completed. The primary mycelium is usually formed from the germination of a haploid basidiospore and it is monokaryotic, and the cells are uninucleate. During initial stages however the primary mycelium may be uninucleate. But this is a short lived stage. The secondary mycelium originates from the primary mycelium as a result of sexual reproduction, only plasmogamy taking place. Much of the life cycle is spent in the dikaryotic secondary mycelium stage. At the completion of life cycle tertiary mycelium is produced as a result of karyogamy. Soon reduction division takes place in this, producing haploid cells (basidiospores) which develop into monokaryotic mycelium.

Clamp connections

This is an interesting structure frequently observed in the hyphae of secondary mycelium. This is seen particularly in narrow hyphae at the time of cell division. The function of clamp, connections is probably to act as bypass and to allow the easy migration of nuclei from one end to the other when the narrow lumen of the hyphae may obstruct their movement. In broad hyphae where nuclear migration is easy, no clamp connections have been found.

Kniep (1915) first discovered clamp connections. Their formation is as follows.

When a uninucleate cell is about to divide, its two nuclei lie a short distance from each other. A hook like clamp appears between the two nuclei. One of the nuclei enters the clamp while the other remains in the main cell. The two undergo division with a specific spidle orientation. In one nucleus it is parallel and in the other it is oblique. After the division four nuclei are formed. Meanwhile clamp gradually bends over the cell and the cell wall dissolves at the point of contact making a passage. After the nuclear division one nucleus will be left in the clamp and it migrates through it to the other end of the cell. In the other nucleus that divides in the cell one of the daughter nuclei gradually migrates to the other end of the cell. After this migration one can see two groups of two nuclei (both of course belonging to different sexes). A cross wall is laid down between the two groups of nuclei dividing the cell into two.

Electron microscopic studies of the cross septum of beasidiomycetes has shown it to be a parenthesis [()] like structure appearing like a barrel with open ends. This type of a septum is known are *Dolipore septum* or *Parenthosome*. It is difficult to evaluate the significance of the septal apparatus.

Some mycologists compare the clamp connections to the crozier formation in Ascomycetes.

Asexual reproduction

Asexual reproduction takes place by a variety of methods such as fragmentation., budding, conida, arthrospores (oidia) etc.

Sexual reproduction

Takes place by two methods

1. Spermatization
2. Somatogamy

Stages of sexual reproduction

There are four stages namely

1. Plasmogamy
2. Dikaryotization
3. Karyogamy
4. Reduction division.

Life cycle pattern. Formation of basidiospores: The primary mycelium starts its life with the germination of a basidiospore. This gives rise to secondary mycelium. In this at definite points special club shaped hyphae develop where karyogamy takes place. These are called basidia. Generally two types of basidia are met with in Basdiomycetes.

1. Phragmobasidium
2. Holobasidium.

1. **Structure and development of a phragmobasidium**. This is found in primitive Basdiomycetes like uredinales and ustilaginales. Here the basidium is septate and develops from a spore called teleutospores (chlamydospores). The spore puts forth a tube into which the diploid nucleus migrates. The spore is formed on secondary mycelium and will be bimucleate in the beginning. Later karyogamy takes place. This tube is the basidium. The diploid nucleus undergoes reduction division and produces four nuclei. The basidium now divides into four uninucleate cells. From each one of the cells a short stalk is produced called *Sterigma* (plural:Sterigmata). The nucleus in each cell migrates to the tip of the stigma which swells to form a spore into which the nucleus migrates. This is the basidiospore. Each basidium produces four basidiospores. At maturity the basidiospores are liberated from the sterigmata. They germinate and develop into a new individual.

2. **Structure and development of a holobasidium**. Holbasidium is an unseptate basidium and it is formed in higher Basdiomycetes. The terminal cell of a dikaryotic hypha develops into a club shaped struc-

ture called the basidium. Karyogamy takes place in this basidium and a diploid nucleus is formed. In the meantime the basidium develops four sterigmata at the distal end into the tips of which the nuclei migrate. The tip of each sterigmat swells to form a basidiospore.

The basidiospores show a great variety in structure and form. Typically they are uninucleate and unicellular. They may be globose, and elongated or saucage shaped, colourless or pigmented. In many cases the pigments are green, yellow, orange, ochre, pink, brown or black coloured.

Liberation of spores. Basidiospores are discharged with a force from the sterigma. When the spores mature, a droplet of water begins to form at its basal end which gradually enlarges. When it reaches a certain volume the basidiosopres is shot off the sterigma, taking the droplet with it. The sterigma may or may not collapse after the spore is discharged.

Fruiting body. In primitive Basdiomycetes the basidia are naked and the spores are discharged as soon as they are formed. There is no protective covering for basidia. In higher basidiomycetes however basidia are produced in highly organised fruiting bodies of various types, their fructifications corresponding to ascocarp of the highr Ascomycetes are termed basidiocaprs (Gr.*basidium* = small base + *karps* = fruits). Basidiocarps may be thin and crust like gelatinous, cartilagenous, papery, fleshy, spongy, corky, woody or of any texture. In size they vary from microscopic to 3 meters. Fructifications of Basdiomycetes are the most familiar among fungi. Mushrooms, puffballs, earthstars, stinkhorns are all examples of basidiocarps of the fungi that bear them.

Basidiocarps may be open from the very first exposing the basidia (Hynenomycetes) or they may open at a late stage at maturiry releasing all the basidiospores (Gasteromycetes).

Inside the basidiocarp, the basidia are formed in definite layers called hymenial layers which are comparable to the layers of Asci in Ascomycetes. Intermingled with the basidia, sterile hairs (paraphyses) are formed. Some mycologists regard them as immature basidia. In some however, paraphyses are larger than basidia and are known as cystidia (Sing.cystidium, Gr.*kystis* = bladder).

Sexual compatibility. About 90% of Basidiomycetes are heterothallic in that sexuality is differentiated at the genetic level though they may be morphologically alike. This is particularly true in some members which sexually reproduce by somatogamy. Somatogamous reproduction wherein two vegetative hyphae fuse apparently seems to be a situation where sexuality is lost. But this is not the case as evidenced by heterothallism. In about 37% of heterothallic species sexual differentiation is bipolar that is governed by a set of factors (Aa). In the remaining 63% it is tetrapolar ie., sexuality is governed by two sets of factors (AaBb).

Classification

The nature of basidium and fructification are of great aid in the classification of Basidiomycetes.

Gwynnevaughan and Barnes (1926) clasisfy Basdiomycetes into 3 sub classes namely

1. Hemibasidiomycetes, 2. Protobasidiomycetes, 3. Autobasidiomycetes.

In hemibasidiomycetes number of basidiospores are indefinite. It includes an order ustilaginales (smuts).

In protobasidiomycetes basidiospores are definite in number, usually four, and basidium is a phragmobasidium. This includes 3 orders.

Uredinales (smuts)

Auricularies and

Tremellales.

In Autobasidiomycetes basidiospores are four in number, basidium is a holobasidium. This includes two orders. Hymenomycetales and Gasteromycetales.

Majority of modern mycologists classify Basidiomycetes into 2 sub classes:

Subclass 1. - Heterobasidiomycetideae (Hemibasidiomycetedeae)

Subclass 2 - Homobasidiomycetideae (Eubasidiomycetedeae)

Subclass - heterobasidomycetideae is primtive. Here no basidiocarp is formed and the basidium is septate. This includes 3 orders.

Ustilaginales, Uredinales and Teremellales.

Subclass - Homobasidiomycetideae is advanced. Here basidiocarp is formed and basidium is unseptate. Depending on the nature of the basidiocarp this is divided into two series.

1. Hymenomycetes and 2. Gasteromycetes.

In Hymenomycetes the basidiocarp is a closed structure. It breaks open only at maturity releasing the basidiospores. This includes four orders.

a. Hymenogastrales, b. Nidulariales, c. Lycoperdales, d. Sclerodermales.

USTILAGO

Common species : *U.maydis, U.scitaminea, U.hordei, U.nuda, U.kolleri, U.avenae* etc.

Occurrence : Majority of the species of *Ustilago* are parasites causing diseases known as smuts. The grains get destroyed and a sooty mass of spores is produced.

Study : 1. Diseases caused and symptoms of the disease

2. Vegetative structure

3. Reproduction

4. Systematic position and identification

Diseases caused and symptoms of the disease

1. *Loose smut of wheat.* This disease is caused by *U.tritici*. The disease is prevalent throughout India. The symptoms are as follows:

a. The disease symptoms appear at the onset of ear development

b. The ovarian wall breaks open releasing the black chlamydospores

c. All the ovaries of the spikelet except the awns get converted into smut spores

d. The smut sorus is covered by a thin membrane in the beginning, but it soon ruptures releasing the dark chlamydospores. The spores are wind dispersed

e. As the spores are dispersed and there is no membrane covering the spores for a long time, the disease is said to be the loose smut

f. After the release of spores, the barren rachis remain exposed

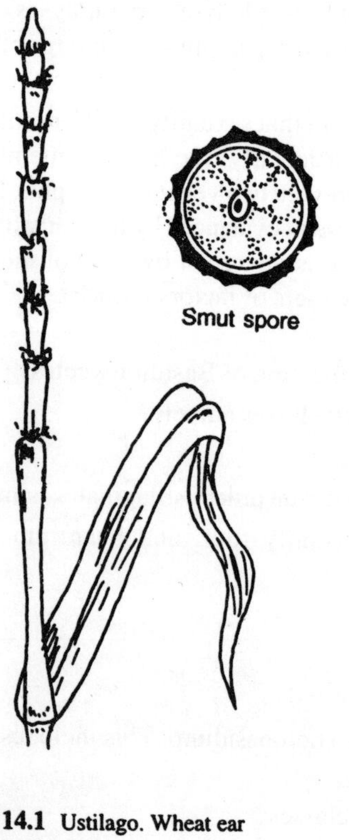

Fig. 14.1 Ustilago. Wheat ear showing loose smut caused by Ustilago tritici

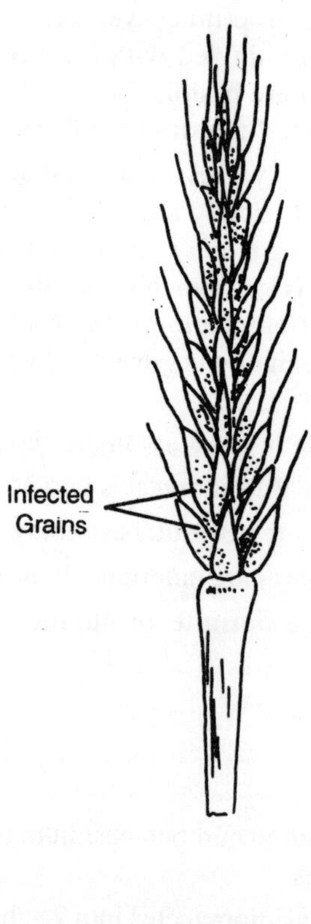

Barley ear showing covered smut caused by Ustilago hordei

g. The spores have smooth endosporium and a thick echinulate exosporium.

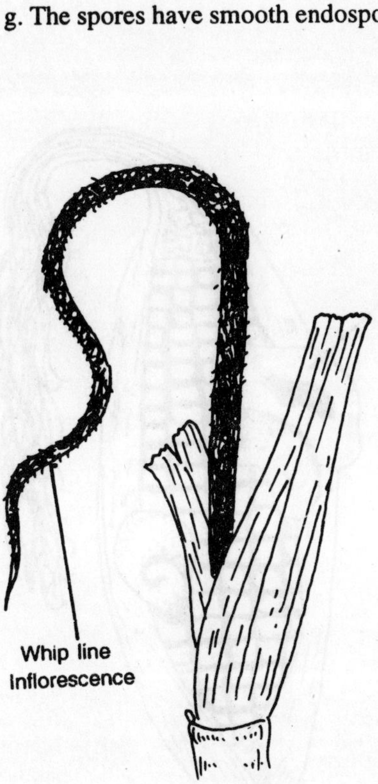

Fig. 14.2 Ustilago. Whip smut of sugarcane caused by Ustilago scitaminea

2. *Covered smut of barley*. This disease is caused by *U.hordei* and is cosmpolitan in distribution. The disease symptoms are:

a. Disease manifestation is at the time when the ears are coming out of the boot

b. No grain remains free of the disease in an ear. All of them turn black. Each grain gets converted into a smut sorus

c. Awns of the flower remaind disease free. Initially the sorus has a membranous covering, but it soon ruptures to release the spores

d. In the chlamydospores, both the exosporium and endosporium are smooth. The endosporium however is thin.

3. *Loose smut of barley*: This is caused by *U.nuda*

a. The grains turn black due to accumulation of spores

b. Spores are discharged when the ear is still attached to the plant.

4. *Whip smut of sugarcane*. This is caused by *U.scitaminea*. Due to the disease the stem tip gets converted into 3-4 feet long dark coloured whip like structure. The main shoot tip as wll as laterals also get converted into smut whips.

5. *Covered smut of oats*. The disease is caused by *U.kolleri*. In this the spores are not exposed they are covered and come out only during the threshing of the grains.

Vegetative structure

1. The plant body is a mycelium. It is septate and much branched.

2. There are two types of mycelia - primary mycelium and secondary mycelium.

3. The cells of the primary mycelium are uninucleate while those of secondary mycelium are dikaryotic ie., (they have two, one male and one female nuclei).

Reproduction

1. Reproductive parts, ie., spores are visible only in the ears of the host plants. These are chlamydospores (also called smut spores, brand spores and teleutospores).

2. The characteristic spores chlamydospores are produced on the secondary mycelium as a result of sexual reproduction.

3. The chlamydospores have a two layered wall. The outer wall called exosporium is thick, pigmented either smooth or ornamental. The inner endosporium is smooth may be thin or thick.

Specimen given in the laboratory

1. Infected ears of the host plants like wheat, maize, oats etc. Mount the spores by crushing the smut sorus. Study the symptoms.

Diagrams to be drawn:

1. Habit sketch of infected ears of the host.

2. A few chlamydospores.

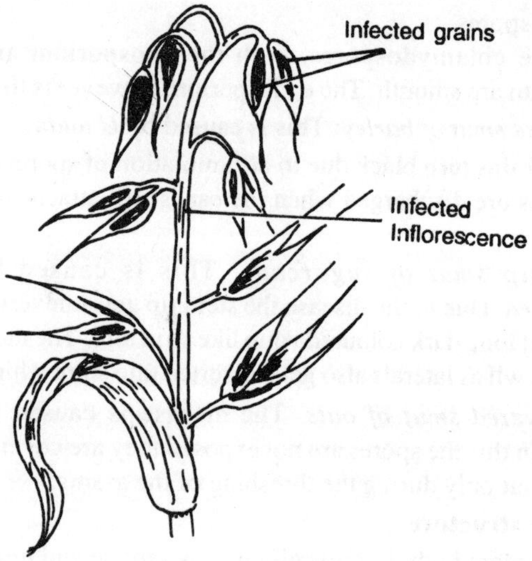

Fig. 14.3 Ustilago. Covered smut of oat caused by Ustilago kolleri

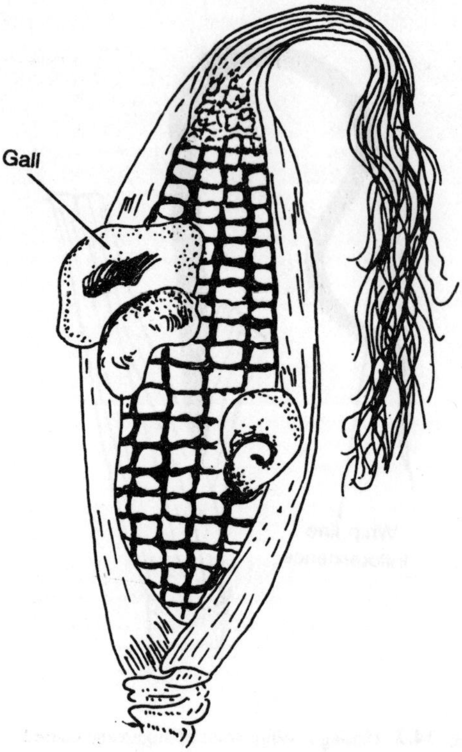

Maize smut caused by Ustilago maydis

Systematic position and identification

A. 1. Chlorophyll absent
   2. Reserve food is glycogen ... *Fungi*

B. 1. Mycelia septate
   2. Basidium and basidiospores formed ... *Basdiomycetes*

C. 1. Basidia divided into 4 cells
   2. No basidiocarp...*Protobasidiomycetes*

D. 1. Disease symptoms visible only in the ears.
   2. Characteristic smut spores formed ... *Ustilaginales*

E. 1. Basidiospores sessile
   2. Basidium fragmented ... *Ustilaginaceae*

F. 1. Chlamydospores black, and separate into single cells (not in pairs)
   2. Sorus dusty when mature ... *Ustilago.*

SPHACEOLTHECA

Common species : *S.sorghi, S.cruenta, S.reliana* etc.

Occurrence : Ther genus *Sphacelotheca* is one of the most notorious among smuts and infects a number of hosts like sorghum, barley, maize etc., causing a great loss to the crop.

Study : 1. Disease caused and symptoms.

2. Vegetative structure
3. Reproduction
4. Systematic position and identification

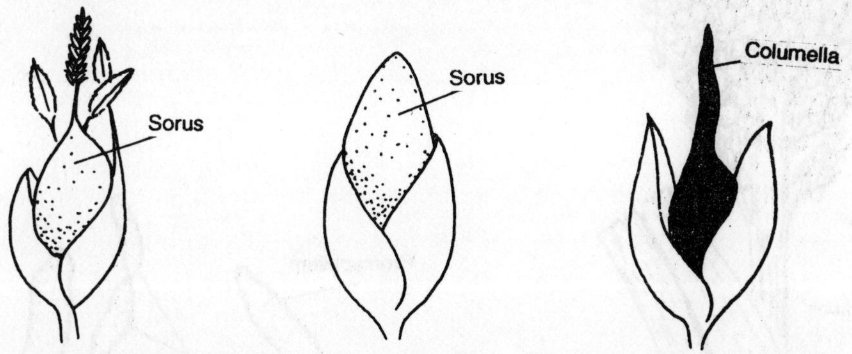

Fig. 14.4 Sphacelotheca . An infected flower A grain containing sorus A sorus showing columella

Diseases caused and symptoms :

A. *Grain smut of sorghum* : The disease is caused by *S.sorghii* and is the most destructive disease of sorghum in India.

The disease symptoms are as follows:

1. The disease manifests itself when the ears are emerging out of the boot.
2. Smut sorus is formed in the grain and each sorus has a central columella made up of a hard tissue.
3. Surrounding each columella are found a number of spores.
4. The sori measure in length from 1/6" to 1/2" in length.
5. Within the spore sac sterile cells are also found intermixed with the spores.
6. Spores are globose to oval, dark brown in colour, smooth walled and 5-9μ in diameter.
7. The smut is externally seed borne.

B. Loose smut of sorghum: This disease is not as common as the grain smut in our country.

The disease symptoms are as follows : The disease is caused by *S.cruenta.*

1. Infected plants are shorter with thin stalks.
2. The ears are produced earlier than in the healthy plants.
3. All the spikelets of the ear get the disease.
4. The covering membrane of the sorus ruptures early releasing the powdery mass of dark spores.
5. The columella in the spore sac persists even after the dissemination of the spores.
6. The wall of the sorus is made up of loosely arranged cells.

7. The spores are round or elliptical darker brown than in grain smut.

Fig. 14.15 Ear of Sorghum infected by Sphacelotheca sorghi

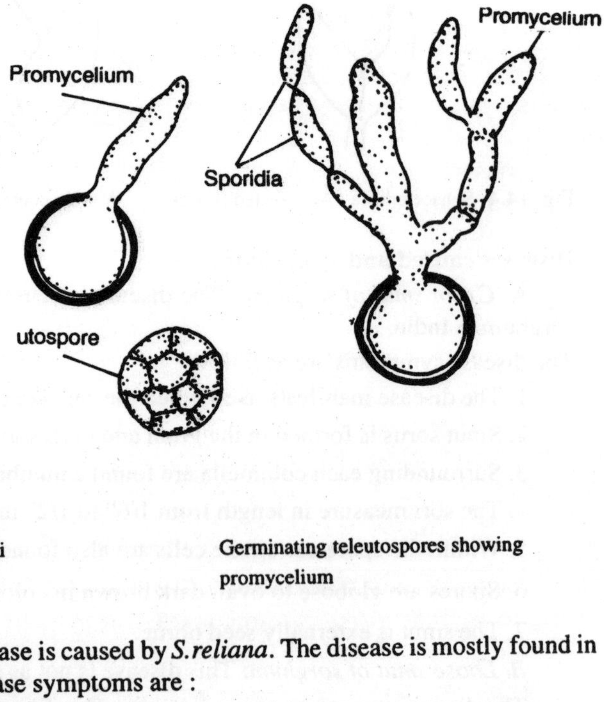

Germinating teleutospores showing promycelium

C. Head smut of sorghum and maize : The disease is caused by *S.reliana*. The disease is mostly found in Kashmir, but present in other states also. The disease symptoms are :

1. In the infected plant the ear is completely damaged.

2. In maize large sori of the smut replace the tassel.

3. In such cases floral bracts grow out into leafy bracts sometimes into small shoots.

4. In sorghum, the entire inflorescence is converted into a big sorus about 4" long and 2" wide.

5. At the early stages a white membrane encloses the sorus.

6. A network of black fibres traverses the spore mass and remains adhering even after the liberation of the spores.

7. Spores are reddish brown to black, finely echinate and irregular to spherical in shape.

Vegetative structure

1. The plant body of *Sphacelotheca* is a septate branched mycelium.
2. The crosswall is porous and allows the movement of cytoplasm from one cell to another.
3. The mycelium is intercellular and hyaline.
4. Haustoria developed from the mycelium help in the absorption of food from the host tissues.
5. The secondary mycelium is dikaryotic and produced from the primary mycelium as a result of sexual reproduction.

Reproduction

1. The cells of the secondary mycelium in the ovaries isolate themselves, acquire thick walls and transform themselves into chlamydospores.
2. The chlamydospores also called teleutospores or brand spores are dark coloured and dikaryotic.
3. Due to the infection, the grain will turn into a compact mass of fungal tissue called sorus. The sori may be long or short.
4. The tough wall of the sorus may entirely be a fungal tissue or may also have the host tissue.
5. Inside the sorus is a central hard columella extending from the base to apex.
6. The columella represents the host tissue and is composed of parenchyma and vascular elements as in *S.sorghii.*
7. The spores are round or oval on liberation they germinate to give rise to basidia and basidiospores.

Specimens given in the laboratory

1. Infected ears of maize, sorghum etc.

Diagrams to be drawn:

1. Infected ears
2. Sours with columella
3. Chlamydospores

Systematic position and identification

A. 1. Chlorophyll absent
   2. Reserve food is glycogen ... *Fungi*

B. 1. Mycelium septate
   2. Basidia and basidiospores produced ... *Basidiomycetes*

C. 1. Basidium is a phargmolbasidium
   2. Basidiocarp is abent ... *Protobasidiomycetes*

D. 1. Disease symptoms muscle only on the ears
   2. Characteristic smut spores are formed ... *Ustilaginales*

E. 1. Basidiospores sessile
   2. Basidium fragmented ... *Ustilaginacae*

F. 1. Chlamydospores entangle in a mass due to the presence of black fibres.
   2. Spores are reddish brown to black finely echinate... *Sphacelotheca.*

PUCCINIA

Common species : *P.graminia, P.recondita, P.stiformis*

Occurrence : The genus *Puccinia* of which 700 species have been reported is an obligate parasite infecting the members of the family Gramineae like wheat, barley, oats, rye etc. *Puccinia* causes three types of rust diseases on wheat, namely.

1. *P.graminis* - Black or stem rust
2. *P. stiformis* (= *P.gimarum*) yellow or striped rust.
3. *P.triticina* Brown or orange leaf rust.

The appearance of red, brown or deep red patches on the leaves are the characteristic symptoms :

Study : 1. Life history
2. Disease symptoms of major types of rusts
3. Systematic position and identification

Specimens given in the laboratory

1. Infected leaves of barberry
2. Infected leaves of wheat
3. Pycnial stage (slide)
4. Aecial stage (slide)
5. Teleutosours stage (slide)

Life history :

1. *Puccinia granminis* is a heteroecious and macrocylic rust
2. It requires two hosts to complete the life cycle hence known as heteroecious.
3. The two hosts are primary host or permanent host wheat - *Triticum* and secondary host or alternate host *Barberry* - Berberis.
4. The rust is said to be macrocyclic because it produces more than one kind of binucleate spores on its secondary mycelium.
5. The life history can be studied in two distinct phases based on the host.

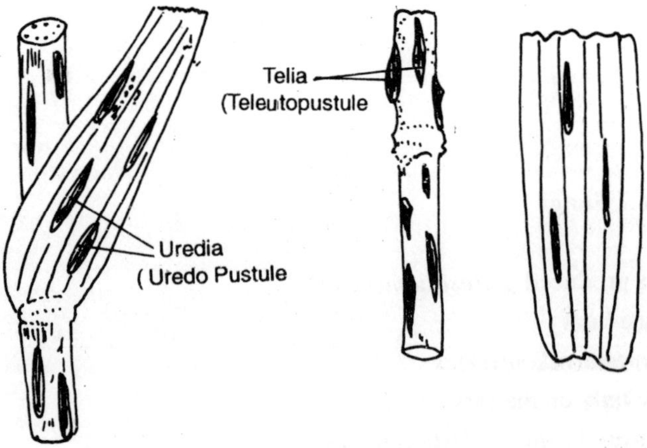

Fig. 14.16 Puccinia graminis. Wheat shoot showing uredosori of Puccinia

Wheat stem and leaf showing teleutosori

Stages of life cycle on Barberry :

1. The barberry leaves are first infected by basidiospores of the rust.
2. The germinating basidiospore produces the primary mycelium in between the mesophyll cells of the host leaf.

3. The primary mycelium is septate, much branched with uninucleate cells.

4. The cross walls are swollen (Dolipore septum) and allow the streaming of cytoplasm from cell to cell.

5. The primary mycelium after a certain period of growth reproduces sexually by forming sex organs in cup like structures called spermagonial cups or pycnial cups.

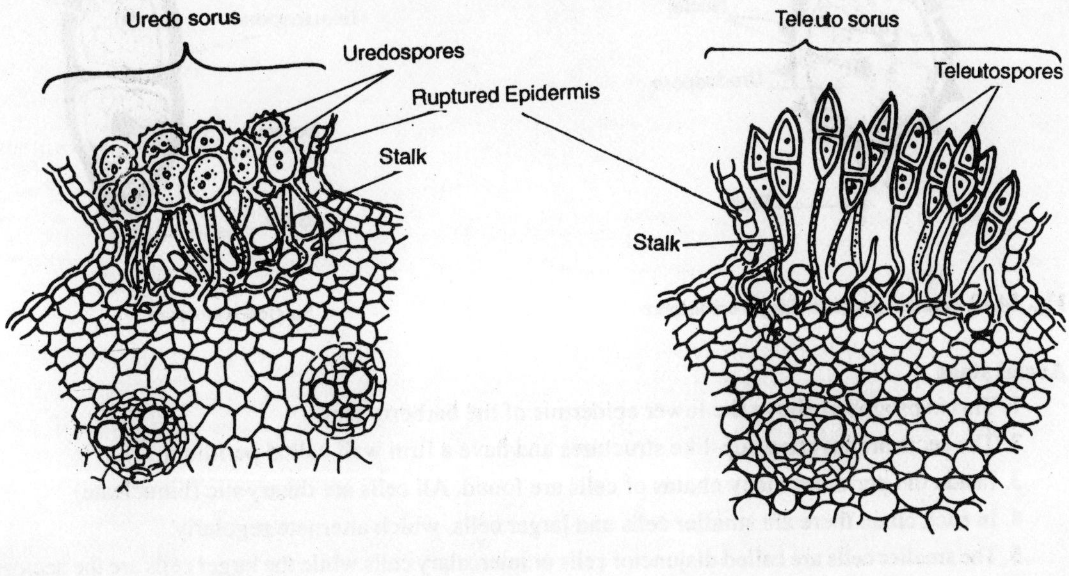

Fig. 14.17 Puccinia graminis. T.S. of wheat stem through uredosorus

T.S. of wheat stem through teleutosorus of Puccinia graminis

6. The formation of these structures is externally manifested by the appearance of black patches on the upper surface of the barberry leaves.

7. A cross section of the leaf passing through an infection patch will show the details of reproductiive structures.

8. Below the upper epidermis of the barberry leaf small flask shaped structures are seen.

These are the spermagonial cupscor pyenia.

9. Each spermagonial cup has an opening called ostiole, through which male sex cells are liberated.

10. Lining the basal region of the spermagonial cup are found many erect, unbranched hyphae called spermatiophores. Each spermahopore cuts off large number of unicelllar, uninucleate spermatia from the tips.

11. Spermatia are the male gametes. They are non flagellate.

12. Some elongated branched hyphae also arise in the spermagonial cups. These are the female sex organs called receptive hyphae. The receptive hyphae come out of the spermagonial cup through the ostiole.

13. The basal cell of the receptive hypha is fertile.

14. The spermagonial cup or the infected host cells produce a sweet juice. Insects visiting the leaves to get the nectar will help to transfer spermatia from one spermagonial cup to the other.

15. As a result of sexual reproduction which is by spermatization, secondary mycelium is produced. This reproduces by means of Aecia or Aecidia.

(Note the name Pycnia for the spermagonia is an old name given when the sexual nature was not known similarly spermatia were called pycnospores).

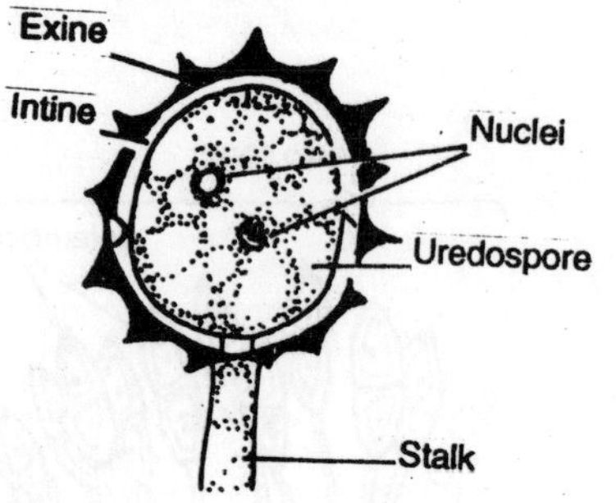

Fig. 14.18 Puccinia graminis. Single uredospore

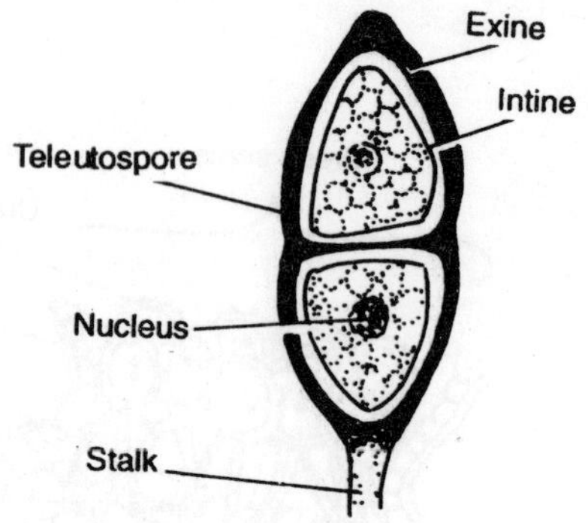

Single teteutospore

Aecial stage

1. This is produced above the lower epidermis of the barberry leaf.
2. The aecia are inverted cup-like structures and have a firm wall called peridium.
3. Inside the peridium many chains of cells are found. All cells are dikaryotic (binucleate)
4. In each chain there are smaller cells and larger cells, which alternate regularly.
5. The smaller cells are called disjunctor cells or intercalary cells while the larger cells are the aeciospores.
6. Aeciospores are produced in large numbers and exert a pressure on the lower epidermis of the barberry leaf.
7. The lower epidermis of the barberry leaf ruptures and the aeciospores are released.
8. The aeciospores carry the infection to the primary host viz., the wheat plant.

Stages of life cycle on wheat :

1. The aeciospores produce two characteristic sori on the wheat plant. These are uredosorus and teleutosours.

Uredosorus stage :

1. This is produced on the wheat by the germination of aeciospores.
2. The formation of uredosorus on the leaves and stem is externally manifested by the appearance of red patches.
3. The details of the uredosours can be studied by taking a cross section of the leaf passing through a red patch.
4. A little below the epidermis, in between the mesophyll cells is found the secondary mycelium.
5. The secondary mycelium is much branched, septate with dikaryotic cells.
6. The uredospores arise in aggregates from the secondary mycelia.
7. Each uredospore has a long stalk and a single oval or eliptical cell. The cell is dikaryotic.
8. The cell wall is thick and two layered. The outerwall is echinulate. The wall has red pigments hence the appearance of red patches on the leaves of wheat.

9. The uredospores are called summer spores as they are produced in the beginning of summer. They are also called repeating spores because a uredospore can infect another wheat plant and produce a uredosorus.

Note : While the initial infection to a wheat plant may come from an aeciospore, the rapid spreading of the disease is due to uredospores.

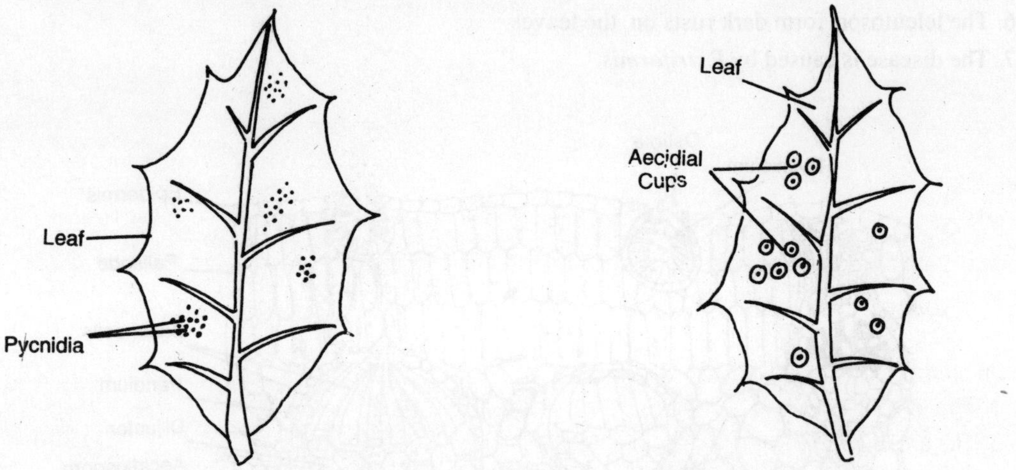

Fig. 14.19 Puccinia graminis. Barberry leaf showing pycnidia on upper surface   Barberry leaf showing aceidial cups on lower surface

10. At the end of summer, the rust produces the next stage ie., the teleutosorus in dried leaves, stubble etc.

11. The presence of the teleutosorus is manifested by the appearance of dark brown patches on the leaves and stem.

12. Each teleutosorus represents the aggregation of teleutospores. The teleutospores are also called winter spores as they are produced at the beginning of winter.

13. Each teleutospore has a long stalk and two terminal cells. Each cell is dikaryotic. The cell wall is thick, brown pigmented, but does not have ornamentaion.

14. Teleutospores on germination produce basidia and basidiospores.

Disease symptoms of some major rust disease caused by Puccinia :

Black rust of wheat :

1. The disease symptoms manifest themselves at the time of maturity of grains.
2. Long, brown coloured pustules on leaf, leaf sheath and stem indicate the presence of infection.
3. The pustules are more in number on the stem. The pustules have uredosori internally.
4. The pustules are 1/4" long and on maturity put out a mass of uredospores.
5. Teleutosori are darker than uredosori, they are also produced in the stems and leaves.
6. The disease causing agent is *P.graminis* var *tritici.*

Stripe rust of wheat :

1. In this disease leaves and inflorescence are completely destroyed.

2. The disease makes its appearance before the grains are mature.

3. When the disease is not intense uredosori.

4. The deep green colour of the leaves gives way to yellow or dull green stripes. Many pustules are found in each stripe.

5. Teleutosori appear little late; when these are being formed black stripes appear on the ventral surface of the leaves.

6. The teleutosori form dark rusts on the leaves.

7. The disease is caused by P.*striformis*.

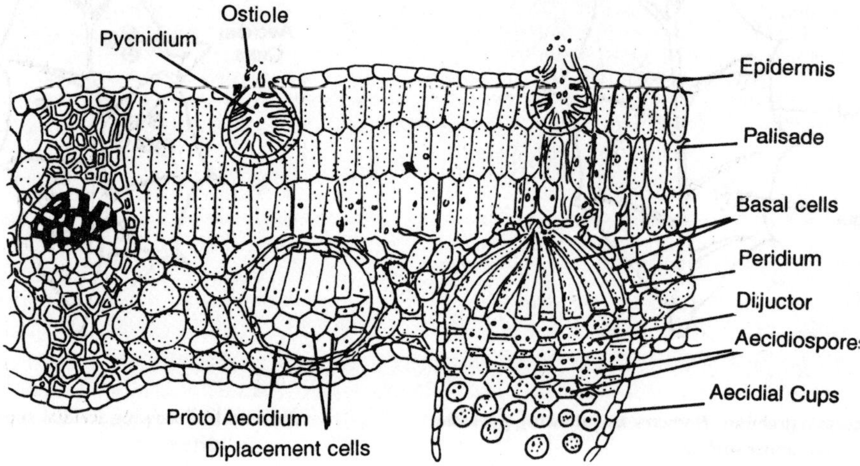

Fig. 14.20 Puccinia graminis. V.S. of barberry leaf showing pycnidia and aecidial cups.

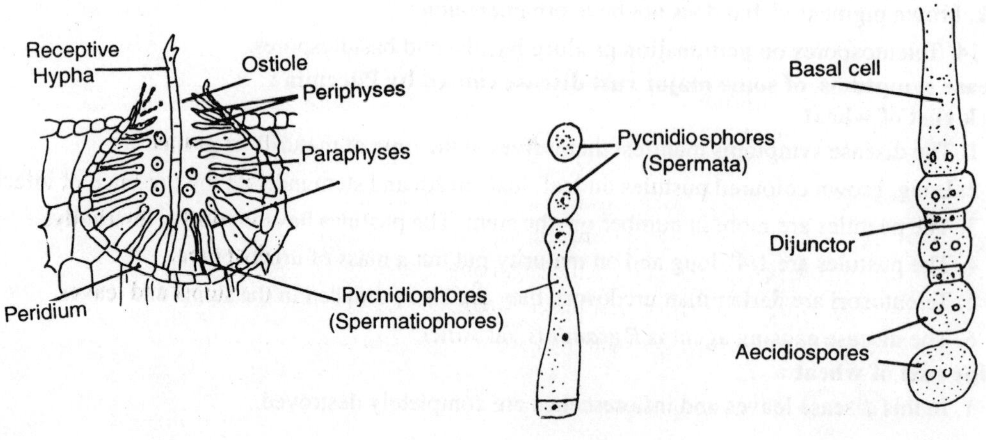

Fig. 14.21 Puccinia graminis. V.S. of pycnidium Single spermatiophore Chain of aecidiospores

Brown or orange rust wheat :

1. The uredosori appear mostly on the leaves
2. The epidermis on the leaves ruptures and deep organe coloured uredosori get exposed.
3. Teleutosori appear quite later and are formed on the under surface of the leaf.
4. The teleutosorus is divided into a number of chambers due to the presence of paraphyses.

Diagrams to be drawn :

1. Habit of infected barberry leaf.
2. Section of the barberry leaf to show spermagonia and aecia.
3. Habit of infected wheat leaf.
4. Uredosours.
5. Teleutosorus.

Systematic position and identification

A. 1. Chlorophyll absent
   2. Reserve food material is glycogen ... *Fungi*

B. 1. Mycelia septate
   2. Basidia and Basidiospores formed ...*Basidiomycetes*

C. 1. Basidiocarp absent
   2. Basidia septate ...*Protobasidiomycetes*

D. 1. Rusty appearance of pustules
   2. Obligate parasites ... *Uredinales*

E. 1. Basidiospores formed laterally
   2. Teleutospore loosely fascicled or more than one celled ... *Pucciniaceae.*

F. 1. Fungus is heteroecious.
   2. Hosts are wheat and barley
   3. Fungus is marrocytic ... *Puccinia.*

UROMYCES

Common species : *U.jasmini, U.fabae, U.pisi* etc.

Occurrence : The species of *Uromyces* may be heteroecious or autoecious. *U.pisi* is heteroecious while *U.fabae* is autoecious. Lentils, peas, jasmins and infected by the various epcies of *Uromyces*.

Study : 1. Disease symptoms and disease caused.
2. Study of life history
3. Systematic position and identification

Diseases and symptoms :

1. Rust of Bengal gram :
   a. The disease is caused by *U.cicer arietini*
   b. Disease symptcms appear when the plants are four months old
   c. Leaves aggregate with the appearance of known patches on them.

The patches which are small in the beginning coalse and become larger.

Disease symptoms may appear on stem and either surface of the leaves.

The spermagonial and aeial stages of this fungus are not known uredospores are rounded with echinulate wall.

Teleutospores are deep brown.

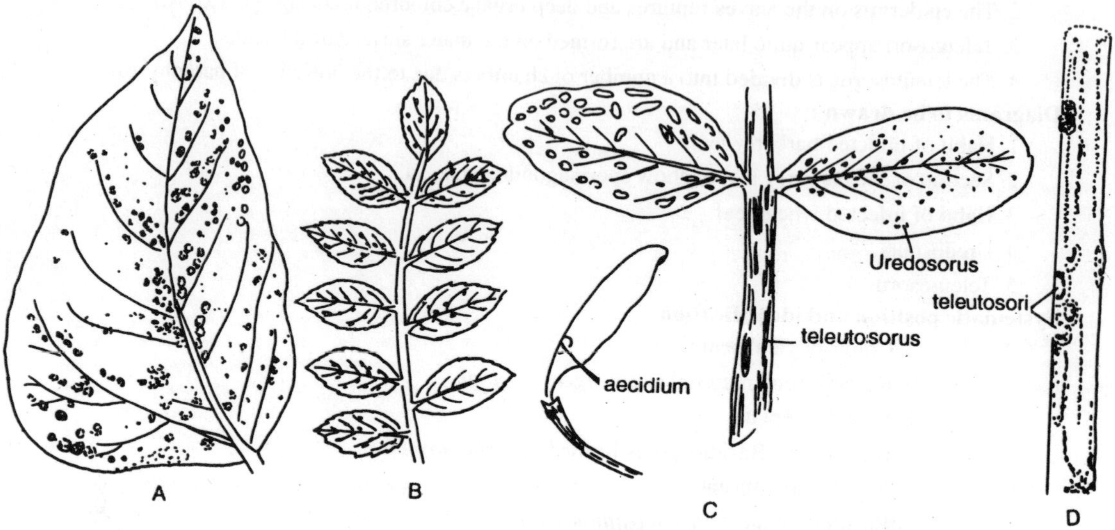

Fig. 14.22 Uromyces A. U.appendiculatus on Vigna leaf. B. U.cicer-arietini on gram leaf. C.U.fabae uredosorus and teleutosorus on leaf and stem, and aecidium on pod. D. U.decoratus (on sunnhemp); teleutosori on stem.

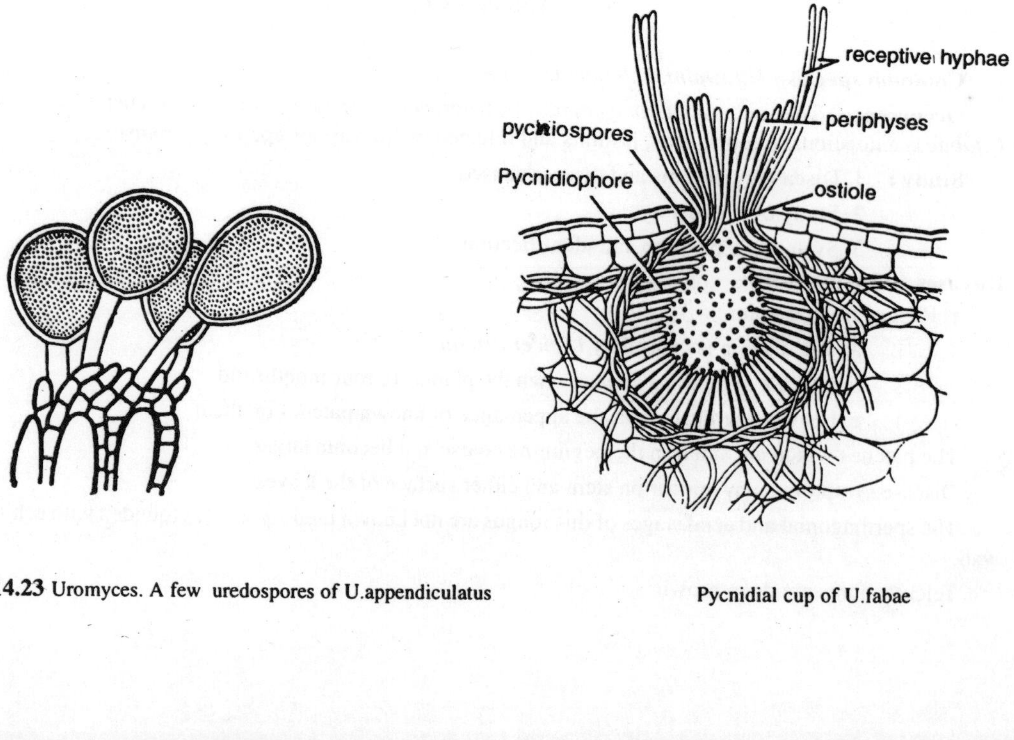

Fig. 14.23 Uromyces. A few uredospores of U.appendiculatus

Pycnidial cup of U.fabae

Rust of peas and lentils.

1. The disease is caused by *U.fabae.*
2. Aecia appear fist on the leaves. *Pycnia* are rare. Sori appear on all green parts of the plant.
3. Even before the formation of aecia, leaves, petioles, tendrils, stem and fruits turn yellow first and later brown.
4. Uredosori appear on either surfaces of the leaves
5. The fungus is autoecious, completing all the stages of the life cycle in a single host
6. Uredospores rounded or oval and light brown in colour.

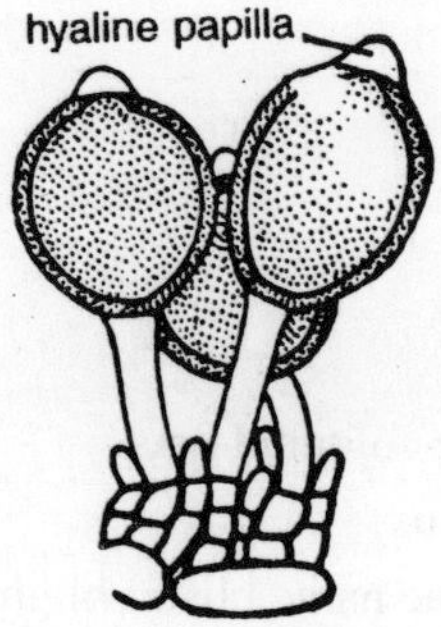

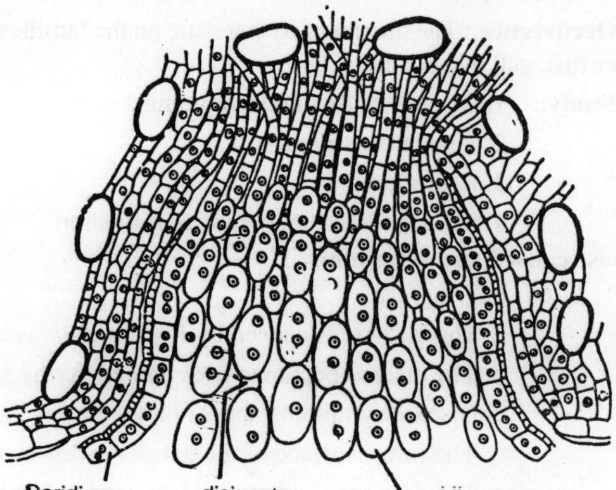

Fig. 14.24 Uromyces. A few teleutospores of U.appendiculatus Aecidial cup of U.erythronii

Life history :

1. Aecia and pycnia appear below the surface of the leaf.
2. Uredia and Telia appear on the same host or on a different host (in heteroecious).
3. Aecia are cup-shaped and dikaryotic.
4. Uredo and Teleutospores are brown coloured and single celled. They have one or more germ pores.

Diagrams to be drawn:

1. Habit of infected leaves.
2. Sectional view of Aecia, pycnia, uredenia and telia.

Systematic position and identification

A. 1. Chlorophyll absent
   2. Reserve food is glycogen ... *Fungi*

B. 1. Mycelium septate
   2. Basidia and basidiospores formed ... *Basdiomycetes*

C. 1. Fructifications absent
   2. Basidia septate... *Protobasidiomycetes*

D. 1. Rusty appearance of pustules
2. Obligate parasites ... *Uredinales*

E. 1. Basidiospores formed laterally
2. Teleutospores one or two celled free or loosely fascicIded ... *Pucuniaceae*

F. 1. Teleutospores one celled
2. Teleutosori do not have a preidium ... *Uromyces.*

EXOBASIDIUM

Common species : *E.vexans, E.vaccinii* etc.

Occurrence : The members are parasitic on the families Ericaceae, Theaceae etc and cause diseases like blister rust, gall disease etc.

Study : 1. Disease caused and symptoms
2. Vegetative structure
3. Reproduction
4. Systematic position and identification

Disease caused and symptoms

1. *Blister blight of tea :*
   a. This disease is caused by *Exobasidium vexans*
   b. The disease becomes wide spread during rainy season
   c. Disease symptoms appear on the leaves in the form of purple coloured dots
   d. The lower surface of the dots have dowry growth of the fungus
   e. The dots appear depressed above and swollen below hence the name blister blight
   f. The leaves drop down and the plant collapses due to the disease
2. *Gall disease :*
   a. The disease is caused by *E.vaccinii*
   b. Members of Ericaceae like cornberries and blackberries get the disease
   c. Both leaves and twigs get the infection.

Vegetative structure

1. The plant body of the fungus is an intercellular much branched, septate mycelium consisting of tubular hyphae.
2. The hyphae produce haustoria to absorb nutrition from the host cells.
3. The mycelium is suuepidermal or deep in the mesophyll tissue or cells.

Reproduction :

1. The fungus reproduces sexually by means of conidia. The conidiospores come out of the host leaf through the epidermis.
2. The conidia are single celled, oval in shape uninucleate and germinate to give rise to a new individual.

Sexual reproduction

1. Sexual reproduction results in the formation of basidia which are aseptate.
2. The basidia come out through the epidermis of the host leaf and bear 4-8 basidiospores at their tip.

3. The basidiospores may be unicellular or may have 2-3 cells by the development of septa.

4. The basidiospores may undergo budding to increase their number or may directly form a new individual.

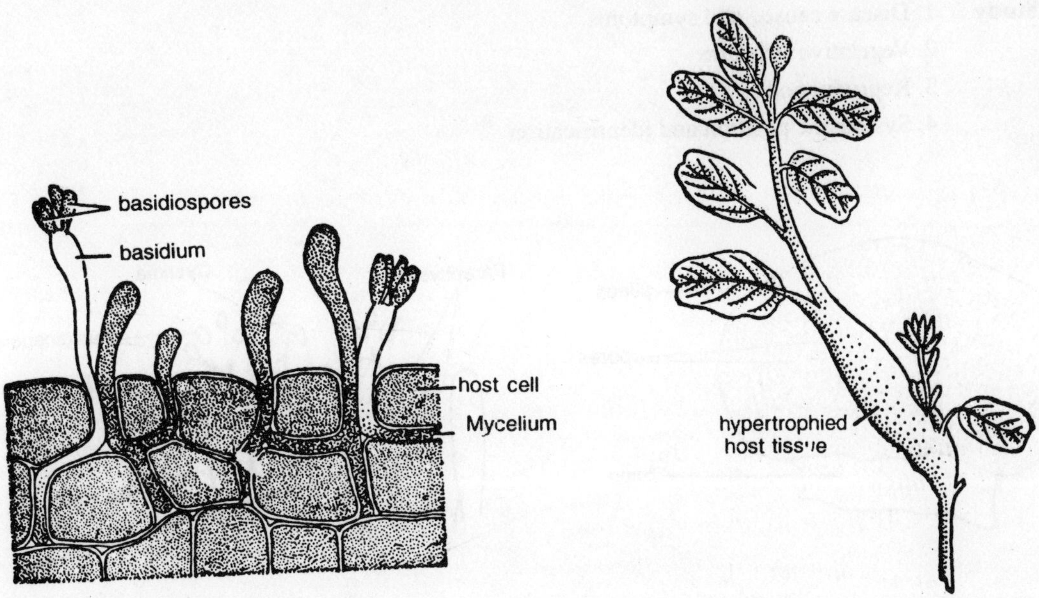

Fig. 14.25 Exobasidium

Diagrams to be drawn:

1. Habit of infected leaves.
2. Sectional view of leaf to show conidia.
3. Sectional view of leaf to show basidia.

Systematic position and identification

A. 1. Chlorophyll absent
   2. Reserve food is glycogen ... *Fungi*

B. 1. Mycelium septate
   2. Basidia and basidiospores formed ... *Basdiomycetes*

C. 1. Basidium aseptate ... *Holobasidiomycetes*

D. 1. No basidiocarp
   2. Basidia superficial on the host ... *Exobasidiales*

E. 1. Basidiospores become multicellular due to septation
   2. Basidia club shaped ... *Exobasdiaceae*

F. 1. Basidia come out in between the epidermal cells.
   2. Paraphyses absent ... *Exobasidium.*

POLYPORUS

Common species : *P.sulphurseus, P.versicolar, P.betulinus* etc.

Occurrence : The species of *Polyporus* are parasitic on a variety of forest trees causing decay and destruction. Oak, Sissoo, Albizzia etc get the attack of the fungus.

Study : 1. Disease caused and symptoms
2. Vegetative structure
3. Reproduction
4. Systematic position and identification

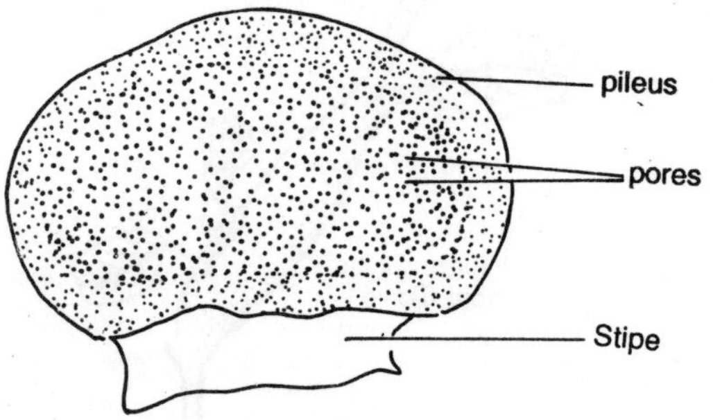

Fig. 14.26 Polyporus. fruiting body (the basidiocarp)

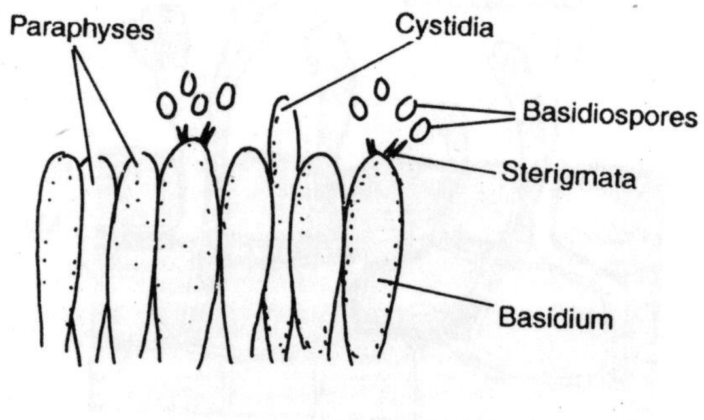

Hymenium showing basidia and basidiospores

Disease caused and symptoms

1. *Wood rot of oaks* : This disease is caused by *P.sulphureus* wherein the wood gets decayed. *P.cinnabearinus* also causes the wood rot oaks.

2. *Heart rot of elms* : The heart wood and sapwood get the attack of the fungus; *P.squamosas elm,* many other trees also get the attack.

3. *Wood rot*: Various tree species are attacked by *P.versicolar.*

4. *Rot of birch* : Species of *Betula* get infected by *P.betulinus.*

Vegetative structure

1. The plant body of the fungus is a mycelium. The mycelium is septate and much branched.

2. The mycelium has uninucleate cells (primary mycelium) in the beginning. Later the cell becomes dikaryotic (secondary mycelium). In many hyphae of secondary mycelium clamp connections are seen.

In cases of severe infection by the fungus the hypha may even penetrate the heart wood.

A. Reproduction

1. Sexual reproduction in the primary mycelium leads to the development of the fruiting body, the basidiocarp.

2. It is this basidiocarp which is seen above the substratum. The basidiocarp is built up lay the secondary mycelium.

3. The basidiocarp has a short stalk and an expanded sheath and disc shaped structure viz., the pileus.

4. The mature basidiocarp commonly called the bracket fungus is woody, corky or leathery and becomes hard as it matures.

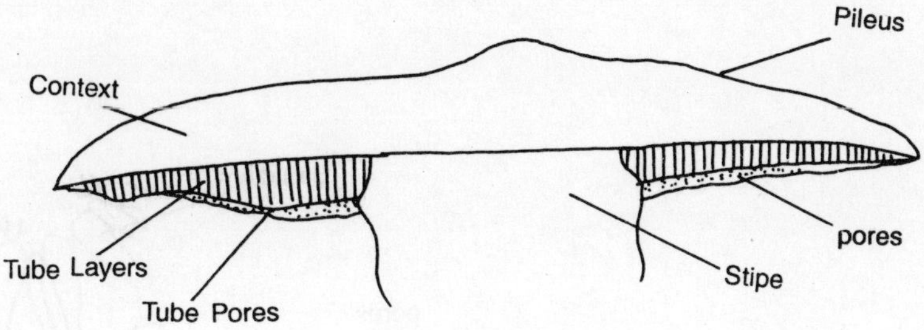

Fig. 14.27 Polyporus. Basidiocarps growing on tree trunk V.S. of basidiocarp (diagrammatic)

5. The stalk or stipe is 2-4" long, somewhat cylindrical and deep or dark brown in colour.

6. The pileus is also light brown coloured and its upper surface is smooth and slightly wavy.

7. Many concentric rings are found at the margin of the pileus. The basidiocarp is perennial and the concentric rings represent the growth rings.

8. The lower surface of the pileus is smooth or flat and has numerous fine pores.

9. Each pore actually represents the opening of a tube, the inner surface of which has basidia.

10. In a sectional view the basidiocarp shows the following regions viz., cortex, trama, pores and hymenium.

11. The cortex represents the outer fibrous part of the basidiocarp made up of compact thick walled hyphae.

12. The trams has loosely arranged branched hyphae.

13. Extending vertically down from below the cortex are several tubes which open out on the central surface in the form of pores.

14. The hymenium constitutes the inner lining of the pores. It consists of basidia and cystidia.

15. Cystidia are regards as sterial basidia. They do not bear basidiospores.

16. The basidia which project into the cavity are aseptate (holobasidia).

17. At the tip of each basidium are 4 sterigmata bearing a basidiocarp each.

18. The basidiospores are released into the pores from where they get dispersed.

19. The spore output is enormous, millions of spores per basidicoarp are produced.

20. The basidiospores are oval, and on germination develop into the primary mycelium.

Specimens given in the laboratory

1. Basidiocarp
2. Section of the pileus (slide)

Diagrams to be drawn

1. Habit of basidiocarp
2. Section of the pileus
3. A pore enlarged to show the basidia.

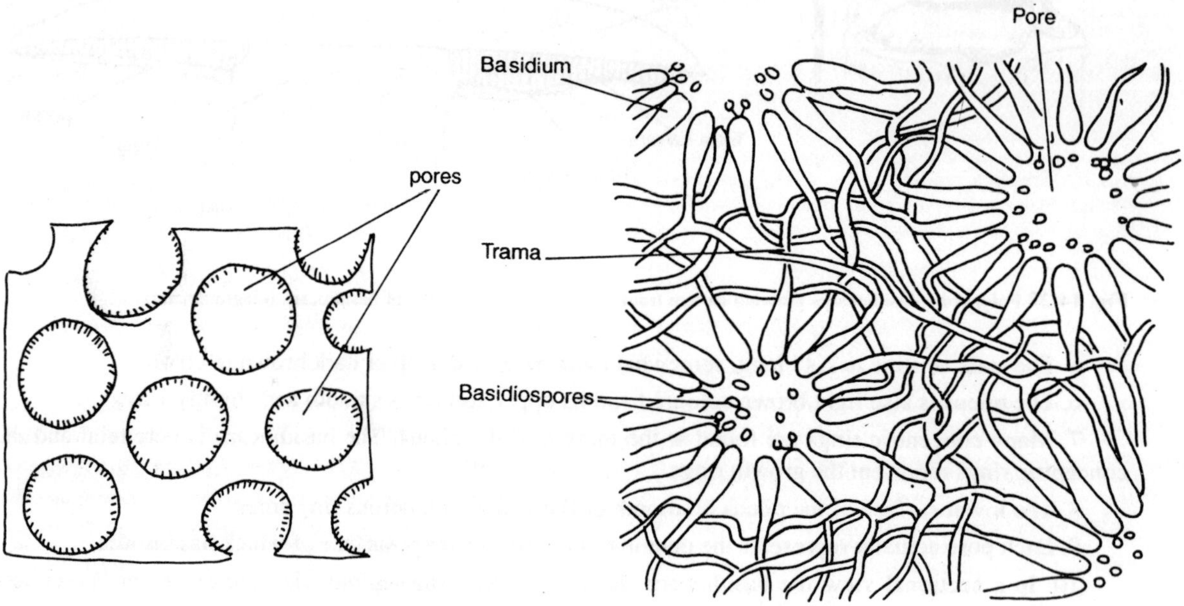

Fig. 14.28 Polyporus. Surface view of pileus showing pores

Cross-section of a portion of pileus

Systematic position and identification

A. 1. Chlorophyll absent
   2. Reserve food is glycogen ... *Fungi*

B. 1. Mycelium septate
   2. Basidia and basidiospores are formed ... *Basdiomycetes*

C. 1. Basidiocarp formed
   2. Basidia aseptate ... *Eubasidiomycetes*

D. 1. Basidiospores released in instalments they are not completely covered till maturity ... *Hymenomycetes.*

E. 1. Fructification macroscopic
   2. Hymenial layers protected ... *Agaricales*

F. 1. Hymenial layer opens out into tubes
   2. Numerous pores present on the undersurface of the pileus ... *Polyporaceae*

G. 1. Ventral surface of the pileus porous and different in colour from the upper surface.
   2. Mature basidiocarp hard and woody ... *Polyporus*

AGARICUS

Common species : *A.campestris bisporous*

Occurrence : Commonly called mushroom, the genus *Agaricus* is a widely distributed saprophytic, fungus growing on dead logs, timber, on soil with plenty of leaf littler.

Study : 1. The vegetative mycelium
2. Basidiocarp
3. Structure of the gill
4. Systematic position and identification

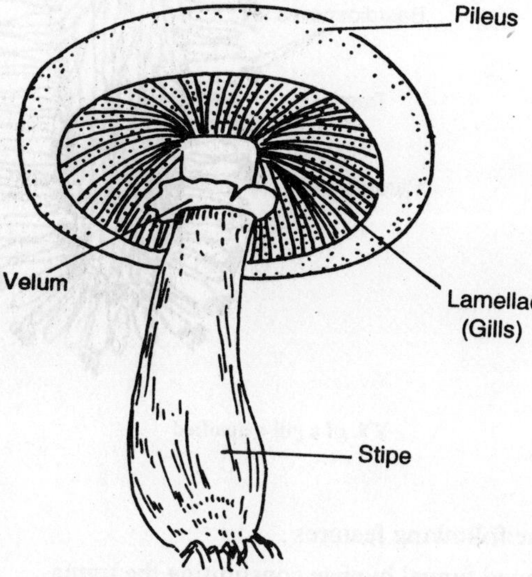

Fig. 14.29 Agaricus. A basidiocarp

Vegetative mycelium

1. The actual plant body of *Agaricus* is a subterranean mycelium

2. The mycelium is extensively branched with anastomosing hyphae. The network of hyphae are called spawns.

3. The hyphae are septate tubular and hyaline.

4. The cytoplasm is granular containing oil globules and other reserve food materials.

5. The mycelium is perennial and survives for several years producing basidiocarps which come above the ground.

The basidiocarp

1. The basidiocarp is formed by the underlying mat of the hyphae after dikaryotization.

2. The young basidiocarp, when it emerges out of the ground will be button shaped, hence it is known as the button stage.

3. The button appears as a small globose whitish structure above the ground.

4. In a long sectional view, the button shows a slightly expanded top portion separated from the lower elongated portion by means of a constriction.

5. A little above the constriction are found two chambers - the lamellar cavities.

6. The margin of the pileus is attached to the stipe with the help of a thin membrane called veil or velum.

Mature basidiocarp:

1. The mature basidiocarp of *Agaricus* is a small umbrella like structure.

2. It has a long, dull white coloured, cylindrical stalk called the stipe.

3. The stipe bears at its tip an expanded portion called the pileus .

4. Midway on the stipe is found the annulus representing the velum, which gets torn as the pileus expands.

5. The upper surface of the pileus is smooth or undulated.

6. The under surface of the pileus has got a number or plates of lamellae radiating from the centre. These lamellae are called gills.

7. The gills that extend right from the centre to the margin of the pileus are called primary gills while those that start midway and reach the margin are called secondary gills.

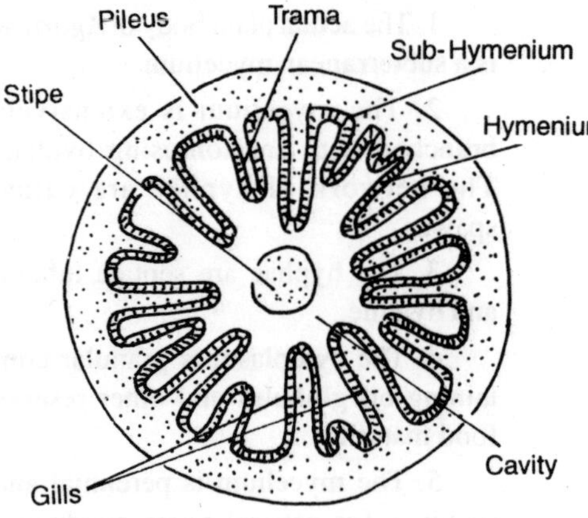

Fig. 14.30 Agaricus. Cross-section of pileus

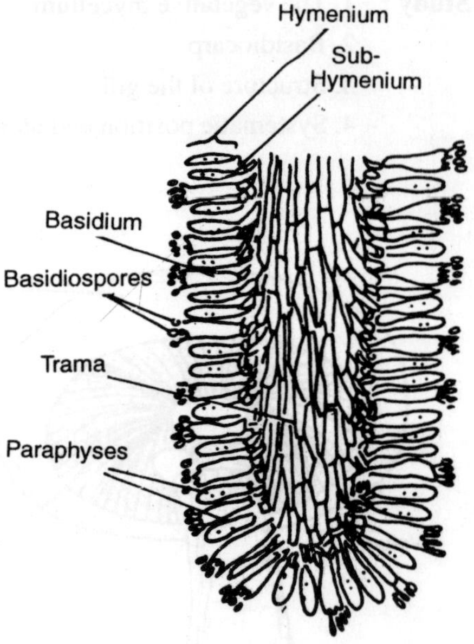

V.S. of a gill magnified

Structure of a gill :

A vertical transverse section of the gill shows the following features :

1. At the central region are present loosely arranged fungal hyphae constituting the trama

2. On either side of the trama are found two to three layers of compactly arranged hyphae which look like rounded cells. This is called the subhymenium.

3. External to the sub-hymenium and arranged at right angles to the flat surface of the gill are found basidia constituting the hymenium.

4. The hymenium also has cystidia, the sterile hyphae intermingled with the basidia.

5. The basidia are club shaped and aseptate. They are called holobasidia.

6. The basidiospores are small, and oval in shape. They are wind dispersed and on the availability of a suitable substratum, they germinate to give rise to the monokaryotic primary mycelium of a new individual.

Specimens given in the laboratory

1. Basidiocarp
2. Section of a gill (slide)

Diagrams to be drawn :

1. Basidiocarp
2. Section of a gill
3. A few basidia enlarged

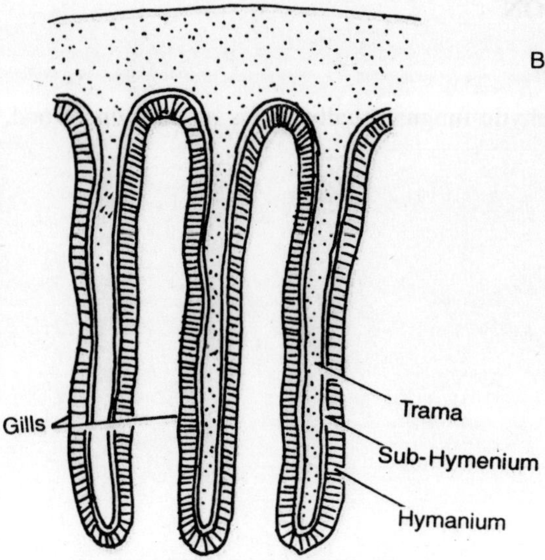

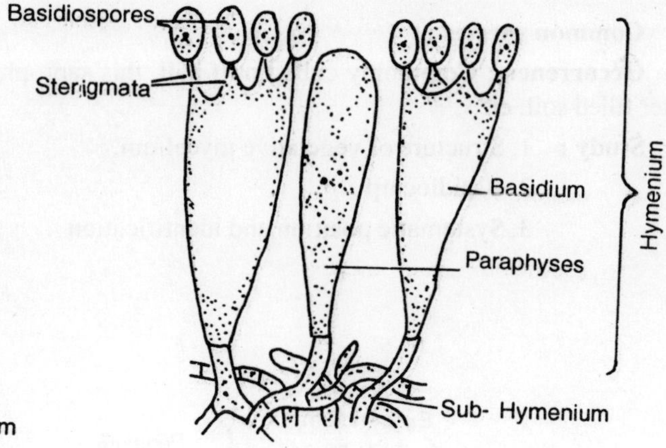

Fig. 14.31 Agaricus. V.S. of a portion of pileus

A portion of hymenium enlarged

Systematic position and identification

A 1. Chlorophyll absent
2. Reserve food is glycogen ... *Fungi*

B. 1. Mycelium is septate
2. Basidia and basidiospores produced ... *Basdiomycetes*

C. 1. Basidiocarp produced
2. Basidium aseptate ... *Eubasidiomycetes*

D. 1. Basidiocarp exposed from the beginning
2. Basidiospores released in instalments ... *Hymenomycetes*

E. 1. Gills present
2. Basidiocarp macroscopic ... *Agaricales*

F. 1. Basidiocarp soft and fleshy
2. Gills narrow and flattened ... *Agaricaceae*

G 1. Gills radiating from the centre of the pileus.
2. Annulus present
3. Primary and secondary gills present
4. Stipe easily seperates from the pileus... *Agaricus*

LYCOPERDON

Common species :

Occurrence : Commonly called puff ball, this saprophytic fungus usually grows on decaying wood, litter filled soil, etc.

Study : 1. Structure of vegetative mycelium.

2. Basidiocarp

3. Systematic position and identification

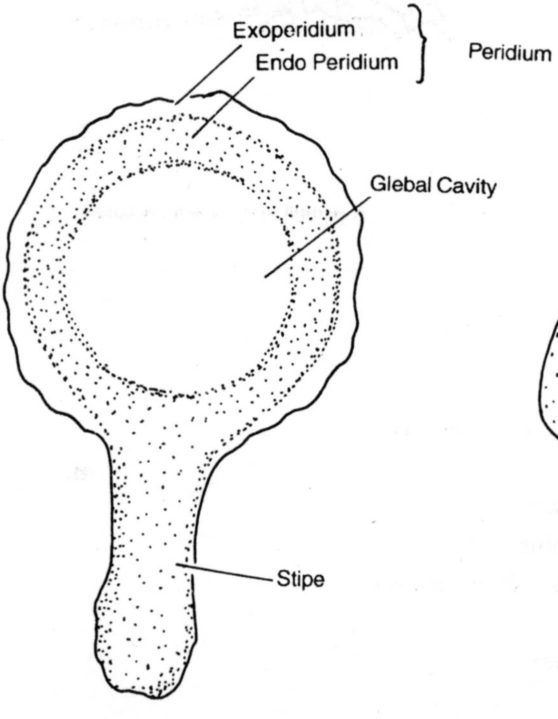

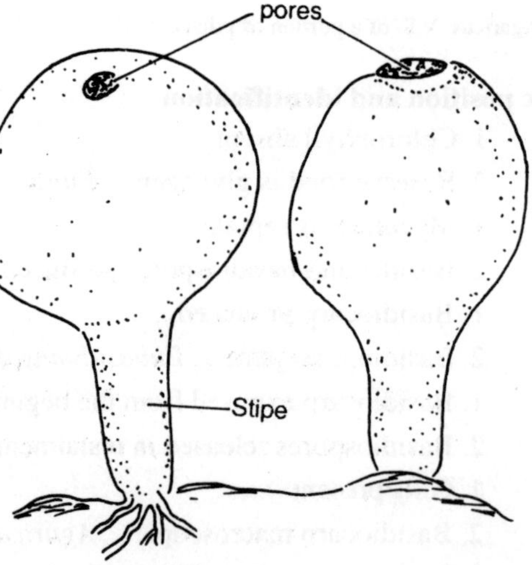

Fig. 14.32 Lycoperdon.The basidiocarps

L.S. of young basidiocarp of Lycoperdon

Vegetative structure

1. The plant body is a subterranean mycelium.

2. The primary mycelium consists of short uninucleate cells and is short lived.

3. The primary mycelium soon develops into dikaryotic mycelium by hyphal fusions. Clamp connections are noticed on the hyphae of the secondary mycelium.

4. The hyphae of secondary mycelium form knots called rhizomorphs.

5. A transverse section of the rhizomorph reveals an outer cortical zone, an inner subcortical zone and a central medulla.

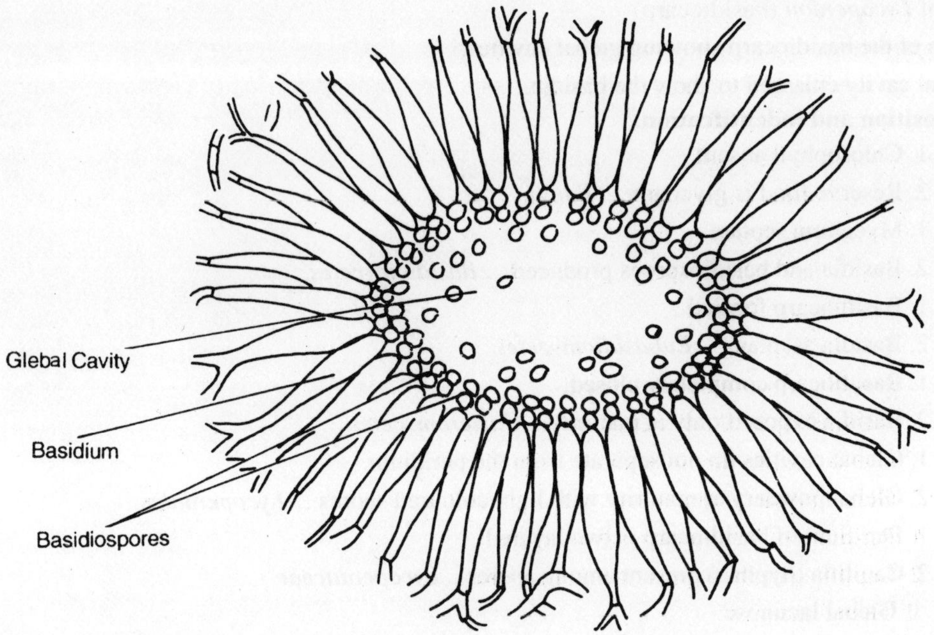

Fig. 14.33 Lycoperdon. Glebal cavity showing basidia and basidiospores

Basidiocarp

1. The basidiocarp arises like a tiny outgrowth from the central region of the rhizomorph.

2. During early stages, the young basidiocarp is homogenous, ie., hyphae are uniformly distributed.

3. A fairly mature basidiocarp is round or pyriform and is attached either laterally or terminally to the rhizomorph.

4. The basidiocarp may range in size from 1-6".

5. The basal portion of the basidiocarp remains sterile.

6. In the upper portion of the basidiocarp, a little later differentiation takes place into outer periderm and inner gleba.

7. The periderm has two layers viz., the outerperidium and the inner peridium.

8. The inner region has many small cavities called glebal cavities.

9. The inner lining of the cavities has many loosely interwoven hyphae with broken ends.

10. The hyphae are made up of binucleate cells and eventually they form the basidia.

11. In a nearly mature baisidum, the baisdia project into the cavities.

12. The basidia are holobasidia and each one produces four basidiospores.

13. All the basidiospores in a basidisocarp are discharged simultaneously in a snuffy mass, whenever the fructification breakes open.

14. The basidiospores germinate to give rise to a new individual, on falling upon a suitable substratum.

Specimens given in the laboratory

1. Basidiocarp

2. Section showing glebal cavities (slides)

Diagrams to be drawn :

1. Habit of *Lycoperdon* (basidiocarp)
2. Section of the basidiocarp showing glebal cavities.
3. A glebal cavity enlarged to show the basidia.

Systematic position and indentifcation.

A. 1. Chlorophyll absent
2. Reserve food is glycogen ... *Fungi.*

B. 1. Mycelium septate
2. Basidia and basidiosproes produced ... *Basidiomycetes.*

C. 1. Basidiocarp formed.
2. Basidia aseptate ... *Eubasidiomycetes.*

D. 1. Basidiocarp completely closed.
2. Basidia exposed only at maturity... *Gasteromycetes*

E. 1. Glebal cavities do not separate from the peridium.
2. Glebal powdery at maturity with light coloured spores ... *Lycoperdales.*

F. 1. Peridium of basidiocarp is two layered.
2. Capilitia (hyphae) present among spores... *Lycoperdaceae*

G. 1. Glebal lacunose
2. Globose or pyriform basidiocarp.
3. Peridium breaks open at maturity... *Lycoperdon.*

CHAPTER 15

DEUTEROMYCETES

Deuteromycetes are called imperfect fungi for they are known only in their asexual stage. The sexual stage or the perfect stage is lacking in them or has not been discovered. These fungi possess septate hyphae that reproduce asexually by means of conidia. Many of these are saprophytes but a great majority of them are parasites causing disease to plants and animals and human beings and hence are of great economic importance.

According to Alexopoulos, the conidial stages of Deuteromycetes resemble those of Ascomycetes and can be regarded as ascomycetes whose ascigerous stage has been dropped during the course of evolution. There have been instances in cases, if ascigerous stage is found they can be placed in Ascomycetes. In some cases, the perfect stage discovered have placed these fungi in Basidiomycetes. Hence the class Deuteromycetes is an artificial one to include genera in which the life cycles are not fully known to classify them.

Alexopoulos defines fungi imperfecti as "one may consider the fungi imperfecti, therefore as conidiale stages of Ascomycetes or more readily Basidiomycetes whose sexual stages have not been discovered or no longer exist". The parasexual cycle discovered in some imperfect fungi indicates that some of them at least never had a perfect stage. There are some fungi which have never produced any spores at any stage.

Somatic structure

The thallus of fungi imperfects consists of well developed septate branched mycelia. The cells are usually multinucleate. The septa are perforated permitting the streaming of cytoplasm.

Deuteromycetes asexually reproduce by conidia which are borne on conidiophores. The conidiophores themselves may be simple isolated structures or are produced in special structures like acervlus, cynnema, pyridium etc. In addition to this, some Deuteromycetes specially animal and human pathogens produce other types of spores called microconidia, blastospores or arthospores. Microconidia as the very name indicates are small conidia, while blastospores are asexual cells produced as a result of budding or directly from a hypha. Arthrospores are produced by the disjoining and isolation of cells. They are also known as oidia.

Some of the Deuteromycetes have parasexuality. This type of sexuality conferred only some advantages of sexual reproduction. Parasexual cycle may be defined as a cycle in which plasmogamy, karyogamy and haploidization (meiosis) takes place but not at specified time or at specified points in the life cycle of an organism.

Classification. Deuteromycetes do not fit into the system of classification followed for other fungi which are mainly based on the characters of the sexual stage. Hence an artificial, but a practical system of classifying Deutromycetes has been followed in which there are many *form* orders, form families, *form* genea etc. The idea of including the prefix *form* is to indicate the members so grouped need not be related. If few morphological characters agree, the members are grouped together. The chief characters used in the classification of Deuteromycetes are the type of fructifications, and the shape, colour and septation of conidia. To date, between 15,000 to 24,000 species belonging to nearly 1200 genera have been included in Deuteromycetes. The form class Deuteromycetes is divided into the following orders.

1. **Sphaeropsidales**. This order includes 500 genera and 12,500 species. Conidia are borne in Pycnidial cavities.

2. **Melanconials**. This order comprises 80 genera and 12,000 species which are mainly parasitic conidia are borne in acervuli, which are sub epidermal or sub cuticular in the host.

3. **Moniliales**. The order consists of nearly 10,000 speices. Many of the monilales are of immense importance to human beings. It is tŏ this group that most of the human pathogens belong. Conidiophores may be simple or branched.

4. **Mycelia sterilia**. This is a conglomerate assemblage of about 20 genera and 350 species. They do not produce any spores.

In this book, the genera :

Colletotrichum, Pyricularia, Alternaria, Fusiarium and *Cercospora* are discussed.

COLLECTORICHUM

Common species : *C.falcatum, C.gossypi, C.circinans* etc.

Occurrence : The genus is a faculative parasite attacking a variety of flowering plants. It causes many diseases known as anthracnose, red rot etc. The plants infected are sugarcane, mango, chillies etc.

Study : 1. The disease caused and symptoms

2. Vegetative structure

3. Reproduction

4. Systematic position and identification

Diseases caused and symptoms:

A. *Red rot of sugarcane*. The disease is caused by *C.falcatum*. The perfect stage of the fungus has been included under Ascomycetes.

1. Disease infects all parts of the plant. Disease symptoms appear when the sugar synthesis is about to begin in the stem.

2. The leaves at the apex get decolourised and drop down. This is followed by the collapse of the entire shoot apex.

3. Anatomically the diseased stems reveal long stripes of red colour with white across.

4. Red patches also appear on the leaves.

B. Fruit rot and die back of chillies. The disease is caused by Colletotrichum capsici.

1. The disease symptoms are of two types- fruit rot and die back.

2. Fruit rot is noticed in ripe red coloured fruits.

3. Black spot first appears on the fruit and then enlarges to form a stripe.

4. Diseased fruits lose their healthy red colour and turn light yellow. Ultimately the fruit decays.

5. Rotten fruits show conidiospores and sclerotia.

In die black disease, the shoot apex starts decaying and it proceeds down. As a result of this the branch or entire plant collapses.

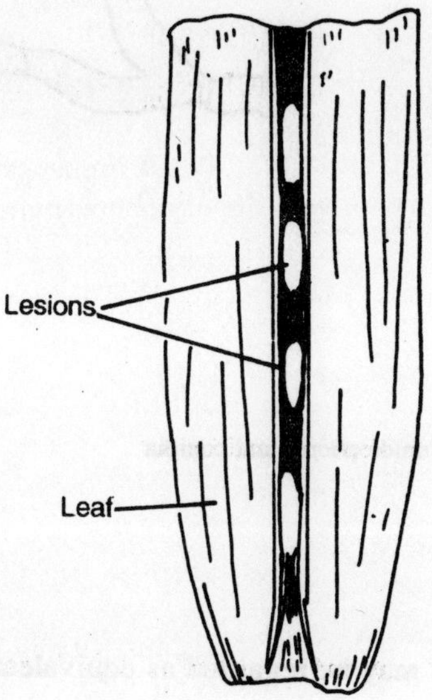

Fig. 15.1 Colletotrichum. A portion of sugarcane leaf showing red rot disease

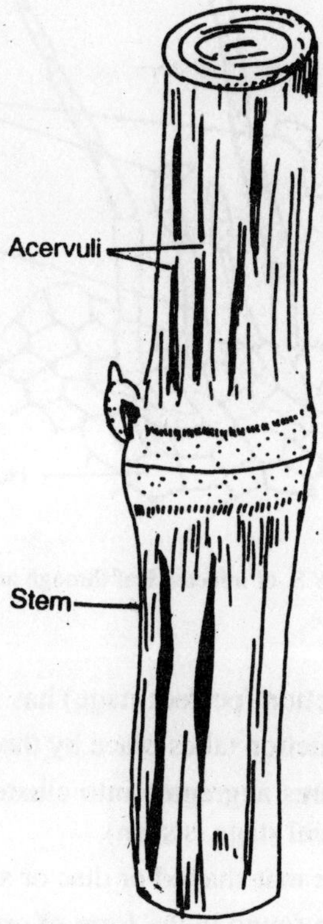

A portion of sugarcane stem showing red rot

Diseased branches have water soaked leisons. Acervuli of the fungus are found in these areas.

C. Black spot disease of mango. The disease is caused by *C.gleosporioides*. The perfect stage of this fungus has been included in Ascomycetes under the name *Glomeralla singulata*.

1. The disease symptoms appear on the skin of the ripening fruits, in the form of slightly depressed patches of different shapes.
2. The skin may crack open and the pulp gets decayed.
3. Disease spots may also appear on the leaves and branches.
4. Tissues of the infected, region decay and die. Tips of branches also drop down due to the disease.

D. Anthracnose of cotton. The disease causing agent is *C.gossypi* and *C.lindeminanthianum*.

E. Smudge of Onions. The pathogen is *C.circinans*.

Vegetative structure.

1. The plant body of the fungus is a septate much branched mycelium.
2. The hyphae are endophytic and inter or intercellular. They are branched and have oil globules as reserve food.

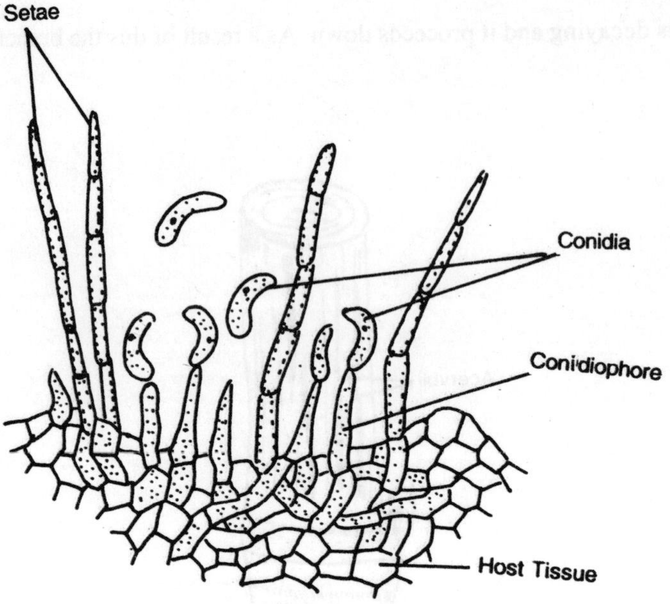

Fig. 15.2 Colletotrichum. V.S. of infected leaf through acervulus

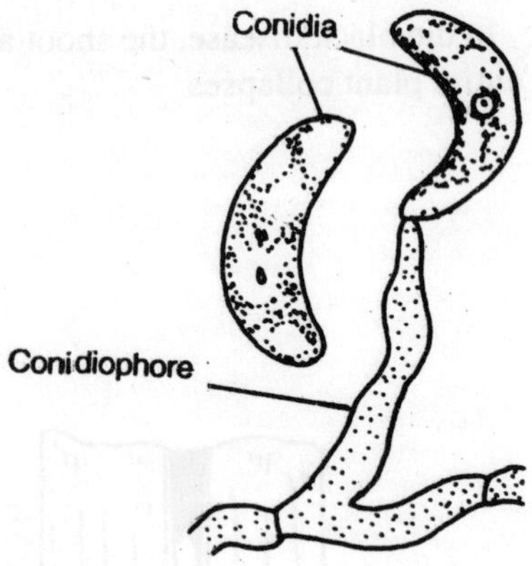

Conidiophores and conidia

Reproduction :

1. Sexual reproduction (perfect stage) has not been reported.
2. Asexual reproduction takes place by the formation of conidia.
3. The conidiospores aggregate into clusters called acervuli, which may be regarded as equivalent to fructifications (no sexual stage is seen).
4. The acervuli are mat shaped or disc or saucer shaped.
5. The acervuli are found in the form of crusts on the host.
6. Within the acervulus, the conidophores are arranged in a palisade like manner.
7. The conidiophores are erect and unbranched. At their tips, they cut off conidia.
8. Each conidium is unicellular, faceate with round ends.
9. In between the conidiophores, sterile hairs called setae are present.

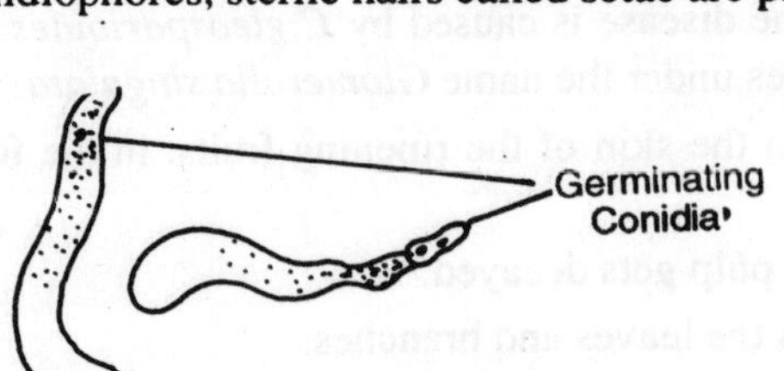

Fig. 15.3 Colletotrichum. Conidia

Diagrams to be drawn:

1. Habit sketch of infected host.
2. Section showing Acervulus.
3. Conidia enlarged.

Systematic position and identification

A. 1. Chlorophyll absent

2. Reserve food is glycogen ... *Fungi*

B. 1. Perfect stage absent

2. Only asexual stages (conidial) are seen ... *Deuteromycetes*

C 1. Conidia found in Acervulus ... *Melanconiales*

D 1. Acervulus has conidiophores and seta ... *Melanconiaceae*

E 1. Red stripes on the host

2. Seta found in acervulus ... *Colletotrichim.*

PYRICULARIA

Common species : *P.oryzae*

Occurrence : The fungus is a parasite on many flowering plants and specially on paddy. The disease is extensive in cold regions.

Study : 1. Diseases caused and symptoms

2. Vegetative structure

3. Reproduction

4. Systematic position and identification

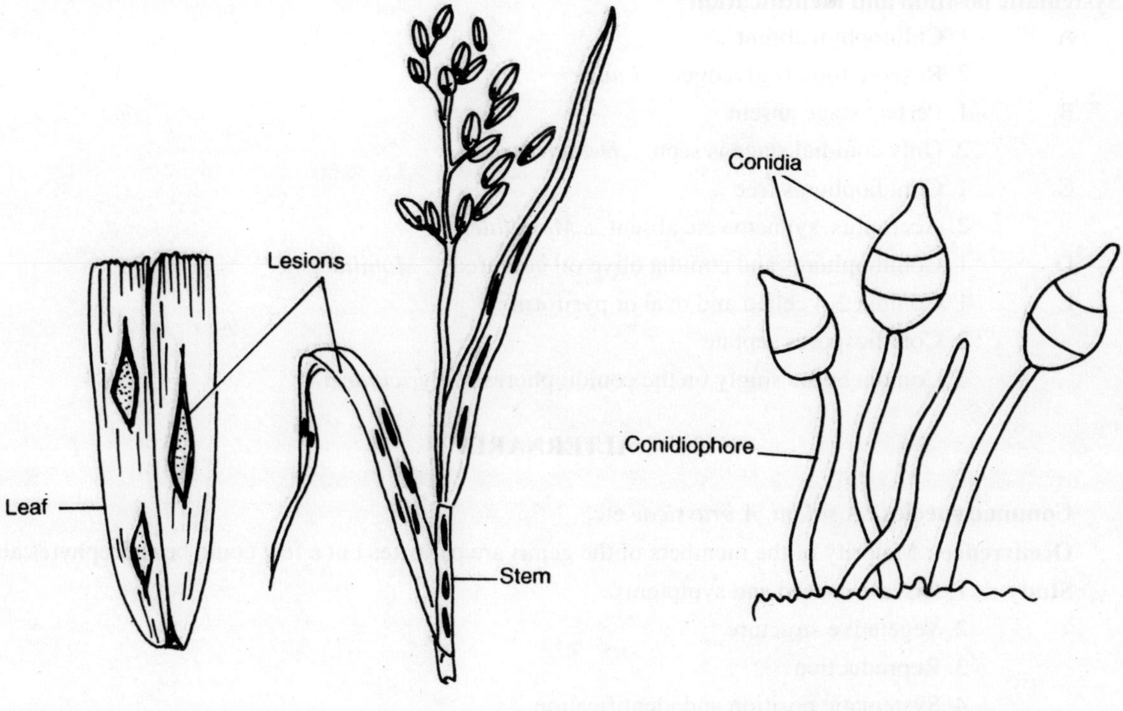

Fig. 15.4 Pyricularia. Rice leaf lesions Rice inflorescence showing lesions Conidiophores and conidia of Pyricularia

Disease caused and symptoms

A. Blast or rotten neck disease of Rice. The disease is caused by *P.oryzae* causing loss in yield as much as 75%.

1. The disease symptoms are visible mostly on the lamina. However leaf sheath, petiole, stem and calyx may also exhibit symptoms of the disease.

2. Small bluish blisters appear on the leaves. These may be rounded or elongated. The centre of the lesions are grayish green in colour.

3. The blisters have water soaked appearance with brown margin.

4. The characteristic symptom of the disease is the rotting of the peduncle just below the flowers. Due to this, the ear collapses, hence the name rotten neck disease.

Vegetative structure :

1. The plant body of the fungus is a septate and much branched mycelium.
2. The hyphae are olive oil coloured and have a number of cross walls.
3. The mycelium may be intercellular or intercellular.

Reproduction :

1. Perfect stage has not been reported.
2. Asexual reproduction is by conidia borne on conidiophores.
3. Conida are small oval in shape and made up of three cells.
4. Conidiophores are simple, unbranched and septate with 2-4 cross walls; colour grayish yellow.

Diagrams to be drawn

1. Habit sketch of infected leaves and inflorescence.
2. Conidiophores and conidia enlarged.

Systematic position and identification

A. 1. Chlorophyll absent
2. Reserve food is glycogen... *Fungi*

B. 1. Perfect stage absent
2. Only conidial stage is seen ... *Deuteromycetes*

C. 1. Conidiophores free
2. Acervulus, synnema etc absent ... *Moniliales.*

D. 1. Conidiophores and conidia olive oil coloured ... *Moniliaceae*

E 1. Conidia 2-4 celled and oval or pyriform.
2. Conidiospores septate
3. Conidia borne singly on the conidiophores ... *Pyricularia.*

ALTERNARIA

Common species : *A.solani, A.brassicae* etc.

Occurrence : Majority of the members of the genus are parasites but a few could be saprophytes also.

Study : 1. Disease caused and symptoms.
2. Vegetative structure
3. Reproduction
4. Systematic position and identification

Diseases caused and symptoms :

1. *Early blight of potato* : This disease is caused by *A.solani* and is widespread in areas wherever potato is cultivated.

a. Light brown patches appear first on the leaves.

b. Leaves on the lower part of the plant get the attack of the disease first.

c. Infected tissues exhibit irregular lines, with the leaf appearing like a target board.

d. Occasionally the disease may spread to the tubers also.

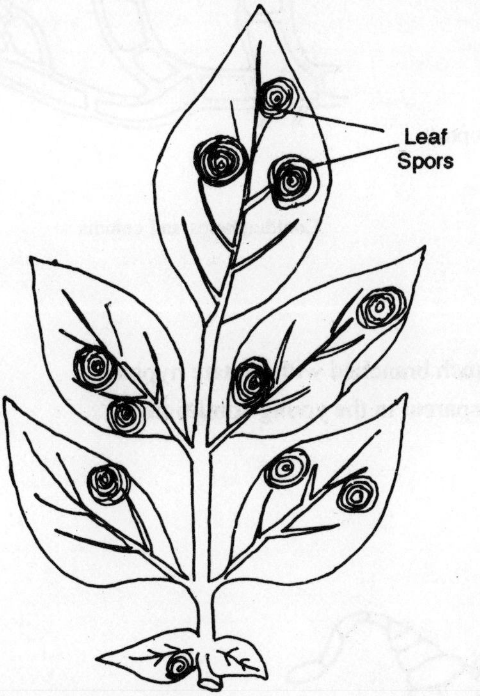

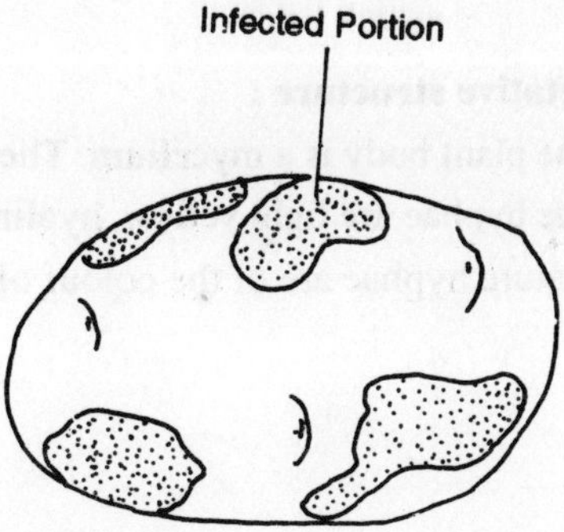

Fig. 15.5 Alternaria. Infected potato leaf showing leaf spots

Infected potato tuber

2. *Alternaria leaf spot of cabbage* : Cruciferous plants like cabbage, mustard, cauliflower, knol khol, raddish etc get the attack of leaf spot. The three species of *Alternaria* viz *A.brassicae*, *A.brassicola* and *A.raphani* cause the leaf spot disease. *A.raphani* is specific to raddish only.

a. The disease symptoms appear in the form of black dots. These are circular with a light blue coloured central region.

b. Many times the spots have concentric lines at the centre.

3. *Leaf blight of wheat* : The disease is caused by *A.triticola*.

a. Disease symptoms first apear on the leaves in the form of scattered oval patches.

b. Gradually, the patches occupy the entire leaf area.

c. The grayish brown patches destroy the chlorophyll completely.

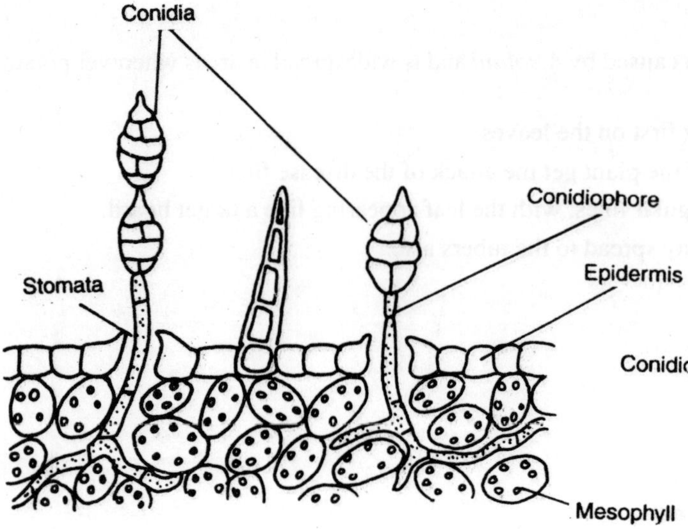

Fig. 15.6 Alternaria. T.S. of host leaf showing conidiophores and conidia

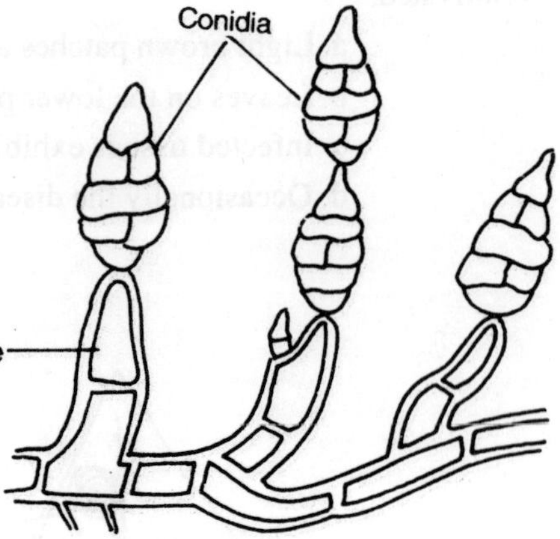

Conidiophores and conidia

Vegetative structure :

1. The plant body is a mycelium. The mycelium is much branched with septate hyphae.
2. The hyphae are light yellow, hyaline and semitransparent in the young condition.
3. Mature hyphae are of the colour of olive oil.

Fig. 15.7 Alternaria. Germinating conidia

Reproduction :

1. Perfect stage of the fungus is not seen.
2. Asexual reproduction takes place by means of conidia.
3. Condiophores are simple, unbranched and septate, they come out of the host leaf through the stomata.
4. Conidia are multicellular and have 5-10 cross walls. In some cases the ends of conidia have beak like projections. They are dictyosporous, ie., have both transverse and longitudinal septa.

Diagrams to be drawn:

1. Habit sketch of infected leaves.
2. Section of the leaf to show conidiophores and conidia.

Systematic position and identification

A. 1. Chlorophyll absent
   2. Reserve food is glycogen... *Fungi*

B. 1. Perfect stage is absent
   2. Reproduction by conidia ... *Deuteromycetes*

C. 1. Conidiophores free.
   2. Acervulus, symnema etc absent ... *Moniliales*

D. 1. Hyphae and reproductive organs semitransparent or olive in colour... *Dematiaceae.*

E 1. Conidiophores single and erect.
   2. Conidia are dictyosporous ... *Alternaria.*

FUSARIUM

Common species : *F.solani, F.lini, F.oxysporum, F.vasinfectum* etc.

Occurrence : The species of *Fusarium* are parasitic and generally cause a disease called *Wilt.*

Study

1. Diseases caused and symptoms.
2. Vegetative structure
3. Reproduction
4. Systematic position and identification.

Diseases caused and symptoms

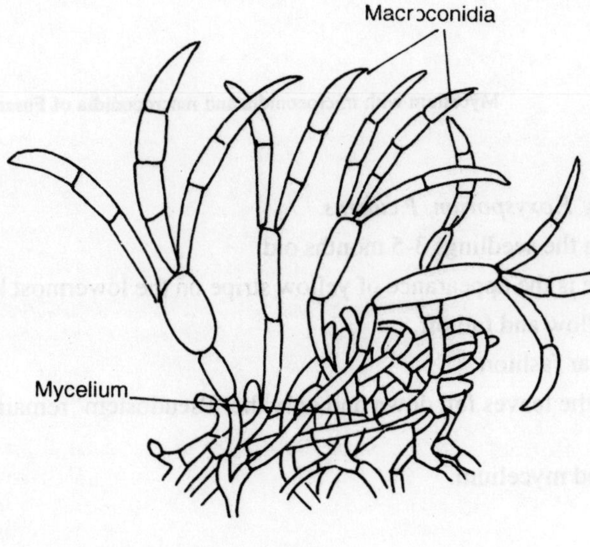

Fig. 15.8 Fusarium. A sporodochium

1. Wilt disease of red gram : The disease causing agent is *F.oxysporum.* The plants of red gram are invariably attacked, wherever they are cultivated.

a. The characteristic symptom of the disease is the wilting of the plant. The plants look as if they are suffering from lack of water, even though there is plenty of water in the soil.

b. The tissues at the basal part of the stem turn dark in color.

2. Wilt disease of cotton : The pathogen is *F.vasinfectum*. The symptoms of the disease are similar to what is seen in the wilt of red gram.

3. Wilt of linseed : *Fusarium lini* is the fungus which causes the disease.

a. Disease symptoms may appear on the host at all stages of growth.

b. Infected seedlings drop down and the cotyledons wither away.

c. In older plants, deep green or brown patches appear on the leaves.

d. Fungus is present in the roots as well as the basal part of the stem.

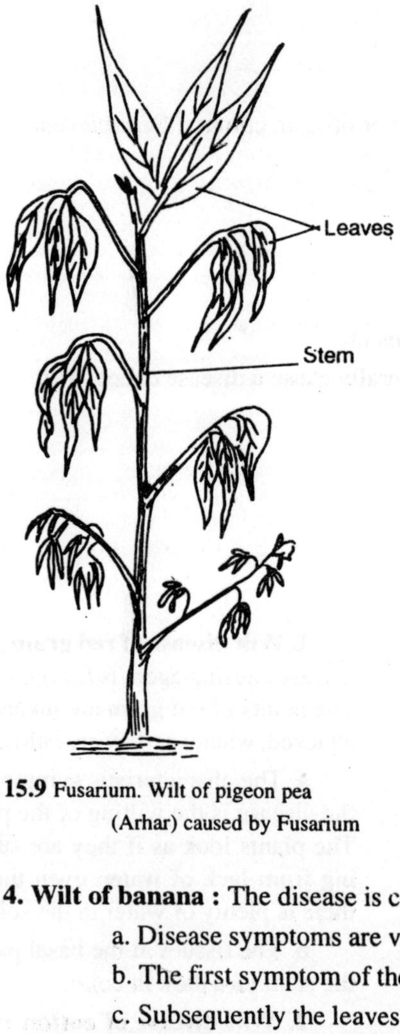

Fig. 15.9 Fusarium. Wilt of pigeon pea (Arhar) caused by Fusarium

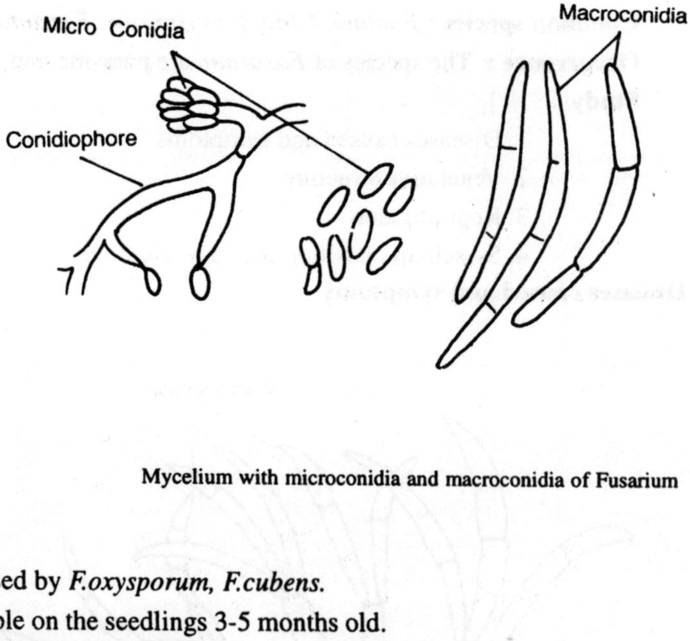

Mycelium with microconidia and macroconidia of Fusarium

4. Wilt of banana : The disease is caused by *F.oxysporum, F.cubens.*

a. Disease symptoms are visible on the seedlings 3-5 months old.

b. The first symptom of the disease is the appearance of yellow stripe on the lowermost leaf.

c. Subsequently the leaves turn yellow and fall off.

d. Leaf sheaths also crack in a linear fashion.

e. When the disease is intense, all the leaves fall down and only the 'Pseudostem' remains.

Vegetative structure

1. The plant body is a septate, much branched mycelium.

Reproduction

1. Prefect stage is not seen.

2. Asexual reproduction is brought about by a variety of methods viz., conidia, (microconidia and macroconidia), chlamydospores, sclerotia etc.

3. Microconidia are small rounded or oval in shape; sometimes crescent shaped also. They are present in small groups and are produced at the tips of branched or unbranched conidiophores.

4. Macroconidia are produced in a fructification called *Sporodochia* which are cushion like. They (conidia) are long, sickle or crescent shaped. The conidia are septate.

5. Chlamydospores are formed by the cells of the hyphae which round off and develop into a thick walled chlamydospore.

6. In some cases, the hyphae aggregate and form small, black compact *Sclerotia.*

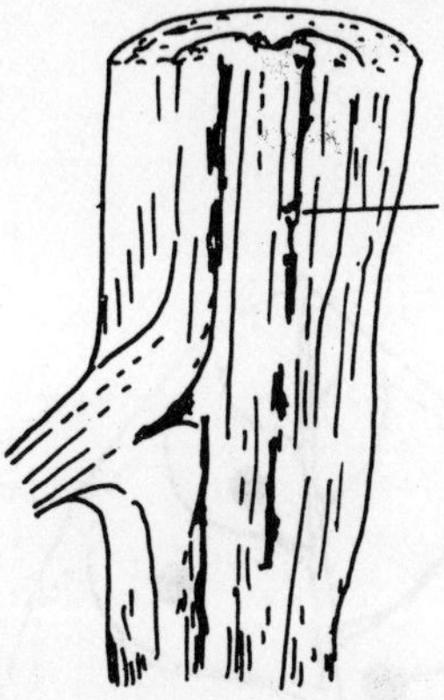

Fig. 15.10 Fusarium. T.S. of infected root (diagrammatic)

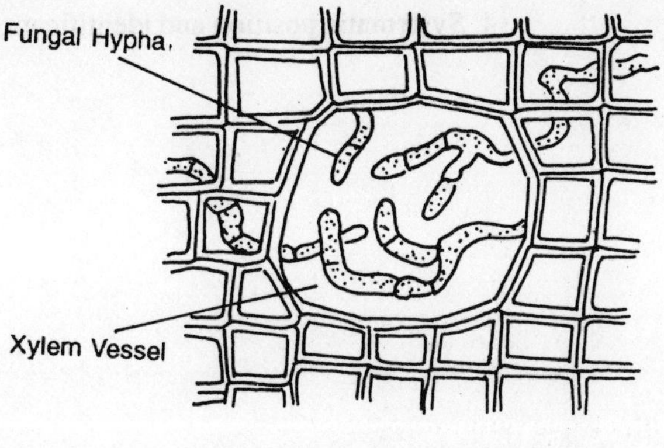

Mycelium in vascular tissue

Specimens given in the laboratory

1. Infected hosts
2. Conidia and conidiophores (slides)

Diagrams to be drawn

1. Habit sketch of infected host
2. Conidiophores and conidia
3. Sclerotia
4. Chlamydospores

Systematic position and identification

A. 1. Chlorophyll absent
   2. Reserve food is glycogen ....*Fungi*

B. 1. Perfect stage absent
   2. Asexual reproduction by conidia ....*Deuteromycetes*

C. 1. Conidiopores generally free
   2. Acervulus, synnema etc., absent ....*Moniliales*

D. 1. Mycelium hyaline
   2. Sporodochia present ....*Tuberculariaceae*

E. 1. Conidia of two types
   2. Hyphae in the vessels of the host ....*Fusarium*

CERCOSPORA

Common species : *C.personata, C.arachnidicola, C.veticola, C.apii* etc.

Occurrence : The various species of *Cercospora* are parasitic pathogens infecting various flowering plants and cause leaf spot diseases. The members attacked include groundnut, chillies, cotton etc. On groundnut, the genus causes a disease called Tikka disease.

Study : 1. Diseases caused and symptoms.
2. Vegetative structure
3. Reproduction
4. Systematic position and identification

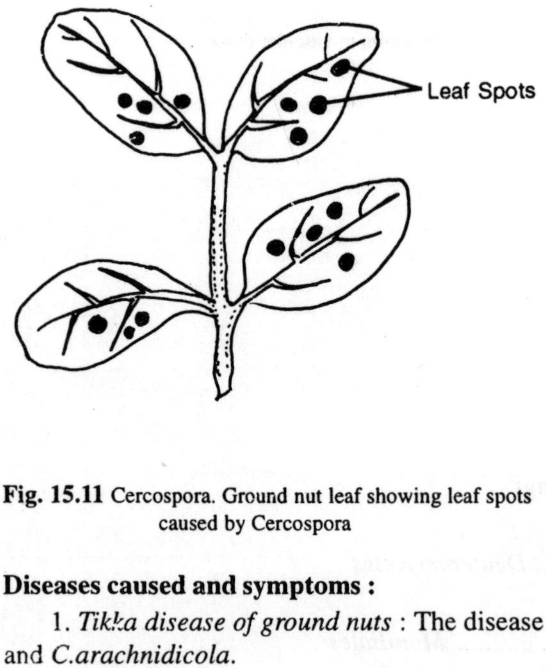

Fig. 15.11 Cercospora. Ground nut leaf showing leaf spots caused by Cercospora

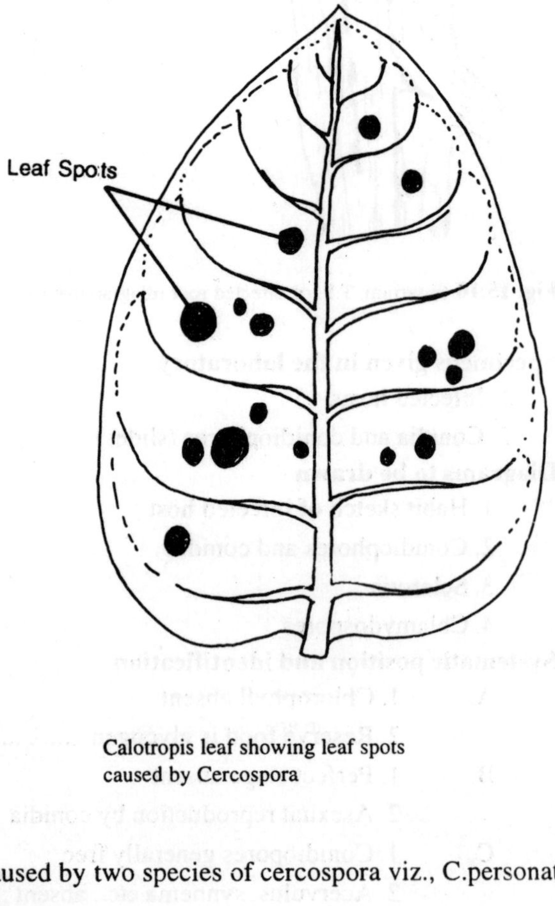

Calotropis leaf showing leaf spots caused by Cercospora

Diseases caused and symptoms :

1. *Tikka disease of ground nuts* : The disease is caused by two species of cercospora viz., C.personata and *C.arachnidicola*.

2. All the aerial parts of the host plant get the infection.

3. As the first symptoms of the disease, light green patches appear on the leaves. The epidermal cells of the leaf lose contact with the mesophyll.

4. In case of infection by *C.arachnidicola*, the patches are somewhat circular with a diameter of 1-10 mm.

5. The patches have a yellow halo right from the beginning.

The patches which are discrete in the beginning later coalsce.

In case of infection by C.personata, the patches are strictly circular with a diameter of 1-6 mm.

There is no yellow halo in case of infection by *C.personata*

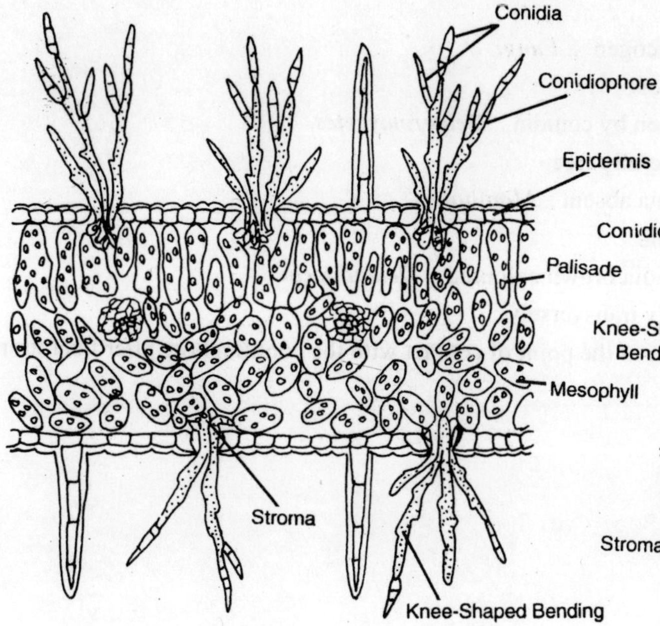

Fig. 15.12 Cercospora. T.S. of host leaf showing conidiophores and conidia of Cercospora

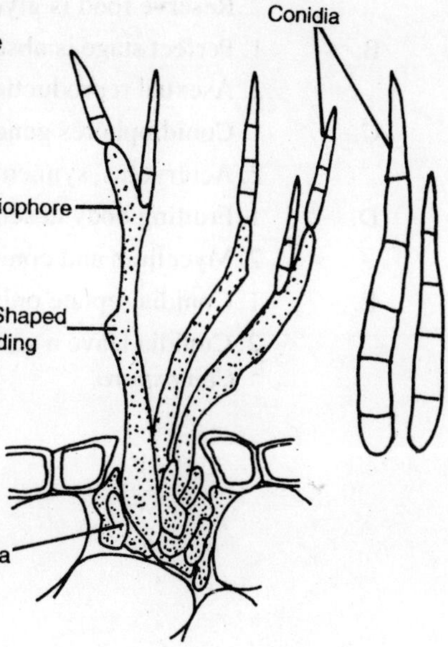

Conidiophores and conidia

Other diseases caused by cercospora :

a. Leaf spot of Dhal - *C.indica*

b. Leaf spot of Cotton - *C.gossypium*

c. Leaf spot of Rice - *C.oryzae.*

Vegetative structure :

1. The plant body is a septate, much branched mycelium.

2. The hyphae are brownish in colour, slender and grow in between the cells of the host.

3. The hyphae produce haustoria as seen in *C.personata.* In *C.arachnidicola*, however the haustoria are absent as the mycelium is intracelluar. C.arachnidicola etc.

4. In *C.personata* the hyphae form stroma.

Reproduction :

1. The perfect (sexual) stage is absent.

2. Asexual reproduction is brought about by means of Conidia.

3. The conidiophores are brown coloured, erect, replete and come out of the host leaf by piercing through the epidermis.

4. Conidia are light yellow in colour, septate with 4 -12 cross walls, they range in length from 0.38 - 108μ.

Diagrams to be drawn :

1. Habit sketch of infected leaves.
2. Section of the leaf showing conidiophores, conidia etc.

Systematic position and identification :

A. 1. Chlorophyll absent
2. Reserve food is glycogen ... *Fungi.*

B. 1. Perfect stage is absent
2. Asexual reproduction by conidia... *Deuteromycetes.*

C. 1. Conidiophores generally free
2. Acrervulus, synnema absent ...*Moniliales*

D. 1. Fruiting body absent.
2. Mycelium and conidia brown coloured ...*Dematiaceae*

E. 1. Conidia septate only transversely.
2. Conidia leave a scar at the point of contact with the conidiophore after liberation ... *Cercospora.*

CHAPTER 16

LICHENS

The light green threads that hang down from the branches of trees, the thick grayish green lathery patches that cover the rocks and clad the trunks of trees, the heavy hard incrustations that form an unsepareable part of the exposed rocks - these are the lichens. Lichens represent one of the ingenious creations of nature in colourizing and occupying any hostile environment. They form a world wide vegetation with 400 genera and nearly 16,000 species.

Occurrence and habit

Lichen thalli have a wide range of tolerance to climatic fluctuations and as such they colonize and settle permanently even on such habitats which are inhospitable to other vegetation. They thrive and multiply in areas where no other vegetation is possible. They can endure and withstand long periods of drought and come to active and normal life at the return of favourable conditions. They are cosmopolitan in distribution.

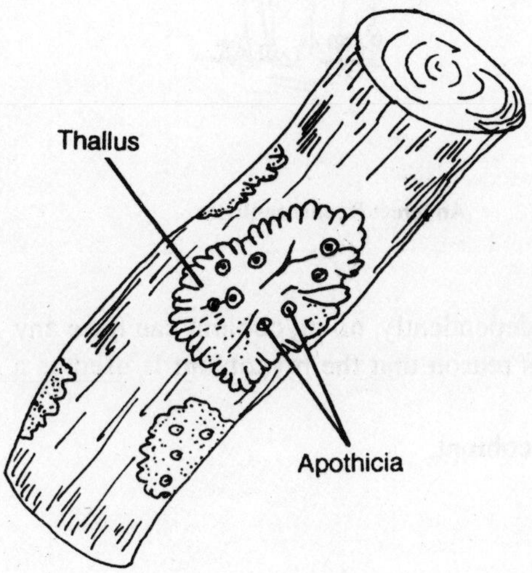

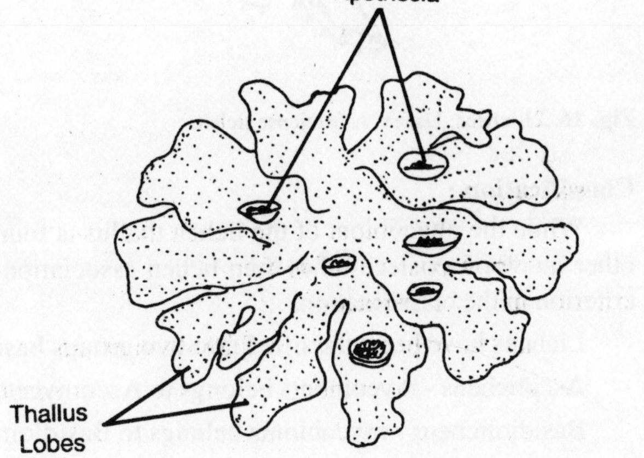

Fig. 16.1 Lichens. Crustose lichen growing on wood   Parmelia. A foliose lichen

Nature and composition :

Lichen is not an individual plant, though the plant body looks like that of an individual organism. Lichens represent an association of two plants of the group Thallophyta. The duality of lichen thallus was first proposed by Schvendner (1867). He identified the lichen thallus as an association of an alga and fungus. In the lichen association the algal partner is called the *phycobiont* and the fungal partner is called the *mycobiont.*

The mycobiont always belongs to the class ascomycetes or basidiomycetes. Of the 400 genera of lichens in about only four, the mycobiont belongs to Basidiomycetes.

The plycobiont belongs to the class cyanophyceae or chlorophyceae. Unlike in the case of mycobiant, phycobionts can be identified up to genus and species for they also exist independently other than the lichen association. The cyanophycean members that have been found as phycoionts are *Nostoc, Gloeocapsa, Rivularia* etc. The chlorophycean algae found in lichens are protococecus, *Cystococcus, Palmella, Trentepohlia* etc.

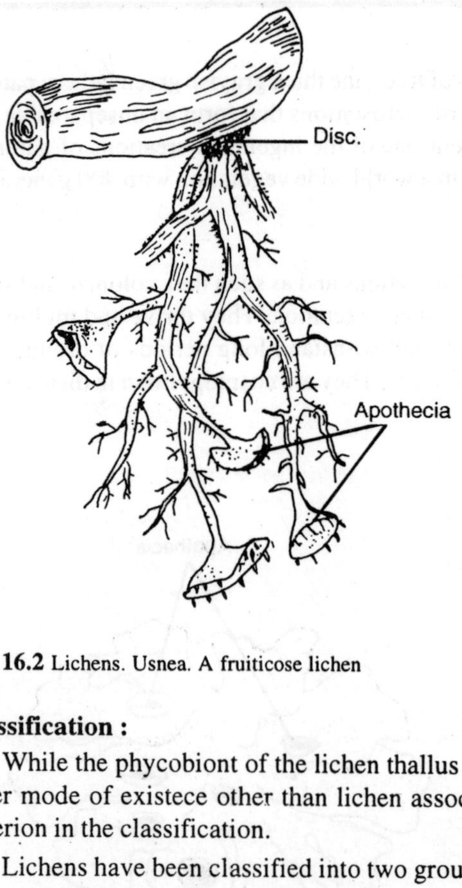

Fig. 16.2 Lichens. Usnea. A fruiticose lichen

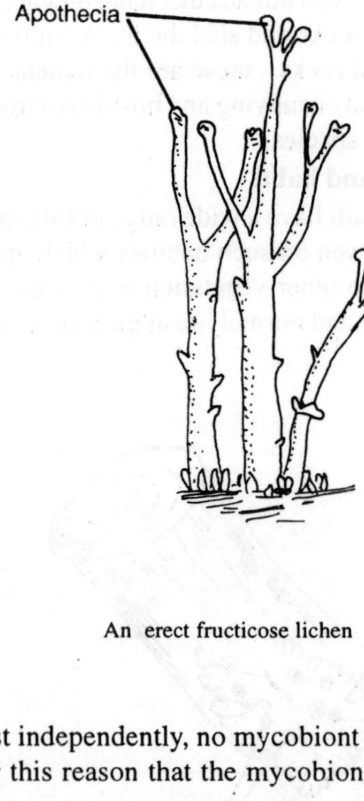

An erect fructicose lichen

Classification :

While the phycobiont of the lichen thallus is found to exist independently, no mycobiont can have any other mode of existece other than lichen association. It is for this reason that the mycobiont is used as a criterion in the classification.

Lichens have been classified into two groups based on the mycobiont.

Ascolichens - mycobionts belongs to Ascomycetes.

Basidiolichens - mycobionts belongs to Basidiomycetes.

There is also a third group called Myxolichens in which the mycobiont belongs to myxomycetes. Myxolichens however do not exist in nature. They have been synthesised in the laboratory.

Study : 1. Different types of Ascolichens
2. Vegetative structure of the thallus
3. Anatomy of the thallus
4. Structure of the Apothecium
5. Structure of other reproductive organs.
6. Structure of Basidiolichens.

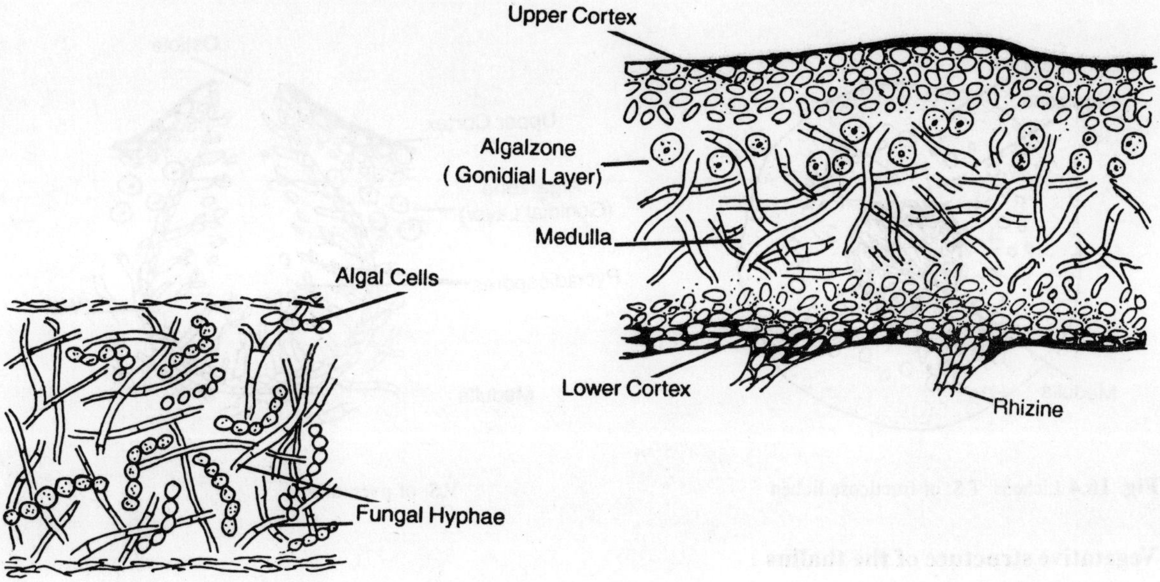

Fig. 16.3 Lichens. V.S. of homoiomerous foliose lichen V.S. of heteromerous foliose lichen

Ascolichens : These are classified into three types based on the nature of the thallus. These are :

Crustose Lichens :

1. They usually grow on rocks or on barks of trees.
2. They form thick or thin incrustations.
3. The thallus is small and dorsiventral
4. Lower surface secretes a gummy substance which sticks the thallus to the substratum and makes it inseparable.
5. Small apothecia are found on the upper surface.

Foliose Lichens :

1. The thallus is leafy and flattened.
2. It is broader than in crustose lichens
3. They usually grow on barks of trees and occasionally or rocks also.
4. The thalli attach themselves to the substratum with the help of rhizoid like outgrowths called rhiznes. Rhizines develop from the ventral surface.
5. The free end of rhizines is expanded into a disc like structure to help in attachment.
6. On the dorsal surface of the thallus are visible many apothecia.

Fruticose lichens :

1. These have cylindrical branched thalli that grow upright or hang down from branches of trees.
2. The plant body is attached to the substratum by dense clusters of hyphae.
3. In some fructicose lichens the basal portion is foliage like with upright cylindrical branches.
4. Apothecia are borne on the upright branches.

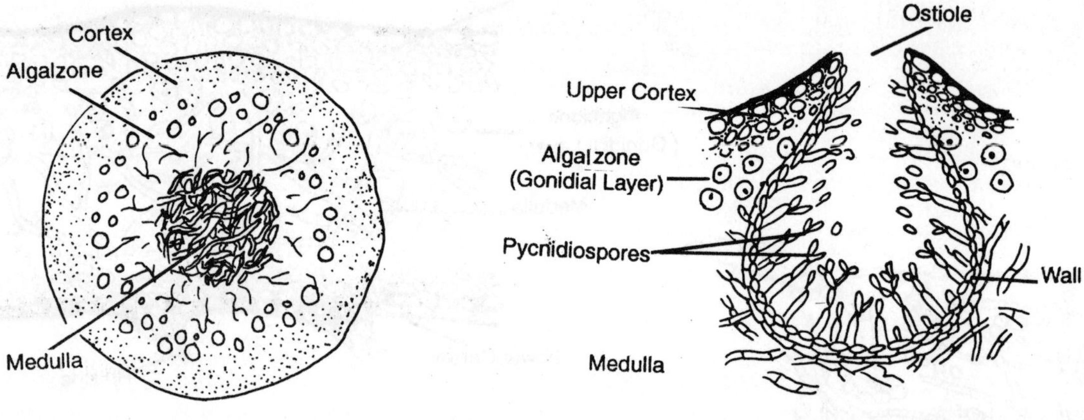

Fig. 16.4 Lichens. T.S. of fructicose lichen V.S. of pycnidium

Vegetative structure of the thallus :

The thallus bears a number of outgrowths or appendages performing various functions. Some of these are :

Breathing pores :

1. These are breaks or openings found in the upper cortex.
2. These pores help in respiration, act as aerating organs.

Cyphellae :

1. Cyphellae are also aerating organs.
2. They are present as circular or oval depressions in the lower cortex.

Isidia :

1. These are small protruberances developing from the dorsal surface of the thallus.
2. The algal partner present in the Isidia is same as in the main thallus.
3. Isidia are supposed to increase the photosynthetic area.

Cephalodia :

1. These are dark coloured outgrowths present on the dorsal surface of the thallus.
2. Each cephalodium has a few algal cells surrounded by fungal hyphae.
3. The algal partner present in cephalodium is different from the one present in the main thallus.
4. Many Lichenologists regard cephalodium as an independent epiphyllous lichen.

Soredia :

1. These are small regular bud like out growth present on the thallus.
2. Soredia may be scattered or present in aggregates called *soralia*.

3. Each soredium has fungal hyphae entangling a few algal cells (same as in the thallus).

4. At maturity the soredium detaches itself from the main thallus and develops into a new lichen.

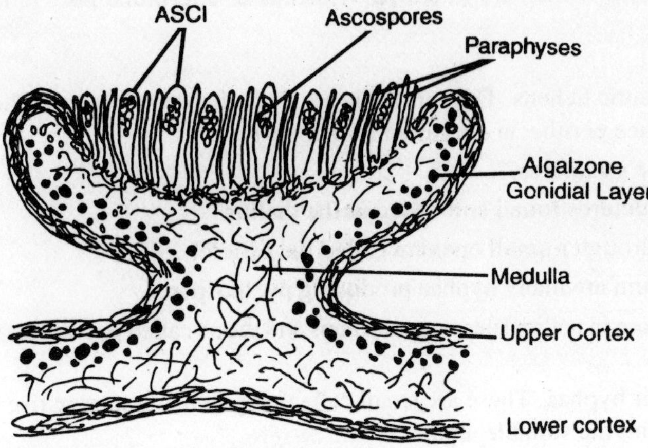

Fig. 16.5 Lichens. V.S. of apothecium

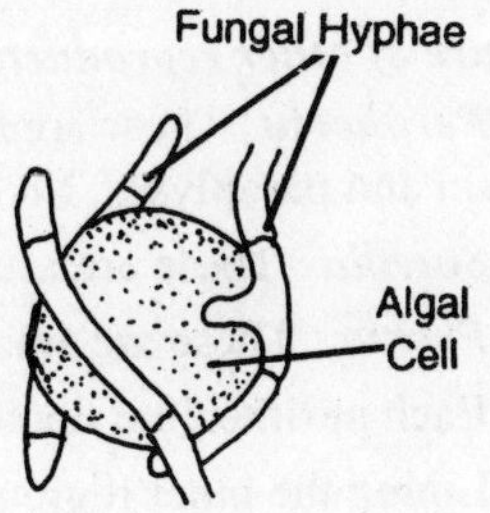

Soredia

Anatomy of the thallus :

1. Anatomically the crustose and foliose lichens are identical.

2. Fruticose lichens are slightly different.

3. A vertical transverse section of either crustose or foliose lichen shows the following features.

4. On the upper and lower surface are found the upper and lower cortex respectively.

5. They are made up of closely interwoven fungal hyphae.

6. Below the upper cortex is the algal zone.

7. This was formerly called the gonidial layer when the dual nature of the lichen thallus was not known.

8. In the algal zone are found algal cells dispersed among the loosely arranged fungal hyphae.

9. In the centre (above the lower cortex) is the medulla, where fungal hyphae are found loosely interwoven.

10. The lower cortex gives rise to rhizoids which help in anchorage.

11. In frutiose lichens the outline of a cross section will be circular.

12. There is an outer cortex, internal to which is the algal zone. In the centre is the medulla.

13. In some lichens the algal cells do not form a definite layer, but are uniformly distributed all over. They are called *Homoimerous,* while those with a restricted algal zone are called *Heteromerous.*

Structure of Apothecium :

1. Sexual reproduction leads to the development of fructifications like apothecium.

2. Apothecium producing lichens are called *Gymnocarpic.*

3. Some lichens produce *Peritherica*, they are called *Pyrenocarpic.*

4. Apothecia are very much similar to the one found in *Peziza* in structure.

5. They are small cup like structures ranging in width from 10 - 25 mm.

6. The inner lining consists of the hymenium having asci and paraphyses.

7. The wall of the apothecium is called the peridium and is made up of closely joined fungal hyphae.

8. Internal to the wall is the algal zone. (In *Lecanorine* type).

9. In some apothecia (*Lecidine* type) algal cells are completely absent.

10. Each ascus - consists of eight ascospores, which are liberated by means of a terminal pore in the ascus.

Structure of other reproductive organs :

1. *Perithecia* : These are found in some aquatic lichens. The perithecia are flask shaped structures. They have asci and paraplyses. Lichens which produce perithecia are called *Pyrenocarpic.*

2. *Soredia* : These are asexual reproductive structures.

3. *Pycnia* : These are small cup shaped structures found embedded in the thallus.

4. Each perithecium open to the exterior through a small opening called the ostiole.

5. Lining the inner region of the perithercium are many hyphae producing pycniospores.

6. Many people however opine, that pycnia are in reality spermagonial cups and the so called pycniospores are spermatia.

Oidia : Some lichens produce *oidia* by their hyphae. These are small cells that germinate to give rise to a new lichen, provided they come in contact with the suitable alga.

***Basidiolichens* :**

1. They are very rare in nature.

2. A typical example of a basidiolichen is *Cora pavonia.*

3. The much lobed thalloid plant body resembles a bracket fungus.

4. The thallus is attached to the tree trunks with the help of rhizines.

5. A vertical section of the thallus reveals three regions.

6. The uppermost part is the *superior layer* consisting of loosely arranged fungal hyphae. The fungus belongs to the class Basidiomycetes, order Agaricales.

7. Beneath the superior layer is found the algal layer having *Chroococcus* or *Scytonema*. The algae are entagled between the hyphae.

8. The *inferior layer* found on the lower face represents a dense matrix of hyphae.

9. The inferior layer bares the sub hymenium and the hymenium. The hymenium has a number of holobasidia arranged like a palisade layer.

10. Each baisidium has four sterigmata, terminating which are found the basidiospores.

11. Occasionally paraphyses are found in between the basidia. They could be immature basidia also.

CHAPTER 17

BACTERIA

Bacteria are one of the smallest among the micro-organisms. Though small in size, they are by no means so in their significance. Bacteria are highly potent organisms. From the point of view of human welfare, they are of greatest significance. In causing such deadly diseases like cholera, influenza, tetanus, leprosy, tuberculosis, typhoid etc, to name only a few, they are man's worst enemies. They are man's best friends also in maintaining N_2 balance in nature, in helping him in various industries (preparation of alcohols and acids, curing of coffee, tea and tobacco) and in various other aspects.

Credit for discovering bacteria goes to Anton Van Leeuwenhock (1976) a dutch scientist. The name of Louis pasteur should be mentioned, for it was he who discovered the fermenting capacity of bacteria. Robert Koch, a German bacteriologist discovered that tuberculosis is caused by bacteria T.J. Burill (1879) an American Scientist discovered that bacteria are responsible for many plant diseases also.

Distribution. Bacteria are omnipresent. They are present in air, milk, water, soil, arctic regions, hot springs, in plants, animals, human beings etc.

Morphology. Bacteria are unicellular organisms. Based on the cell shapes, four morphological forms have been identified in bacteria. These are :

1. Bacillus. The cells are rod shaped and elongated. They are the second largest among bacteria. Eg. *Mycobacterium*.

2. Coccus. The cells are spherical. These are the smallest among the bacteria. Eg. *Streptococcus Staphyloccus* etc.

3. Spirillum. The cell will be in the form of a loose spiral. These forms are the largest among bacteria. Eg. *Rhodospirillum*.

4. Comma. The cell will be in the form of punctuation mark, comma. Eg. *Vibrio cholorae*.

Coccus forms exist in various ways:

Micrococcus - single isolated cells.

Diplococcus - cells present in pairs.

Streptoccus - cells exist in chains.

Staphylococcus - cells exist in sheets.

Locomotion. Except cocci, all the other forms of bacteria may possess organs of locomotion called flagellae. On the basis of number and position of flagella bacterial forms are classified as follows:

Atrichous - no flagella.

Monotrichous - many flagella present at one end.

Lophotrichous- many flagella present at one pole.

Amphitrichous - two tufts of flagella attached at either end of the cell.

Peritrichous - flagella present all round the cell.

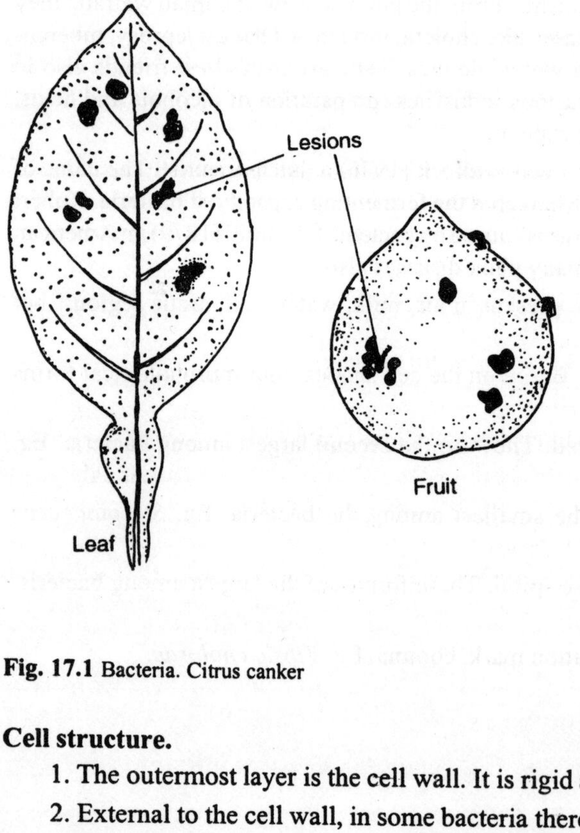

Fig. 17.1 Bacteria. Citrus canker

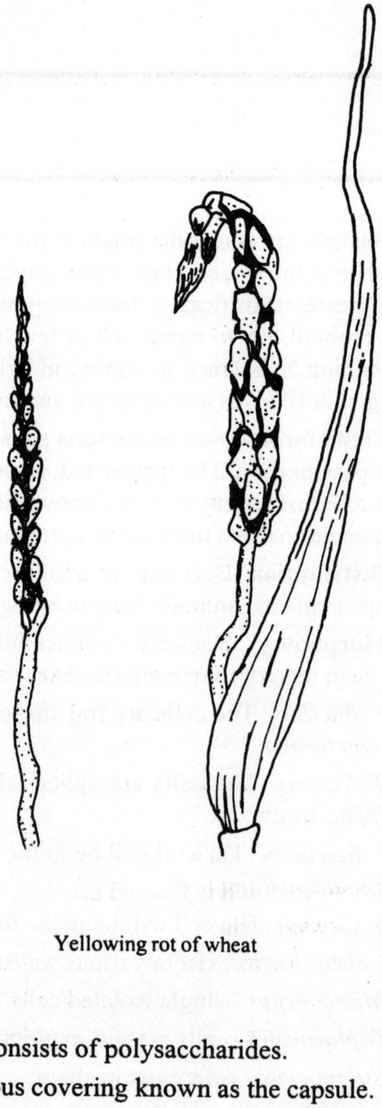

Yellowing rot of wheat

Cell structure.

1. The outermost layer is the cell wall. It is rigid and mainly consists of polysaccharides.
2. External to the cell wall, in some bacteria there is a gelatinous covering known as the capsule. This is composed of polysaccharides and polypeptides.
3. Next to the cell wall, internally is the plasma membrane.
4. The cytoplasm of bacteria may not always have organelles like mitochondria and endoplasmic reticulum.

There is a nucleus, but it lacks a nuclear membrane (prokaryotic).

Ribosomes present in cytoplasm help in protein synthesis.

Reproduction. Bacteria reproduce asexually as well as sexually. Sexual reproduction takes place by genetic recombination and difficult to observe in the laboratory.

Asexual reproduction takes place by means of fission, budding and endospore formation.

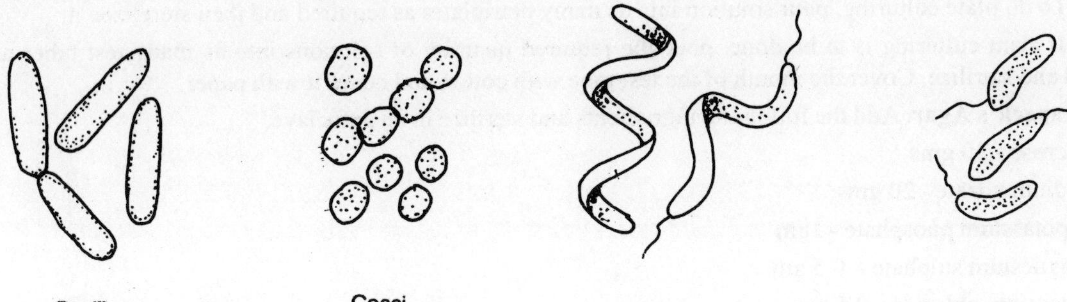

Fig. 17.2 Bacteria. Various shapes of bacterial cells

Fission :

1. Usually bacteria divide by means of binary fission in which the cell divides into two halves.
2. The division is perpendicular to the long axis of the cell.
3. The nucleus elongates and divides into two.
4. The cell wall is laid down in the centre of the cell.

Budding :

1. A protrusion develops from a cell into which a little cytoplasm migrates.
2. The nucleus splits into two and one bit enters the protrusion.
3. The protrusion cuts off from the cell by means of a cross wall. This is the bud.
4. At maturity, the bud separates itself from the parental cell and develops into a new individual.

Endospore :

1. The endospore formation is one of the unique features of bacteria.
2. This is formed to withstand, the adverse environmental conditions.
3. Only one cell is formed per cell. It may lie in the centre or towards one side of the cell.

Culture of bacteria.

For the purpose of study, bacteria, can be cultured in the laboratory using several culture media. Two of them which are simple and easy to prepare are given below:

PDA medium. The medium is prepared as given below. The ingredients are potato, dextrose sugar, agar and water.

Skinned potato - 250 gms

Glucose - 20 gms.

Agar - 15 gms.

Water - 1 litre

1. Cut the peeled potato into small pieces.
2. Boil it in 500 ml of water and filter through a Muslin cloth.
3. Add glucose to this solution. Stir briskly.
4. In the remaining 500 ml water, dissolve agar.
5. Mix both the solutions and make up the volum to 1 litre.
6. Sterilize the solution in an autoclave.
7. To do plate culturing, pour solution into as many petriplates as required and then sterelizee it.
8. If slant culturing is to be done, pour the required quantity of solutions into as many test tubes as needed and sterilize. Cover the mouth of the test tube with cotton and cover it with paper.

Czapeck's Agar. Add the following ingredients and sterilize in an autoclave.

Sucrose - 30 gms

Sodium nitrate - 20 gms

Dipotassium phosphate - 1gm

Magnesium sulphate - 0.5 gm

Potassium chloride - 0.5 gm

Ferrous sulphate - 0.01 gm

Agar - 15.0 gms

Distilled water - 1 litre

The medium can be poured into either petriplates or test tubes as required before sterilizing.

Sources of bacteria for culture:

1. Obtained from diseased plants. For this purpose crush the diseased part of the plant tissue and macerate it. Dissolve this in a little water. This will be the source of bacteria.
2. Bacteria in tap water.
3. Bacteria at the tip of nails.
4. Bacteria in garden soil. Dissolve a little amount of garden soil in water. Allow it to settle. Use the supernatant.

Plating the bacteria :

1. Keep a bunsen burner or spirit lamp burnigng on the table.
2. Wash the table surface with formalin.
3. Keep the petriplates close to the spirit lamp.
4. Take a platinum loop, hold it to the flame, withdraw and dip it in the source of the bacteria.
5. Carefully open the petriplate or the test tube, touch the medium with the platinum loop containing bacteria and close the petriplates or the test tubes immediately.
6. In case you want to culture bacteria present at your nail tip, open the petriplate having the medium, touch it and close the petriplates immediately.
7. In case you want to culture bacteria present in air, open the petriplates containing the sterilized medium for about a second and close it immediately.
8. If you want to culture bacteria present in your breath, open the petriplates, breath on it for a second and close the petriplates immediately.

Culturing:

1. After the required bacteria are plated on the medium keep the petriplates or test tubes either at room temperature or in a hot air oven at 28°C.

2. Within a few days bacterial colonies appear on the medium. These can be stained and identified.

Staining of bacteria:

1. One of the simplest staining procedure is the staining by crystal violet also known as *Gram's stain*.

2. The staining procedure is as follows:

a. Take a clean microslide. Heat it a little on the open flame

b. Take a little of the bacterial colonies and smear it and spread evenly

c. Allow it to dry for a few minutes

d. Add a few drops of the Gram's stain. Leave it for 2 minutes

e. Wash the stain with water and add a few drops of Iodine solution (potassium iodide). Allow it to settle for 2 minutes

f. Wash it thoroughly with alcohol followed by water and observe

g. Alcohol is a decolourising agent and it will remove the stain from some bacteria

h. Bacteria may appear to have blue colour on their cell walls. Such bacteria are called Gram positive bacteria

i Bacteria that have not retained the stain after the application of the decolourising agent are called Gram negative bacteria.

j. The capacity of the bacteria to retain or not to retain the Gram's stain depends on the chemical composition of the cell wall, which is of importance in the classification

k. In medical bacteriology, Grams staining technique is employed to identify pathogenic bacteria.

Plant diseases caused by bacteria :

Bacterial plant diseases may be broadly classified into three categories namely:

1. *Parenchyma diseases*: In this type of disease, the soft tissues decay; patches or pustules appear on various parts of the plant. These are leaf spot disease, canker etc.

2. *Vascular diseases or Bacteriotracheoses*. In infected plants bacteria multiply in large number and accumulate in the lumen of the vascular elements. As a result of the blockade of the lumen translocation, ascent of sap etc will stop; and the plant wilts.

3. *Galls or tumours*. The infection by bacteria brings about hyptertropy on the roots or stems resulting in the development of tumours and galls.

There are about 40 and odd bacterial plant diseases occuring in India. Plants like Rice, wheat, maize, citrus, apple tomato, onion, pea etc infected. Symptoms of a few plant diseases are listed below:

Citrus canker :

1. The disease attacks citrus plants like lime, sweet lime, orange etc.

2. The pathogen is *Pseudomonas citri*.

3. The fruits have brown or dark coloured cruptions are pustules.

4. The pustules are corky and hard.

5. The pathogen is an aerobic, monotrichous bacillus.

6. The disease symptoms appear on all parts of the plant except the roots.

7. Early symptoms appear in the form of brown scabs resulting in the yellowing of leaves.

8. The scabby spots rupture abd enlarge to form corky pustules.

9. Occasionally, the disease may form a girdle around branches. The portion of the branch above the girdle dies.

10. The corky outgrowths on the rind do not spoil the pulp of the fruit.

Blight disease of Rice:

1. The disease infects rice plants (*Oryza sativa*).
2. The disease symptoms may appear on all the aerial parts, but specifically present on the leaves.
3. Lamina and sheathing leaf bases exhibit yellow streaks 5-10 mm long.
4. As the disease increases in intensity the streaks enlarge to form patches.
5. In some instances, sap containing bacteria may exude from the plant.

Brown of potato:

1. The host plant is *Solanum tuberosum* (Potato)
2. The pathogen is *Pseudomous* solanacearum
3. Leaves lose their green colour and turn yellow.
4. The bacteria infect the vascular tissues which turn brown. When a cut in made sap containing bacteria exudes out of the plant.
5. The plants generally collapse before they can produce tubers. Even if tubers are produced, they get discoloured and rot with a brown colour. Sap also exudes out of the tubers.

Leaf spot of Mango:

1. The host plant in *Mangifera indica* (Mango).
2. The pathogen is *Pseudomonas mangiferae indicae.*
3. Water soaked areas appear at leaf tips, which turn brown at later stages.
4. Death of leaf tissue occurs as the lesions enlarge causing necrosis.
5. The disease may infect the fruits, in which case, they crack open and drop down in a premative condition.

CHAPTER 18

VIRUSES

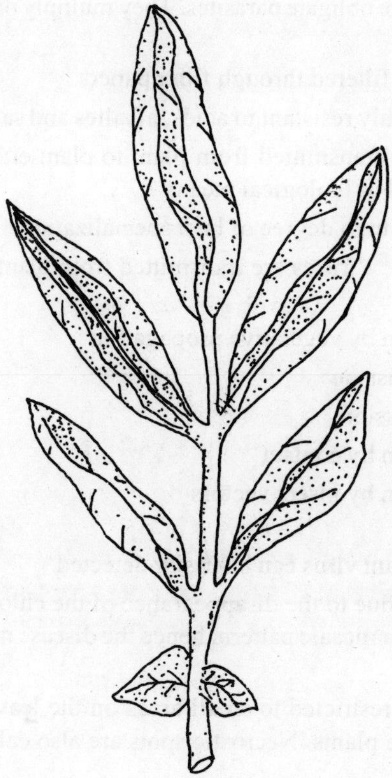

Fig. 18.1 Viruses. Leaf roll of potato

Viruses are the smallest biological entities, that are on the border line between living and non living. They do not possess all the characters of the living beings. They are acellular, but have the capacity to reproduce like the living beings. The most surprising thing about viruses is that they are not destroyed when their organisation is disturbed. They can be crystalized and kept in a bottle like another chemical, and yet when allowed to enter a living host, they start reproducing. This has led to their being described as "chemicals in a test tube but living beings in a host".

The name Virus (= poison) was first coined by Louis pasteur. While studying rabies disease he named the cause as Virus. Four names figure prominently in the discovery of Viruses.

1. Adolf Meyer (1886) set to find out the reason for tobacco mosaic disease came to the conclusion, that no fungus or bacterium causes the disease. But he observed that the sap of the diseased plant was capable of transmitting the disease to healthy plants. He claimed that the sap filtered through a bacterial filter loses the infectivity.

2. Iwanowsky (1892) of Russia proved that even filtered sap retains the capacity of infection. But he did not find out the exact cause of the disease.

3. Beijernick (1896), a Dutch scientist demonstrated that there is a disease causing principle (definitely not a bacterium) present in the sap which beings about the mosaic disease. He called the infective principle *Contaguim vivum fluidum*.

4. Stanley (1935) an American biochemist, crystalized the tobacco mosaic viruse and showed that the crystals when introduced into tobacco plants would bring about the disease.

Types of viruses:

1. *Plant* viruses : These infect the plants and bring about the diseases. RNA is the genetic material in them.

2. *Animal viruses*: These infect animals and human beings brining about the disease. DNA is the genetic material in them.

3. *Bacteriophages*. These infect the bacteria and kill them. DNA is the genetic material.

Structure of a virion:

1. Virion is the techniacl term for viruses.

2. It consists of a shell a coat of protein and encloses the nucleic acid.

3. The nucleoid may be DNA or RNA, but never both.

4. The protein coat is called capsid. It consist of many units capsomeres.

Properties of viruses:

1. They can resist high temperature. Even after being subjected to 80°C, they can infect plants.

2. All viruses are obligate parasites. They multiply only within a living cell.

3. They can be filtered through filter papers.

4. They are highly resistant to acids, alkalies and salts.

5. They can be transmitted from plant to plant either through mechanical or biological means.

6. They have a high degree of host specialization.

Transmission : Viruses are transmitted from plant to plant through

1. Transmission by vegetative propagation
2. Seed transmission
3. Soil transmission
4. Transmission by contact
5. Transmission by insect vectors

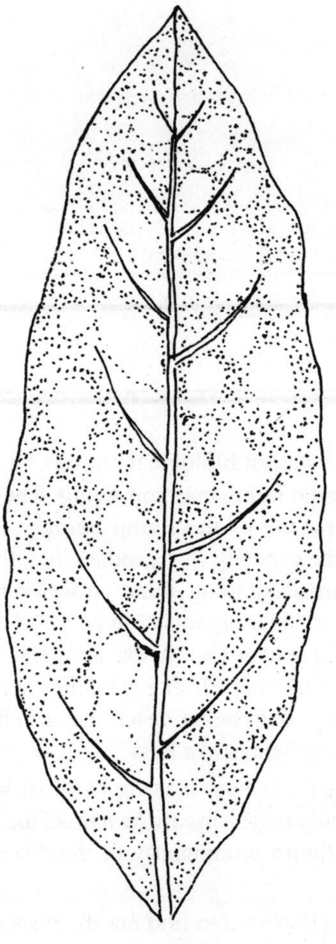

Fig. 18.2 Viruses.Tobacco mosaic

Symptoms of viral plant diseases:

Viruses produce a number of external symptoms, by which a plant virus can be easily detected.

1. **Chlorosis**. This refers to the yellowing of the leaves. This is due to the disappearance of the chlorophyll at places leaving yellow spots. These yellows spots appear like a mosaic pattern, hence the disease may also be called mosaic disease.

2. **Necrosis**. This refers to the death of the cells. It is usually restricted to small areas on the leaves, sometimes it extends over large areas causing complete death of the plants. Necrostic spots are also called lesions.

3. **Distortions.** This is a common symptom of viral diseases. It may be in the form of cushing or ciniling of leaves caused by the slow growth of the veins, sometmes there may be the production of a cluster of thin wiry shoots called Witches broomes.

4. **Vein banding** and **vein clearing**. Vein banding refers to the disappearance of chlorophyll and banding surrounding the veins, where as in vein clearing, chlorophyll disappears along the veins of the leaves.

5. **Yellowing** : When the chlorophyll completely disappears from the infected parts, which consequently turn yellow, the disease is known as yellowing.

Plant disease caused by viruses :

Viruses cause a number of plant diseases. A few of the symptoms are described below:

Tobacco mosaic :

1. The disease is brought about by tobacco mosaic virus (TMV).

2. TMV infects tobacco and many other flowering plants.

3. Typical mosaic pattern of yellow and green areas appear on the leaf.

4. On the young tobacco leaves, the veins may show clearing followed by mottling.

5. Finally dark green spots appear, which develop into blister like areas.

6. The plant may become stunted.

Leaf curl of potato:

1. Young plants with few leaves exhibit disease symptoms prominently.

2. Seed borne infections usually result in the seedling becoming infected. The viruses are systematic (present inside the plant system).

3. Leaf curl virus causes the disease.

4. Pathces of yellow appear on the leaves appear in a scattered fashion.

5. As the growth proceeds, the margins of the leaves curl inwords giving a crinkled look.

6. Yield gets considerably reduced, as the tubers are not of the normal size.

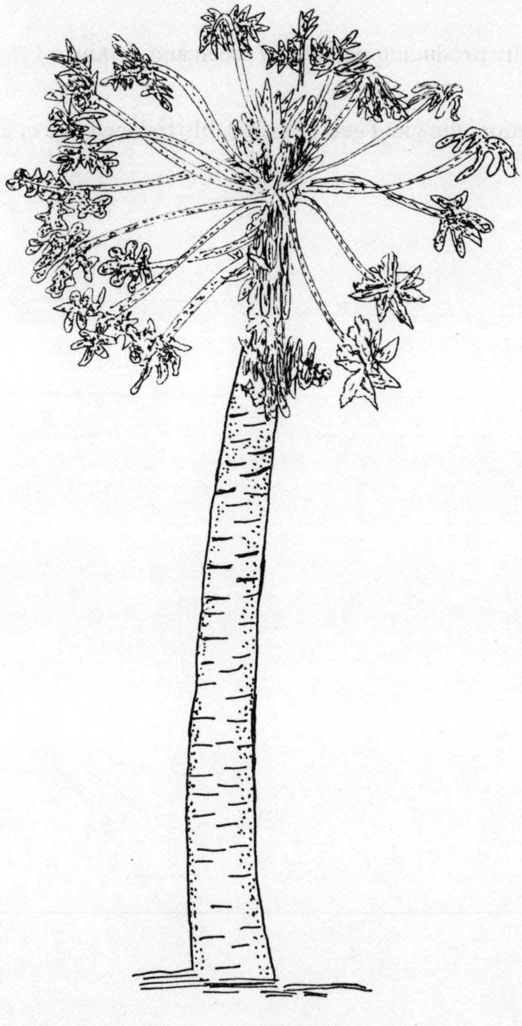

Fig. 18.3 Viruses. Leaf curl of papaya

Leaf curl of papaya :

1. The leaf curl virus causes the disease.

2. The symptoms of the disease appear conspicuously on the leaves.

3. The infected plants show leaves of reduced size. Sometimes malformation also appears on the leaves.

4. At advanced stages of the disease, the leaf margins curl inwards.

5. If the infection appears quite early, the plants may not produce flowers at all.

6. The leaves drop down and ultimately fall down.

7. The plant may survive for a long time even after infection, but it will produce little or no fruits.

Bean mosaic :

1. The host plants infected are beans, double beans belonging to the genus *Phaseolus*.

2. The virus is bean mosaic virus.

3. The disease appears on the leaves characteristically producing patches of green and yellow.

4. All the leaves of the plant get the disease.

5. Eventhough the plant may not die due to the infection, the yield gets considerably reduced due ot lack of photosynthesing pigments.

Leaf curl of Tobacco:

1. This is caused by leaf curl virus.

2. Insect vectors transmit the disease.

3. Leaf gets badly curled and crumpled.

4. Heavily infected leaves drop down.

PART THREE
BRYOPHYTES

CHAPTER 19

GENERAL CHARACTERS AND CLASSIFICATION

Bryophytes include nearly 960 genera and 24,000 species. They represent a unique group of plants intermediate between aquatic thallophytes and terrestrial pteridophytes. They occupy and inhabit a zone which is neither completely land nor completely water. It is for this reason Bryophytes are often regarded as the amphibians of the plant kingdom.

Bryophytes are intermediate between thallophytes and pteridophytes not only in their habitat but even in the structure and organisation of the plant body also. As such one can see the blending of algal and pteridophytic characters in Bryophytes.

Bryophytes abound on habitats which offer moisture. They grow in large numbers on wet soil, damp cliffs of rocks and even on the branches of trees as epiphytes (Porella).

Life cycle pattern

The life cycle of Bryophytes consists of a regular hetermorphic alternation of generations. The gametophyte is the predominant plant body in the life cycle whereas the sporophyte is a short lived generation totally or partially dependent on the gametophyte. In fact whenever we refer to the plant body in Bryophytes, it is the gametophyte that we have in mind.

Structure of Gametophyte

The gametophytes of Bryophytes can be broadly categorised into 2 types namely (1) Thalloid forms and (2) Leafy forms.

Thalloid forms. In this type, the plant body consists of a dorsiventrially flattened structure called the thallus which grows prostrate on the substratum. The thallus is usually dichotomously branched with or without a mid-rib on the dorsal surface. The thallus is green and nutrition is photosynthetic. The thallus is anchored to the soil with the help of root like structures called rhizoids. The rhizoids are unicellular except in Musci. In Hepaticae two types of rhizoids are found namely smooth or simple rhizoids and pegged or tuber-culate rhizoids. In pegged rhizoids the inner layer of the wall is infolded into small projections which is supposed to increase the absorptive area of rhizoids. No such infolding of the wall is seen in smooth rhizoids. Besides rhizoids, multicellular wedge shaped structures are also found in some heptaciae. These are called *scales*. Scales are protective in function.

The thalloid gametophytes anatomically show only parenchyma tissues. While there is no tissue differentiation in some members (*Anthoceros*) in Riccia, Marchantea and others, the internal tissues are distinguishable into upper photosynthetic region and the lower storage region. The cells of the former have a number of chloroplasts while those of the latter are rich in starch grains.

2. **Leafy forms**. In leafy forms the gametophyte consists of a central axis on which are arranged the "leaves". The axis is comparable to the stem. It may be branched or unbranched. It may grow prostate (*Porella*) or erect (*Funaria*). It is anchored to the soil with the help of rhizoids. The rhizoids in Musci are multicellular. The leaves usually arise spirally on the stem. The leaves are usually one cell in thickness but towards the centre may be more than one cell in thickness.

Reproduction

The gametophyte reproduces by 2 methods namely 1. Vegetative propagation and 2. Sexual reproduction

1. **Vegetative propagation** : Vegetative reproduction is an accessory means of reproduction to bring about a rapid increase in the number of gametophytes. In fact it is the vegetative propagation which is mainly responsible for the rapid spreading of the gametophyte in an area. Vegetative propagation is brought about by a variety of methods.

2. **Sexual reproduction** : This is the usual method of reproduction of the gametophyte by which the sporophyte is formed. Sexual reproduction is of oogamous type. Sex organs are the antheridium (male) and the archegonium (female). Gametophytes may be homothallic or heterothallic. Sex organs may be produced directly on the thallus or on special branches set apart for that purpose as in *Marchantia*.

Structure of the Antheridium

Antheridium is usually a globose structure (Hepaticae) or elongated and club shaped (Musci) as in some cases. It usually develops from an initial cell called the antheridial initial. A mature antheridium consists of a stalk and the terminal swollen region in which the antherozoid mother cells which give rise to antherozoids (sperms). The antherozoids of Bryophytes are spirally coiled and biflagellate. At maturity the wall breaks open and the antherozoids are liberated.

Structure of the Archegonium

Archegonium is a flask shaped structure. It also develops from an initial cell called the archegonial initial. It may be stalked or sessile. The basal bulbous region is called the neck. The neck region has a single layered jacket, while the venter is surrounded by jacket more than one cell in thickness. In the venter region is found the egg above which is found a small cell the venter canal cell. In the neck region there is a row of central cells called neck canal cells varying in number from four (Hepaticae) to sixteen (Mosses). The tip of the neck is covered by 4 neck cells or cover cells.

Fertilization

Fertilization is dependent on external moisture. At the time of fertilization the contents of the archegonium except the egg, dissolve forming a mucilagenous mass. Due to the pressure exerted by this, the neck cells open apart and the mucilage oozes out.

The antherozoids liberated from the antheridium are said to be attracted chemotactically by the mucilage towards the archegonium (chemical analysis of the mucilage shows the presence of malic acid besides other constituents). Many antherozoids enter the archegonium. Ultimately one of the antherozoids succeeds in fusing with the egg forming the zygote.

Sporophyte

Sporophytic generation in Bryophytes is not predominant in that it is dependent on the gametophyte either completely or partially. It is short lived and perishes soon after the liberation of the spores. The zygote first develops into embryo which matures later into the sporophyte.

The mature sporophyte may be an undifferentiated sac containing spores as in *Riccia* or may show differentiation to a lesser or higher degree. Usually the sporophyte consists of three parts namely foot, seta and capsule. Foot is the basal portion and is embedded in the gametophyte drawing nutrition from it. Seta is an elongated stalk like structure serving to draw the capsule away from the gametophyte. In the capsule region which is swollen, haploid spores are fomred from spore mother cells. All the spore mother cells may or may not develop into spores. Many of them become sterile bodies known as elaters that serve to dessimate the spores (*Marchantia, Porella*).

The spores on dispersal develop into the gametophyte either directly (Hepaticae) or through the protonemal stage (Mosses).

Apogamy and Apospory

Both these phenomena are of rare occurrence in Bryophytes. Apogamy is the production of a sporophyte without involving fusion of gametes. Any part of the gametophyte develops into a sporophyte. Such sporophytes are haploid and will not survive for long.

Apospory is the production of a gametophyte from a sporophyte without involving spore production ie., circumventing reduction division. Aposporous gametophytes will be diploid. Among Bryophytes, apospory has been recorded in Musci and Anthocerotae (Bornhagen 1926, Rink 1935).

Classification

The name Bryophyta was first coined by Braun in 1864. But he included algae, fungi, lichens and mosses in his group Bryophyta. It was Schimper (1879) who first recognised Bryophyta as a division in the sense as it is used today.

Eichler (1833) classified Bryophytes into two groups namely Hepaticae and Musci. Till recently this classification was followed by majority of botanists. Engler (1892) followed Eichler in the classification divides Marchantia Jungermanniales and Anthcerotales. Many eimnent botanists like Bessey (1911), Fritsch (1929), Wetlsteir (1935), Verdoon and Evan (1989) follow the system of Engler.

Underwood (1894) and Baryex Janezeroslei (1872) however recognised the special features of Anthocerotales and opined that it cannot be included under Hepaticae.

Consequently Gayet (1897) and other biologists emphasised the need to separate Anthocerotales from Hepaticae. Majority of botanists including Campbell (1940), Smith (1955) and Takhtajan (1953) opine that Bryophyta should be divided into 3 classes namely Hepaticae, Anthocerotae and Musci.

Rothmaler (1951) changed the endings of the class names as follows :

Hepaticae to Hepaticopsida.

Anthrocerotae to Anthoceropsida.

Musci to Bryopsida.

This change is in accordance with the International code of Botanical nomenclature. Proskauer (1957) changed the name Anthroceropsida to Anthocerotopsida.

CHAPTER 20

HEPATICOPSIDA

The members of Hepaticopsida are commonly called liver worts (*Hepatica* = liver) because of the lobed nature of the thallus which resembles the liver in outline. This class includes approximately 225 genera and 8500 species (Smith 1955).

Hepaticopsida of temperate climate are mostly restricted to shady terrestrial habitats. Such habitats include swamps and bogs. In tropical climate also their distribution is limited to cool, moist shady areas like the floor of forests.

There is great diversity in the external make up of gametophyte. But essentially the plant body is a dorsiventrally flattened thallus except in some members like *Porella*. The thalloid gametophyte is dichotomously branched with or without a prominent midrib. In leafy gametophytes the leaves are arranged in two or three rows on the axis.

Anatomically, the thalloid gametophyte may or may not show internal differentiation into storage and photosynthetic regions.

Sex organs are formed on the dorsal surface of the thallus developed from a superficial cell. The sporogonium may or may not show differentiation. When differentiated the sporophyte consists of foot, seta and capsule. All the spore mother cells are not fertile. Some of them become sterile and form either nurse cells or elaters, the latter helping in spore dispersal.

The class is divided into 4 orders namely Marchantiales, Spherocarpales, Calobryales and Jungeramanniales.

In this book, *Ricca, Plagiochasma, Marchantia, Porella* and *Pellia* of Hepaticopsida are discussed.

RICCIA

Common species : *R.himalyensis, R.curciata, R.pearsoni, R.fluitans* etc.

Occurrence : The species of *Riccia* occur on damp soil, moist rocks etc. They grow very well in cool and shady places. *R.fluitans* is an aquatic member.

Study : 1. External features of the gametophyte

2. Internal structure of the thallus

3. Methods of vegetative propagation

4. **Structure of the sex organs**
5. **Structure of the sporogonium**

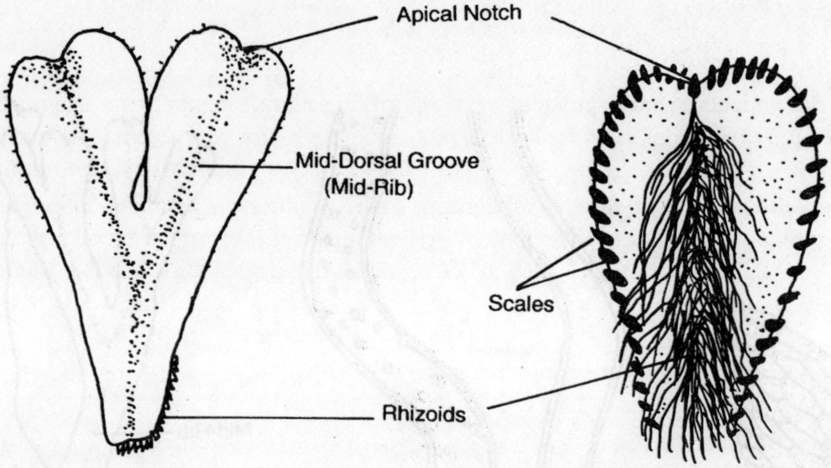

Fig. 20.1 Riccia. Dorsal surface of thallus

Ventral surface of thallus

Practical work to be done

1. Cut the vertical transverse section of thallus
2. Mount a few rhizoids and scales
3. Observe the thallus with a hand lens.

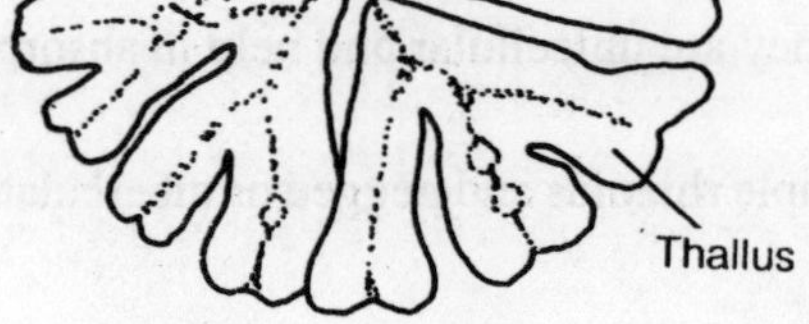

Fig. 20.2 Riccia. Rosette

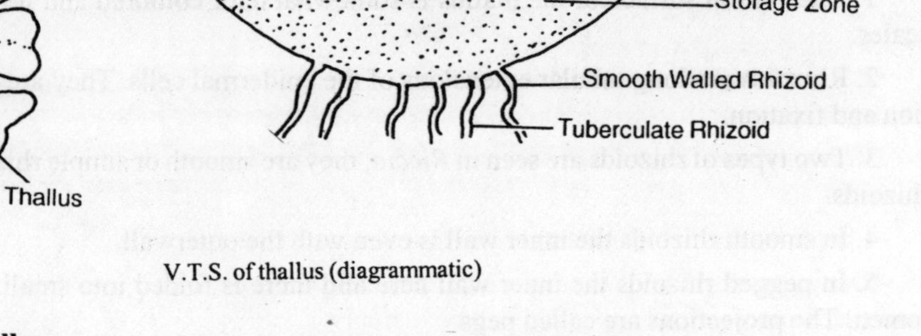

V.T.S. of thallus (diagrammatic)

External features of the thallus

1. The plant body of *Riccia* is a dorsiventrally flattened, deep green coloured thallus.

2. The thalli grow in groups called rosettes. In this, many thalli radiate from a central point to give a circular outline.

3. The thallus is small, thick towards the centre, gradually thinning out towards the margin.
4. The thallus shows typical dichotomous branching.
5. There is an apical notch in which is situated the growing point (apical cell).
6. The thallus can be clearly distinguished into dorsal and ventral surfaces.

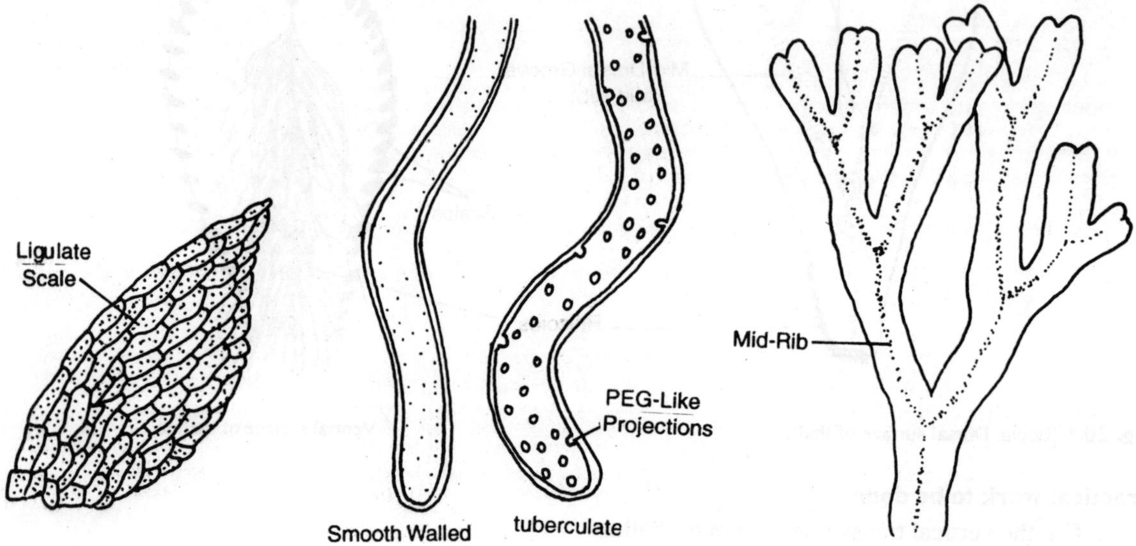

Fig. 20.3 Riccia. Single scale — Rhizoids — Riccia flutians. Thallus

Dorsal surface :

1. This is the upper part facing the sunlight
2. A mid rib is present and it is represented by a median furrow.
3. Here and there small dots are present, embedded in the thallus. They are the sex organs. Sometimes the dots are dark coloured in which case they represent the sporogonium.

Ventral surface

1. The ventral surface of the thallus is somewhat dark coloured and it bears a number of rhizoids and scales.
2. Rhizoids are long tubular extensions of the epidermal cells. They are unicellular and help in absorption and fixation.
3. Two types of rhizoids are seen in *Riccia*, they are smooth or simple rhizoids and pegged or tuberculate rhizoids.
4. In smooth rhizoids the inner wall is even with the outerwall.
5. In pegged rhizoids the inner wall here and there is folded into small projections extending into the lumen. The projections are called pegs.
6. Scales are multinuclear, but one cell thick, wedge shaped structures. They are violet in colour.
7. They are present in two rows along the margin of the thallus, specially near the growing apices.
8. The scales protect the young growing points.

Internal structure of the thallus

A transverse section of the thallus shows the following features.

1. Two distinct regions are noticed internally. These are the photosynthetic region and the storage region.

2. Photosynthetic region is present above ie., towards the dorsal surface.

3. It is made up of many vertical rows of cells called photosynthetic filaments.

4. The photosynthetic filaments are unbranched being made up of cells joined end to end. The cells are chlorophyllous, constituting the chief photosynthetic tissue of the plant.

5. The photosynthetic filaments are arranged parallel with air spaces between them.

6. There is no layer of cells that can be called upper epidermis.

7. The uppermost cell of each photosynthetic filament functions as an epidermal cell. Hence one can say the upper epidermis is discontinuous with air spaces between cells.

8. The storage region consists of irregular closely packed parenchyma cells.

9. The cells generally lack chloroplasts as their main function is storage.

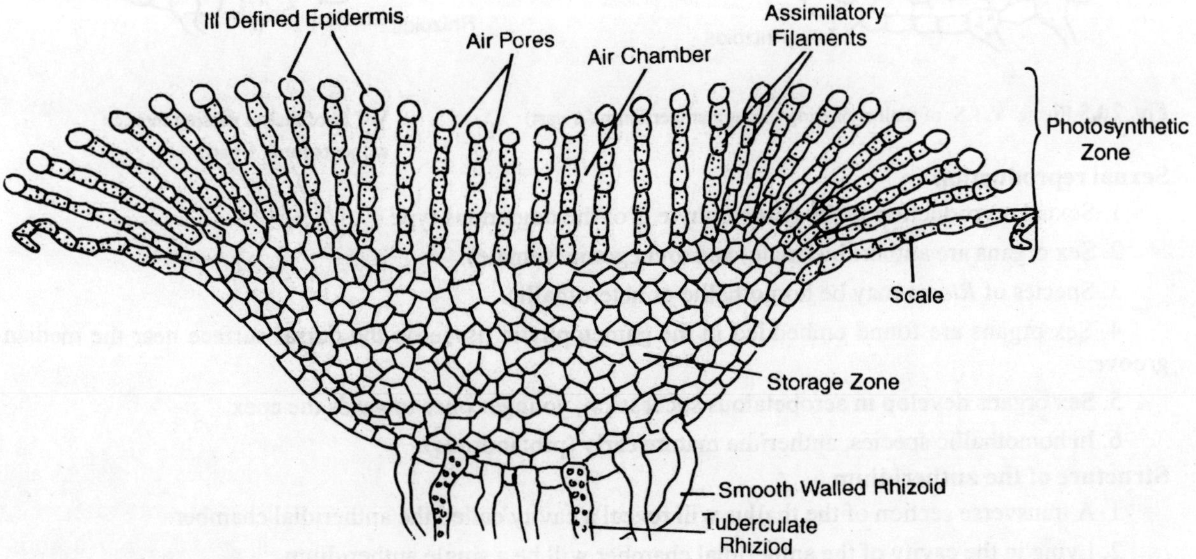

Fig. 20.4 Riccia. V.T.S. of thallus

Vegetative propagation

1. **By progressive death and decay**. Older parts of the thalli die, leading to the separation of dichotomies which develop into new individuals.

2. **Persistent apices.** During unfavorable conditions all the parts of the thallus, except the apex die. The apex survives and develops into a new individual at the return of the favourable condition.

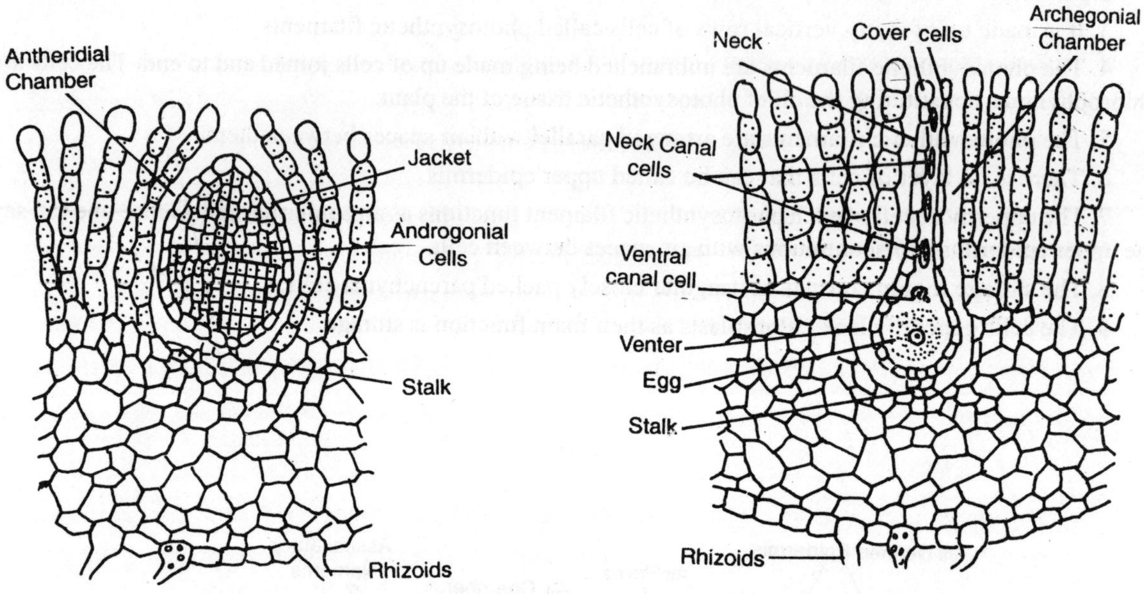

Fig. 20.5 Riccia. V.T.S. of thallus passing through antheridium (A part)

V.T.S. of thallus passing through archegonium (A part)

Sexual reproduction

1. Sexual reproduction in the gametophyte is of the oogamous type.
2. Sex organs are antheridia (male) and archegonia (female)
3. Species of *Riccia* may be homothallic or heterothallic.
4. Sex organs are found embedded in the gametophytic tissue on the dorsal surface near the median groove.
5. Sex organs develop in acropetalous succession, younger ones towards the apex.
6. In homothallic species, antheridia mature early (protandrous).

Structure of the antheridium

1. A transverse section of the thallus will reveal a cavity called the antheridial chamber.
2. Lying in the cavity of the antheridial chamber will be a single antheridium.
3. Each antheridium is a sub globose or pear shaped structure having a multicellular stalk.
4. The upper sub-globose part of the antheridium is called the capsule.
5. The capsule has a single layered wall called the jacket layer.
6. Inside the wall is a mass of cubical cells called androcyte mother cells.
7. Each androcyte mother cell divides diagonally to form two triangular androcytes.
8. Each androcyte metamorphoses itself into a biflagellate coiled antherozoid or sperm.
9. At maturity the antheridium absorbs water and breaks open to release the antherozoids which come out in a mass.

Structure of the archegonium

1. A transverse section of the thallus reveals an archegonial cavity.
2. Each archegonial cavity encloses a single archegonium.
3. The archegonium is a flask shaped structure having a bulbous base and an elongated neck.
4. The wall of the archegonium is one celled thick. It has six rows of cells constituting the wall.
5. At the venter region inside the archegonium is an egg.
6. Above the egg cell is a single venter canal cell.
7. Inside the neck are found four to six neck canal cells.
8. The tip of the neck is covered by four cover cells or lid cells (in a long section however, only 2 cells are seen).
9. At the maturity of the archegonium, all the cells inside, except the egg dissolve, which comes out in the form of a mucilaginous mass.
10. The lid cells open apart and the antherozoids enter inside.
11. Fertilization of the egg results in the formation of a zygote which eventually develops into the sporophyte (sporogonium).

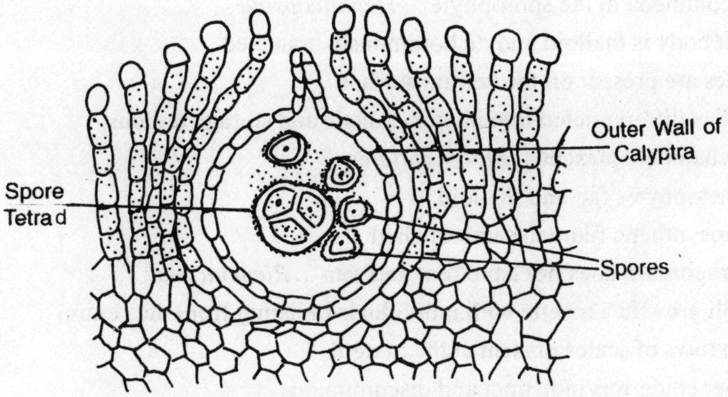

Fig. 20.6 Riccia. V.T.S. of thallus passing through sporogonium (A part)

Structure of the sporogonium

1. The sporogonium is a globose structure and it is completely embedded inside the thallus.
2. It has a small peg like projection at the tip representing the degenerated neck of the archegonium
3. There is no differentiation of the sporophyte into foot, seta and capsule.
4. The sporogonium has a single layered wall called the calyptra.
5. Within the sporogonium are found a number of spore mother cells which undergo reduction division to form tetrads of spores. Tetrads are tetrahedral.
6. Some of the spore mother cells degenerate and form nurse cells, which provide nutrition to the surviving spore mother cells.
7. There are no elaters.

8. All the spores are of the same kind homosporous.

9. The spore will be pyramidal in shape, after separating from the tetrad, later it becomes round.

10. The spore is surrounded by a thick sculptured wall.

11. The wall has 3 layers viz., outer thick cultinized exospore, middle thin mesospore and inner thin endospore.

12. Spores are uninucleate and haploid. They are liberated at the decay of the sporogonium and germinate to develop into new individual gametophytes.

Systematic position and identification

A. 1. Plant body is a parenchymatous gametophyte.
2. It is leafy or thalloid
3. Vasculture and roots absent
4. Sex organs surrounded by a sterile jacket layer
5. Life cycle exhibits heteromorphic alternation of generations with a predominant gametophyte ... *Bryophyta*

B. 1. Thallis are lobed and dorsiventrally flattened or leafy
2. Chloroplasts do not have pyrenoids
3. Scales and two types of rhizoids present
4. No columella in the sporophyte ... *Hepaticopsida*

C. 1. Plant body is thalloid and dichotomously branched
2. Scales are present on the ventral surface
3. Thallus differentiated into photosynthetic and storage regions.
4. Air chambers present ... *Marchantiales*

D. 1. Gametophytes flat and thalloid
2. Photosynthetic filaments unbranched.
3. Sporogonium does not have foot and seta ... *Ricuciaceae*

E. 1. Thalli grow in a rosette with individuals radiating from the centre.
2. Two rows of scales present at the margin
3. Upper epidermis indistinct and discontinuous
4. Sporogonium has only spore mother cells and nurse cells.
5. Elaters are absent
6. Spores have a sculptured, three layered wall ... *Riccia*

PLAGIOCHASMA

Common species : *P.articulatum, P.simlensis, P.cordatum, P.intermedium, P.appendiculatum* etc.

Occurrence : The gametophytic plant occurs in moist, cool and shady places. Majority of the species are frequently seen in mountainous regions. They grow on moist soil, rocks, trunk of trees etc.

Study : 1. External morphology
2. Internal structure
3. Structure of sex organs
4. Structure of sporophyte

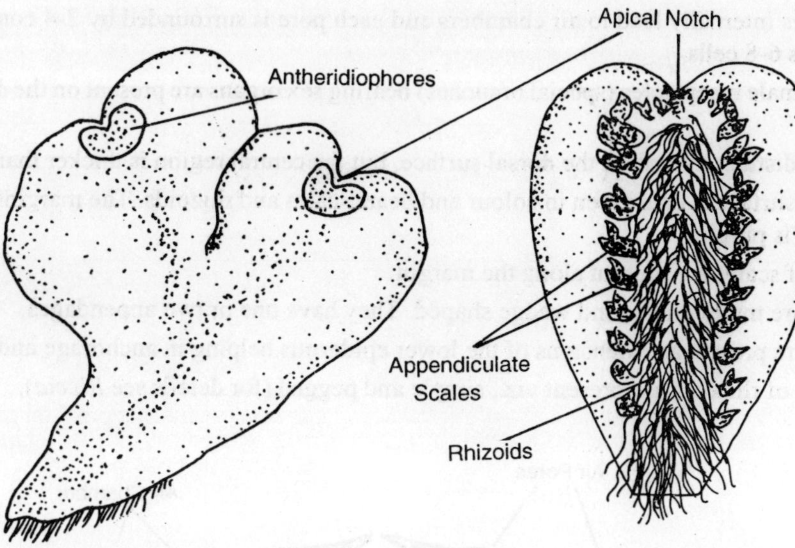

Fig. 20.7 Plagiochasma. Dorsal surface of thallus

Ventral surface of thallus

Practical work to be done

1. Cut a cross section of thallus
2. Section the thallus to show the sex organs
3. Section the thallus to study the sporophyte\

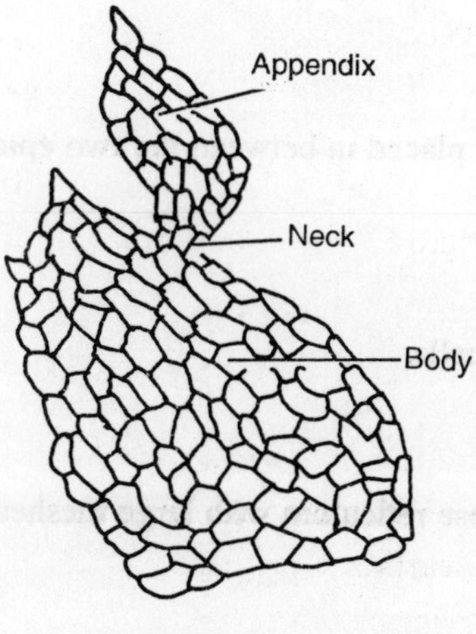

Fig. 20.8 Plagiochasma. An appendiculate scale

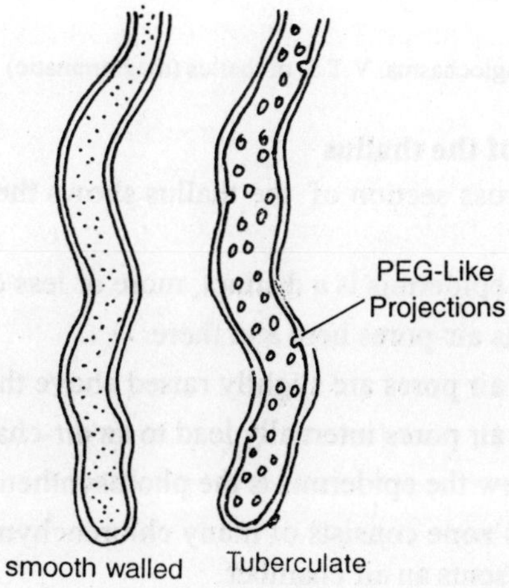

Rhizoids

External morphology

1. The gametophytic plant body is a dorsiventrally flattened thallus
2. The thallus is dichotomously branched and deep green in colour

3. The dorsal surface of the thallus is greener than the ventral surface and has a number of air pores

4. The air pores internally lead to air chambers and each pore is surrounded by 2-4 concentric rings of cells. Each ring has 6-8 cells.

5. Male and female receptacles (special branches) bearing sex organs are present on the dorsal surface of the thallus.

6. There is no distinct mid ribon the dorsal surface, but the central region is thicker than the margins.

7. The ventral surface is pale green in colour and bears scales and rhizoids. The margins of the thalli on the ventral surface is purple.

8. Two rows of scales are present along the margin.

9. The scales are multicellular and wedge shaped. They have one or two appendages.

10. Rhizoids are present as extensions of the lower epidermis helping in anchorage and absorption.

11. Two types of rhizoids are present viz., simple and pegged (for details see *Riccia*).

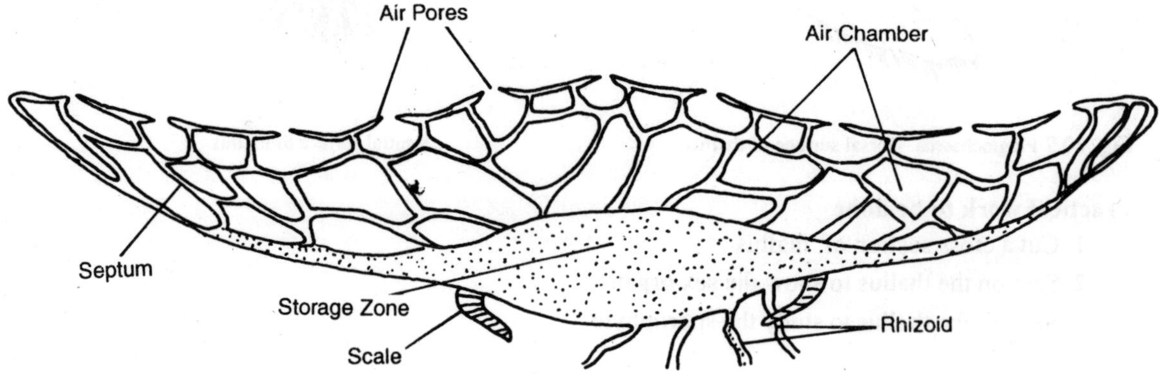

Fig. 20.9 Plagiochasma. V.T.S. of thallus (diagrammatic)

Anatomy of the thallus

1. A cross section of the thallus shows the two distinct regions placed in between the two epidermal layers.

2. The epidermis is a distinct, more or less continuos layer.

3. It has air-pores here and there.

4. The air pores are slightly raised above the level of epidermal cells.

5. The air pores internally lead to an air-chamber.

6. Below the epidermis is the photosynthetic zone.

7. This zone consists of many chlorenchyma cells forming a loose reticulum with large meshes. Each mesh represents an air chamber.

8. Below the photosynthetic region is found the storage region.

9. The storage region consists of closely packed parenchyma cells.

10. The lower epidermis is a single continuos layer of cells. It gives rise to rhizoids and scales.

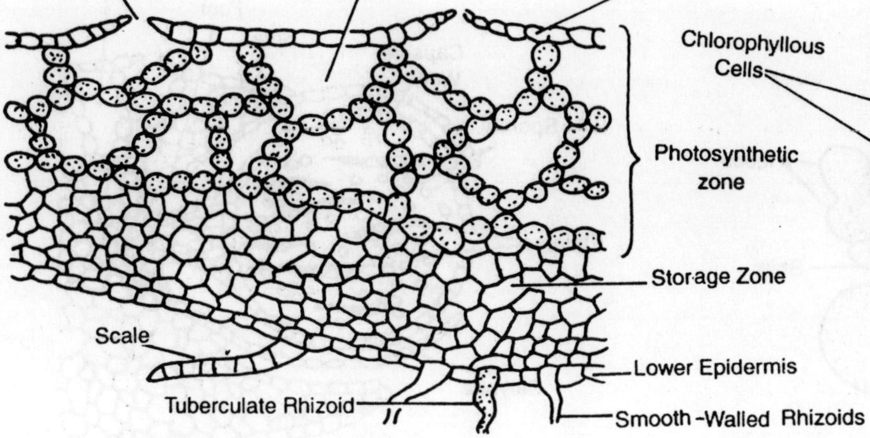

Fig. 20.10 Plagiochasma. V.T. S of thallus (A part)

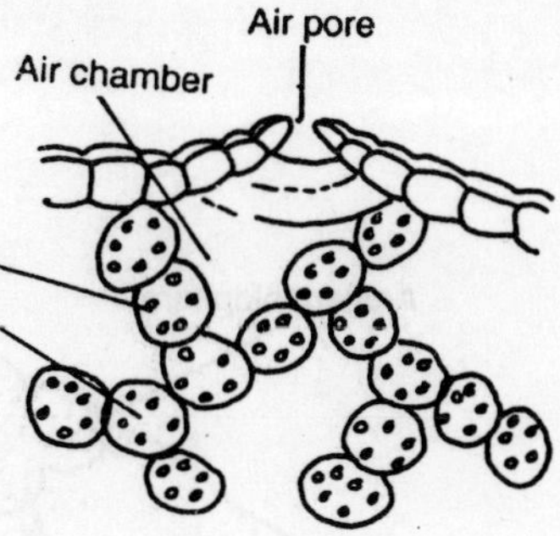

A portion of V.S. of thallus showing air pore

Vegetative propagation

1. Special adventitious branches of the thalli develop at the apex. These separate from the parent plant and envelop into a new individual.

2. By the progressive death and decay of older parts, the dichotomies get separated and develop into new individuals.

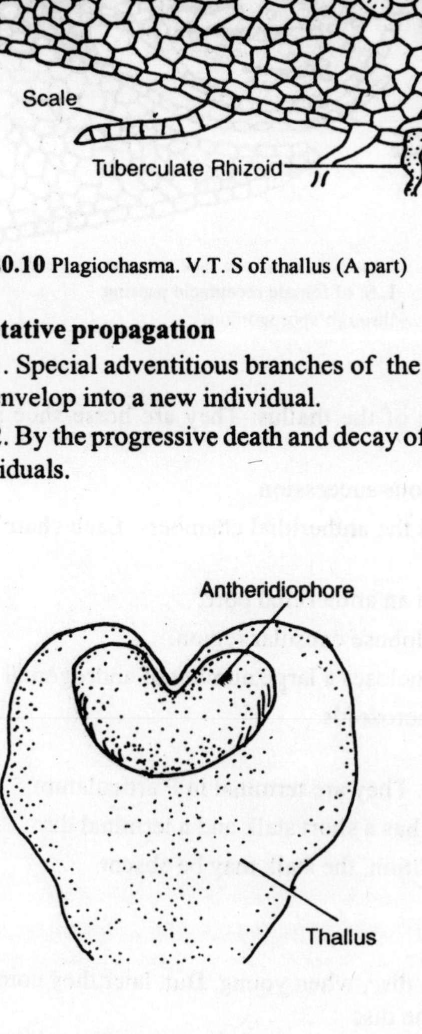

Fig. 20.11 Plagiochasma. Thallus showing male receptacle

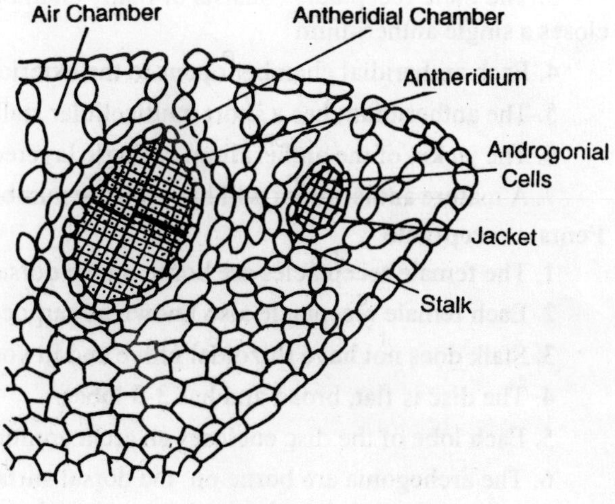

L.S. of male receptacle

Sexual reproduction

1. It is oogamous, sex organs are antheridia and archegonia. Majority of the species are homothallic.

2. The sex organs are found in special structures called receptacles. Consequently there are male and female receptacles.

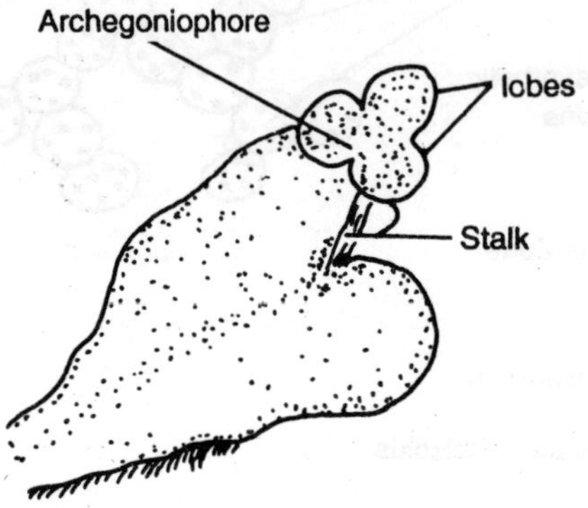

Fig. 20.12 Plagiochasma. Thallus showing female receptacle

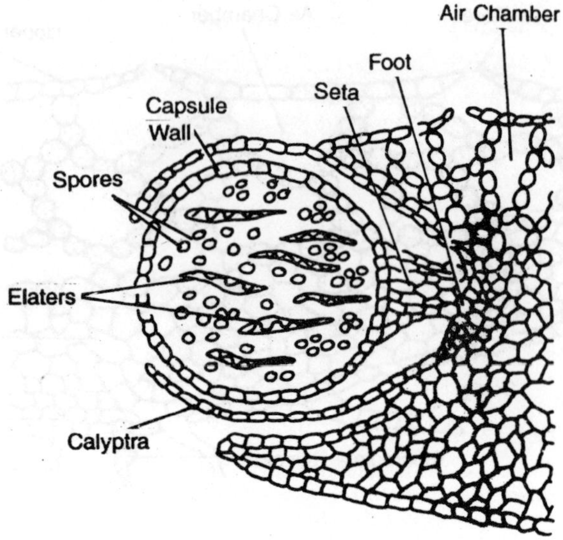

L.S. of female receptacle passing through sporogonium

Male receptacle

1. The male receptacles are borne on the dorsal surface of the thallus. They are horse shoe shaped structures and are sessile.

2. Antheridia develop on the male receptacle in acropetalous succession.

3. The male receptacle consists of many chambers called the antheridial chambers. Each chamber encloses a single antheridium

4. Each antheridial chamber opens to the exterior through an antheridial pore.

5. The antheridium has a short multicellular stalk and a globose capsular region.

6. The jacket of the antheridium is single layered and it encloses a large number of androgonial cells.

7. A mature antheridium will have a large number of antherozoids.

Female receptacle

1. The female receptacles are borne on the dorsal surface. They are terminal in P.articulatum.

2. Each female receptacle also known as carpocephalum has a short stalk and a terminal disc.

3. Stalk does not have rhizoidal grove and in young condition, the stalk may be absent.

4. The disc is flat, broad and has 3-9 lobes.

5. Each lobe of the disc encloses an archegonium

6. The archegonia are borne on the dorsal surface of the disc, when young. But, later they come to lie towards the ventral surface due to the centripetal growth of the disc.

7. In a mature disc, the archegonia will be having their neck pointing down.

8. The overgrowth of the disc surrounding the archegonium will form an involucure.

9. The archegonium is a flask shaped structure having a bulbous basal venter and a narrow neck.

10. The archegonium is similar to that of *Riccia* (see *Riccia* for details). The number of neck canal cells ranges from 6-9.

11. Fertilization is same as in *Riccia*.

Structure of the sporophyte

1. The sporophyte is seen hanging down from the ventral surface of the disc of the female receptacle.

2. The sporophyte is differentiated into three parts viz foot, seta and capsule.

3. Foot is somewhat bulbous and is made up of polygonal parenchyma cells. It helps in anchoring the sporophyte to the gametophytic tissue.

4. The seta is a stalk like structure and it serves to carry the capsule slightly away from the gametophyte.

5. The capsule is the terminal part of the sporophyte. It is somewhat globose in shape.

6. The wall of the capsule is made up of a single layer of cells except at the apeex, where it is double layered to form an operculum or lid.

7. The lid is thrown off at maturiry to facilitate the liberation of spores.

8. Within the sporophyte are found the diploid spore mother cells.

9. Some of the spore mother cells become sterile, elongate and form the elaters.

10. Elaters are spindle shaped structures and have one or two spiral band like thickenings which are hygroscopic. Elaters help in spore dispersal.

11. The remaining spore mother cells undergo reduction division and form tetrahedral tetrads of haploid spores.

12. Spores are sub globose, uninucleate and have a two layered wall. The outer wall exine is thick and ornamented, while the inner wall intine is thin.

13. The spores on liberation and on the availiability of a suitable substratum develop into new individuals of gametophyte.

Diagrams to be drawn

1. Habit sketch
    a. Dorsal surface
    b. Ventral surface
2. Rhizoids and scales (enlarge)
3. Transverse section of the thallus
4. Male and female receptacle
5. Sections through the receptacle to show antheridia and archegonia
6. Section of the female receptacle to show sporophyte
7. A few elates and sproes (enalrged)

Systematic position and identification

A. 1. Plant body is a parenchymatous gametophyte.
2. It is leafy or thalloid
3. Vasculature and roots absent
4. Sex organs surrounded by a jacket layer
5. Life cycle exhibits a heteromorphic alternation of generations with a predominant gametophyte ... *Bryophytes*

B. 1. Thalli are lobed and dorsiventrally flattened or leafy
2. Chloroplasts do not have pyrenoids
3. Scales and 2 types of rhizoids present ... *Hepaticopsida*

C. 1. Plant body is thalloid and dichotomously branched
2. Scales are present on the ventral sruface

3. Thallus differnaitaed into photosynthetic and storage regions.
4. Air chambers present ... *Marchantiactles*

D. 1. Gametophytes flat and thalloid
2. Photosynthetic filaments absent
3. Scales have appendages
4. Air pores simple
5. Capsule has terminal operculum... *Rebauliaceae*

E. 1. Male receptacle horse shoe shaped
2. Capsules have a two lipped envelope of the involucre
3. Ventral margins of the thallus purple in colour ... *Plagiochasma.*

MARCHANTIA

Common species : *M.polymorpha, M.palmata, M.nepalensis* etc

Occurrence : The thallus of *Marchanita* grows luxuriantly in cool and moist climates. The species of *Marchantia* are abundant in hilly regions. In South India, the Nilgiri hill ranges show luxuriant growth of *Marchantia.*

Study : 1. External features of the thallus
2. Internal structure
3. Vegetative propagation
4. Sexual reproduction
5. Structure of the sporophyte

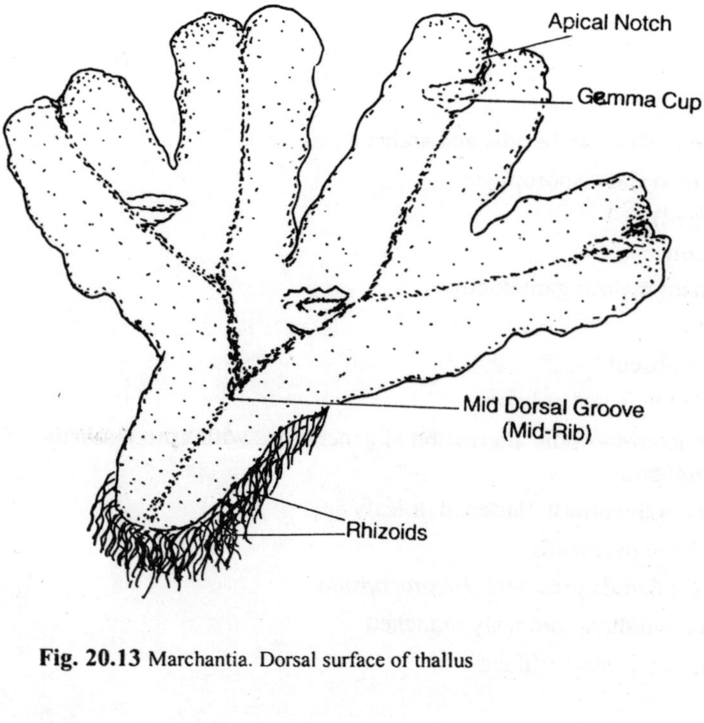

Fig. 20.13 Marchantia. Dorsal surface of thallus

I. External features of the thallus :

1. The plant body of *Marachantia* is a dorsiventrally flattened dichotomously branched thallus.

2. The thallus is grass green in colour with an undulted margin.

3. There is a prominent median midrib on the dorsal surface represented by a groove. On the ventral surface, the mid-rib is represented by a ridge.

4. If the thallus is observed with a hand lens, many pologynoal areas may be seen on the dorsal surface. These are the areoles and demarcate a photosynthetic chamber within.

5. In the centre of each areole is a dot like structure. This represents the air pore or the chimnery pore.
6. The growing point of the thallus is situated in the apical notch.

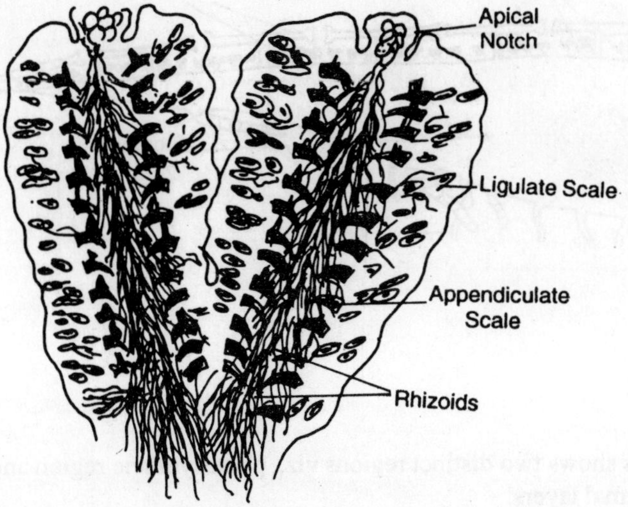

Fig. 20.14 Marchantia. Ventral surface of thallus

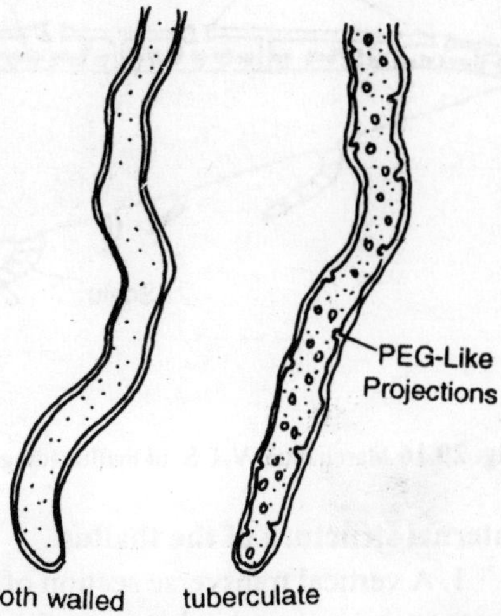

Rhizoids

Reproductive structures present on the dorsal surface:

1. Along the midrib here and there are found small cup shaped structures. These are called the gemmae cups. Inside the gemma cup are found a large number of gemmae.
2. The thalli of *Marchantia* also have special branches to bear the sex organs.
3. The thalli are dioecious. The male thalli bears the antheridiophore and the female thallus bears the archegoniophore.

Ventral surface of the thallus

1. The ventral surface of thallus is pale green in colour.
2. The midrib on the ventral surface is represented by a ridge.
3. Along the mid-rib region on the ventral surface are found scales and rhizoids.
4. Rhizoids are unicellular and develop from the lower epidermal cells. They are of 2 types viz., simple and pegged rhizoids.
5. Scales are multicelluar wedge shaped structures also present on the ventral surface.
6. Four rows of violet coloured scales are present along the ventral margin of the thallus, specially near the growing points.

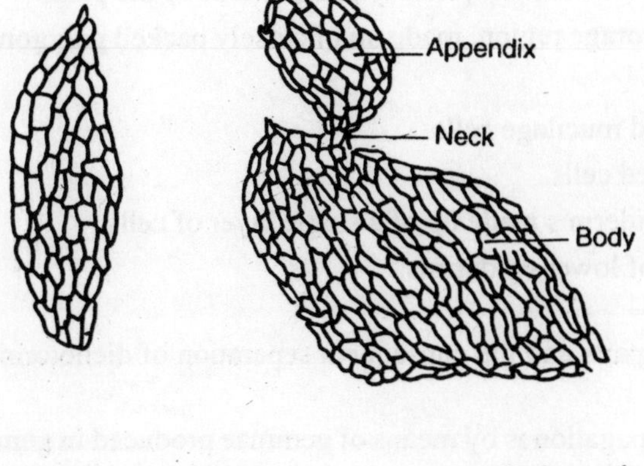

Fig. 20.15 Marchantia. Ligulate scale Appendiculate scale

7. Scales are of two types - *Appendiculate* and *ligulate*. Appendiculate scales have an appendage arising out of a wedge shaped structure, while legulate scales are simple wedge shaped structures.

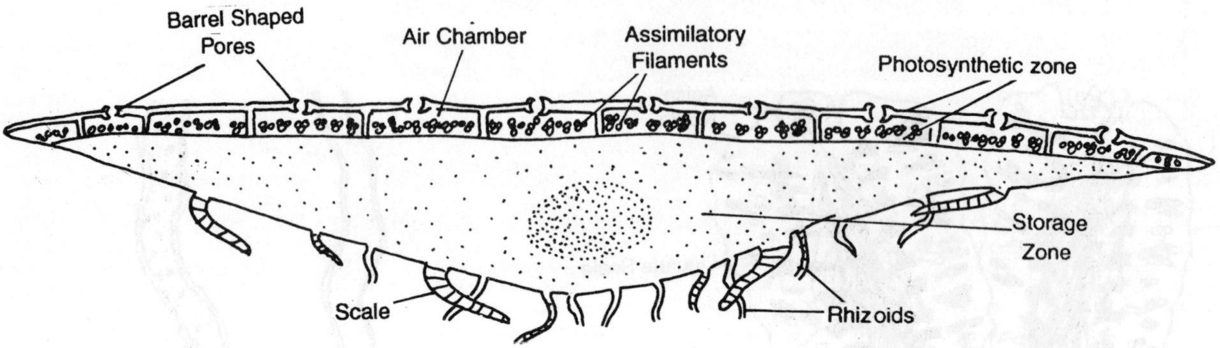

Fig. 20.16 Marchantia. V.T.S. of thallus (diagrammatic)

Internal structure of the thallus

1. A vertical transverse section of the thallus shows two distinct regions viz., photosyntheic region and storage region placed in between the two epidermal layers.

2. The upper epidermis is made up of a single layer of barrel shaped cells.

3. The continuity of the epidermal layer is broken by the presence of air pores. These pores have a rim or border made up of a few concentric rings of cells. As a result, the pores are slightly elevated, looking like a chimney, hence they are also called chimney pores.

4. There will be one chimney pore at the centre of each aereole.

5. Internal to the upper epidermis is the photosynthetic region divided into a number of chambers by means of parenchymatous partitions.

6. It is these chambers that constitute the aereoles superficially.

7. Arising from the floor of each photosynthetic chamber are a number of branched filaments. The cells of these filaments are chlorophyllous and constitute the cheif photosynthetic tissue of the plant.

8. Next to the photosynthetic region is the storage region, made up of closely packed polygonal cells.

9. The cells are rich in starch grains.

10. The storage tissue also consists of oil and mucilage cells.

11. The midrib region consists of thick walled cells.

12. Below the storage region is the lower epidermis made up of a single layer of cells.

13. Scales and rhizoids arise from the cells of lower epidermis.

Vegetative propagation

1. Progressive death and decay of the older parts and the consequent separation of dichotomies brings about vegetative propagation.

2. The commonest method of vegetative propagation is by means of gemmae produced in gemma cups.

3. A transverse section of the thallus passing through a gemma cup reveals the following details.

4. The gemma cup has a fringed margin and is made up of parenchyma cells.

5. Rising from the floor of the gemma cup are found a number of gemmae at different stages of growth.

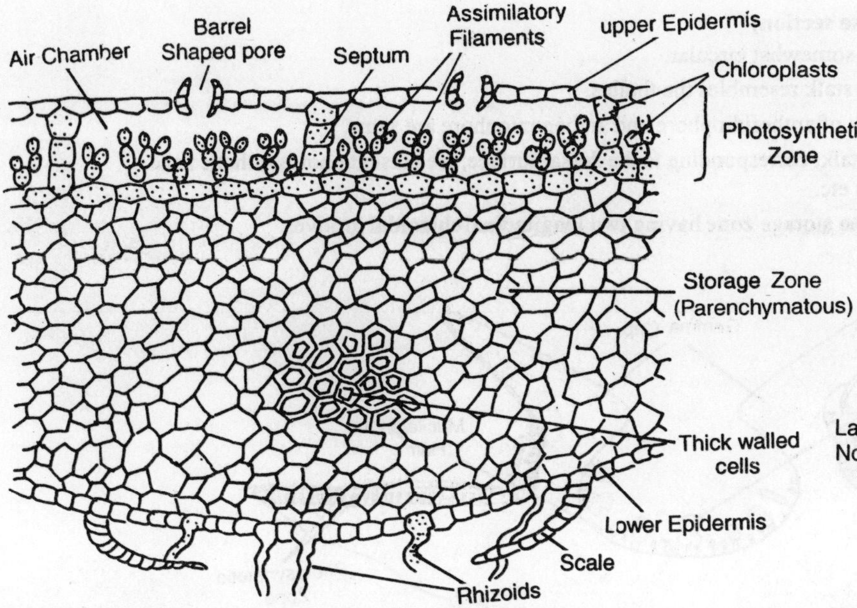

Fig. 20.17 Marchantia. V.T.S. of thallus (A part)

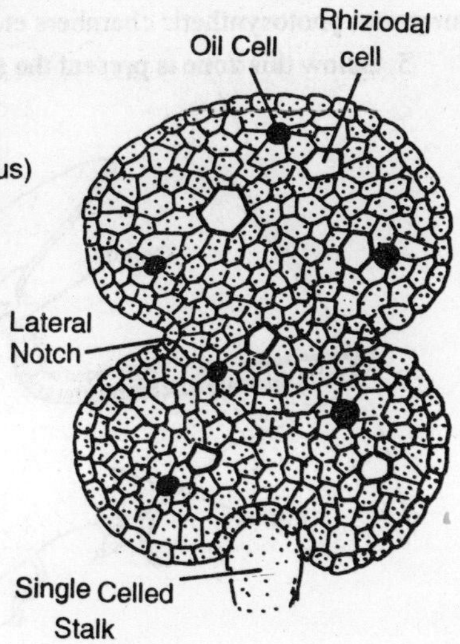

Single gemma

6. Each gemma has a single celled short stalk.

7. The gemma is a dumb bell shaped structure and is made up of parenchyma cells.

8. The cells of gemma have chloroplasts.

9. The gemma is many celled thick at the centre, but one celled thick towards the margin.

10. There are two notches in the gemma situated in the lateral margins. It is at these notches that the growing points are situated.

11. Oil cells and rhizoidal initials are also found in each gemma.

12. A mature gemma separates from the gemma cup and develops into two new individuals. Each growing point of the gemma gives rise to a new thallus.

Sexual reproduction

1. Sexual reproduction is of the oogamous type- sex organs are antheridia (male) and archegonia (female).

2. The sex organs are formed on special branches of the thallus called "Gametophores".

3. The male gametophore is called Antheridiophore and the female gametophore is called Archegoniophore.

4. The antheridiophore and archegoniophores arise at the apical notches on different thalli. Species of *Marchantia* are strictly dioecious.

Structure of an Antheridiophore

1. Each antheridiophore has a cylindrical stalk and a terminal peltate disc.

2. The stalk has two longitudinal grooves called rhizoidal grooves. All along the groove the epidermal cells produce rhizoids.

Anatomy of the stalk - (transverse section)

1. The outline of the section is somewhat circular.
2. In essentials of anatomy the stalk resembles the thallus.
3. Internal structure of the stalk of antheridiophore and archegoniophore are same.
4. On the posterior side of the stalk, corresponding to the dorsal surface, are present photosynthetic zone, air pores, photosynthetic chambers etc.
5. Below this zone is present the storage zone having two longitudinal rhizoidal grooves.

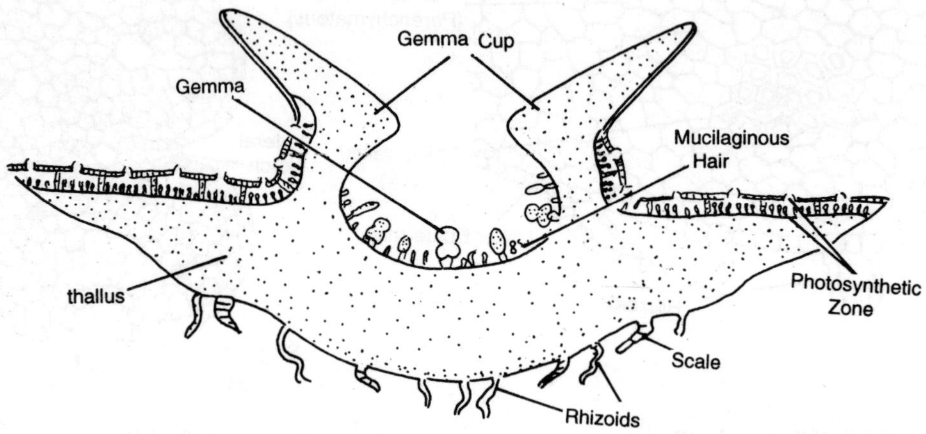

Fig. 20.18 Marchantia. V.T.S. of thallus passing through gemma cup (diagrammatic)

Structure and anatomy of the disc

1. The disc is also called the receptacle. It is irregularly circular in outline and has eight lobes.
2. The dorsal surface of the disc is slightly convex.
3. The antheridia are found embedded in the dorsal surface of the disc.
4. A vertical median longitudinal section of the disc will show the details of anatomy.
5. Anatomically the disc resembles the thallus in having an upper photosynthetic zone and a lower storage zone.
6. There is an upper epidermis, whose continuity is interrupted by chimney pores.
7. The photosynthetic zone consists of photosynthetic chambers.
8. Unlike in the thallus, the photosynthetic chambers alternate with cavities called antheridial chambers.
9. Each antherdial chamber is a flask shaped cavity opening on the dorsal surface of the disc through an opening called the ostiole.
10. Chimney pores and ostioles alternate on the disc.
11. Each antheridial cavity encloses a single antheridium.

Structure of the antheridium

1. The antheridium is a sub globose or oval structure arising from the floor of the antherideal chamber.
2. The antheridium has a short multicellular stalk.
3. The wall of the antheridium is made up of a single layered jacket enclosing the antherozoid mother cells.
4. The antherozoid mother cells eventually form the coiled, biflagellate antherozoids.

5. The antheridia develop in acropetalous succession ie., oldest antheridium at the centre and the youngest near the margin.

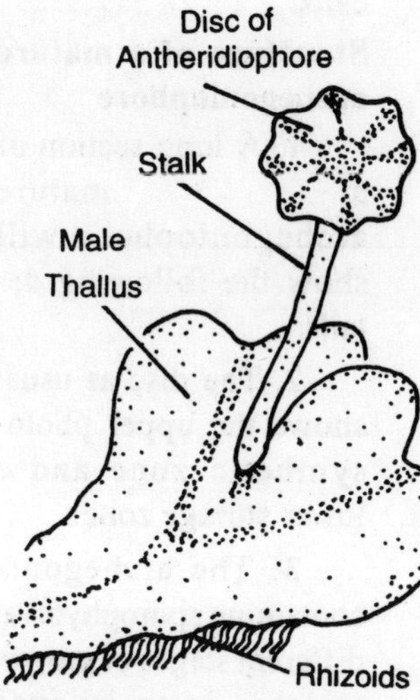

Fig. 20.19 Marchantia. Thallus bearing antheridiophore

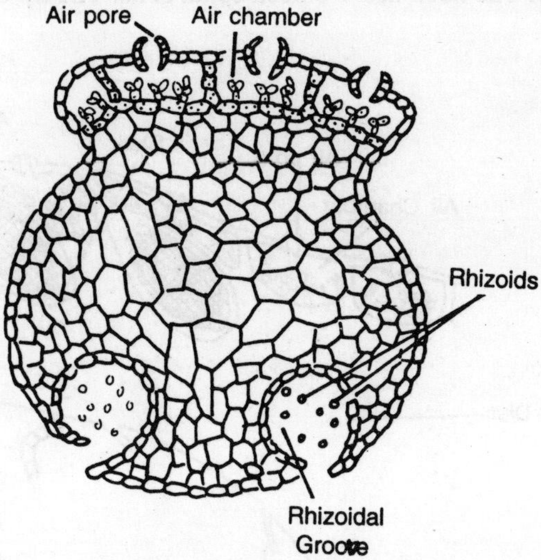

T.S. of stalk

Archegoniophore or the female receptacle

1. The archegoniophore also arises on the dorsal surface of the thallus near the growing point.

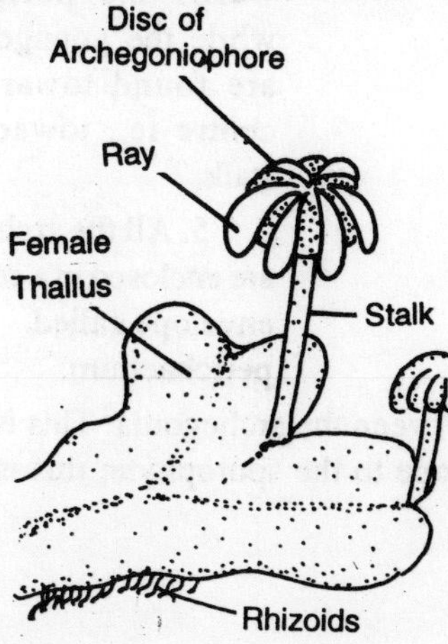

Fig. 20.20 Marchantia. Thallus bearing archegoniophore

2. It has a stalk and a peltate disc.

3. The stalk is similar to that of the antheridiophore in structure and anatomy.

4. The disc is smaller than that of antherideophore and it has got 6-8 finger like projections called rays.

5. The dorsal surface of the disc is convex.

6. Archegonia are found on the dorsal surface of the disc in young conditions, while after fertilization they become inverted and pushed downwards due to the centrifugal growth of the disc. Ulimately the archegonia will be hanging down with their neck pointing downwards.

Long Section of a Young Archegoniophore

1. The disc in its anatomy resembles the thallus in having photosynthetic zone above and storage zone below.

2. Archegonia are found on the dorsal surface of the disc in erect position.

3. They develop in acropetalous succession ie., older archegonia towards the centre and younger archegonia towards the margin.

4. Each lobe of the disc corresponding to a ray has a row of archegonia.

Structure of the archegonium

1. Each archegonium is a flask shaped structure.

2. It has a bulbous base called venter and a narrow elongated part called the neck.

3. The venter region encloses a venter canal cell and an egg cell.

4. The wall of the neck is made up of six rows of cells.

5. The neck has 4-6 neck canal cells. The tip of the neck is covered by means of four cover cells or lid cells.

Structure of a mature archegoniophore

1. A long section of a mature archegoniophore will show the following details.

2. The disc as usual shows the upper photosynthetic zone and a lower storage zone.

3. The archegonia containing sporophytes at different stages of growth are seen in an inverted condition on the ventral surface of the disc.

4. The older archegonia are found towards the periphery, while the younger ones are found towards the centre ie., towards the stalk.

5. All the archegonia are enclosed in a common envelope called perichaetium.

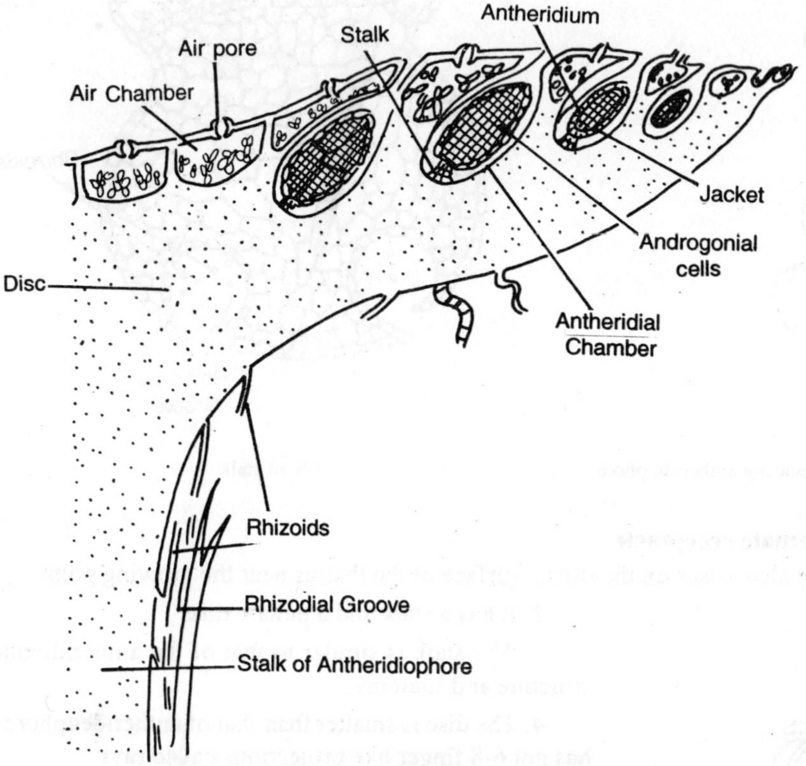

Fig. 20.21 Marchantia. L.S. of antheridiophore (diagrammatic)

6. The ventral surface of the disc produces another envelope in between the archegonia. This is called the perigynium. In addition to this, the venter also provides an envelope to the sporophyte; this is called Calyptra.

Structure of a mature sporophyte

1. The mature sporophyte of *Marchanita* will be in an inverted position hainging down from the ventral surface of the archegonial disc.

2. The sporophyte has three distinct regions viz., foot, seta and capsule.

3. The foot is made up of parenchymatous cells. It is Sagittate (arrow head like) and helps to anchor the sporophyte in the gametophytic tissue.

4. The seta is an elongated stalk. The cells are vertically extended.

5. The terminal part of the sporophyte is the globose capsule.

6. The wall of the capsule is made up of a single layer of cells.

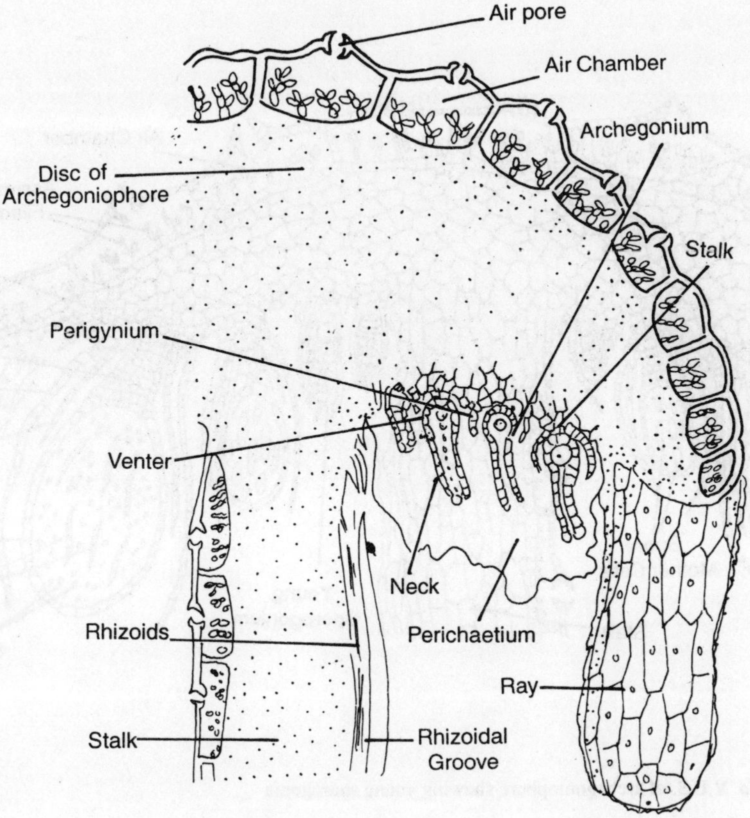

Fig. 20.22 Marchantia. L.S. of archegoniophore (diagrammatic)

7. Within the capsule, in the young condition there will be a number of spore mother cells. Some of them become sterilized to form the elaters, while the rest undergo reduction division to form tetrads of haploid spores.

8. Elaters are spindle shaped structures having special thickenings. They are hygroscopic and help in the dispersal of the spores.

9. The spores are uninulceate and have a double layered wall. The outer layer exine is highly ornamented, while the inner intine is thin. The spores on liberation develop into new individuals.

Diagrams to be drawn

1. Habit sketch - Dorsal and Ventral views
2. Dorsal view enlarged to show gemma cups, antheridiophore and archegoniophore (on different thalli)
3. T.S. of the thallus
4. T.S. of the thallus through a gemma cup (ground plan)
5. T.S. of stalk of antheridiophore
6. Vertical section of antheridiophore.
7. Vertical section of archegoniophore
8. A sporophyte enlarged.

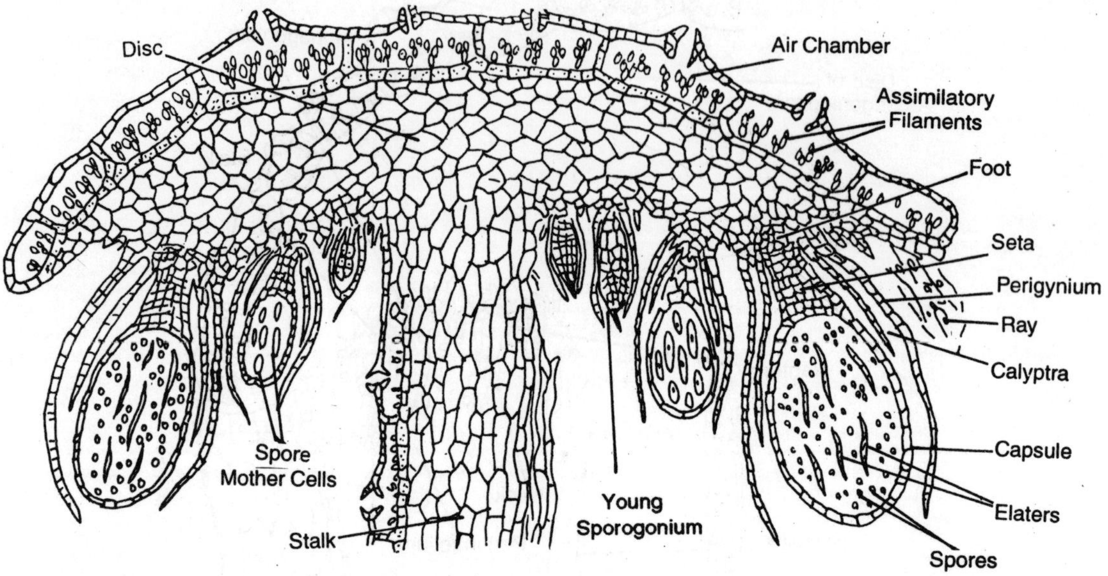

Fig. 20.23 Marchantia. V.L.S. of archegoniophore showing young sporogonia

Systematic position and identification

A. **1. Plant body is a parenchymatous gametophyte.**

2. It is leafy or thalloid.

3. Vasculature and roots absent.

4. Sex organs surrounded by a sterile jacket layer.

5. Life cycle exhibits heteromoprhic alternation of generations with a predominant sporophyte ... *Bryophyta*.

B. **1. Thalli are lobed and dorsiventrally flattened or leafy.**

2. Chloroplasts do not have pyrenoids

3. Scales and two types of rhizoids present

4. No columella in the sporophyte ...*Hepaticopsida*.

C. 1. Plant body thalloid and dichotomously branched

2. Scales are present on the ventral surface.

3. Thallus differentiated into photosynthetic and storage regions

4. Air chambers present ...*Marchantiales*.

D. 1. Sex organs borne on gametophores.

2. Air pores are like chimney.

3. Sterile cells form elaters...*Marchantiaceae*.

E. 1. Four rows of scales presents.

2. Scales are of two types.

3. Aereoles present on the dorsal surface.
4. Photosynthetic filaments branched.
5. Gemma cup with fringed margins, it is not crescent shaped.
6. Mucilage cells present in the storage region of the thallus ...*Marchantia.*

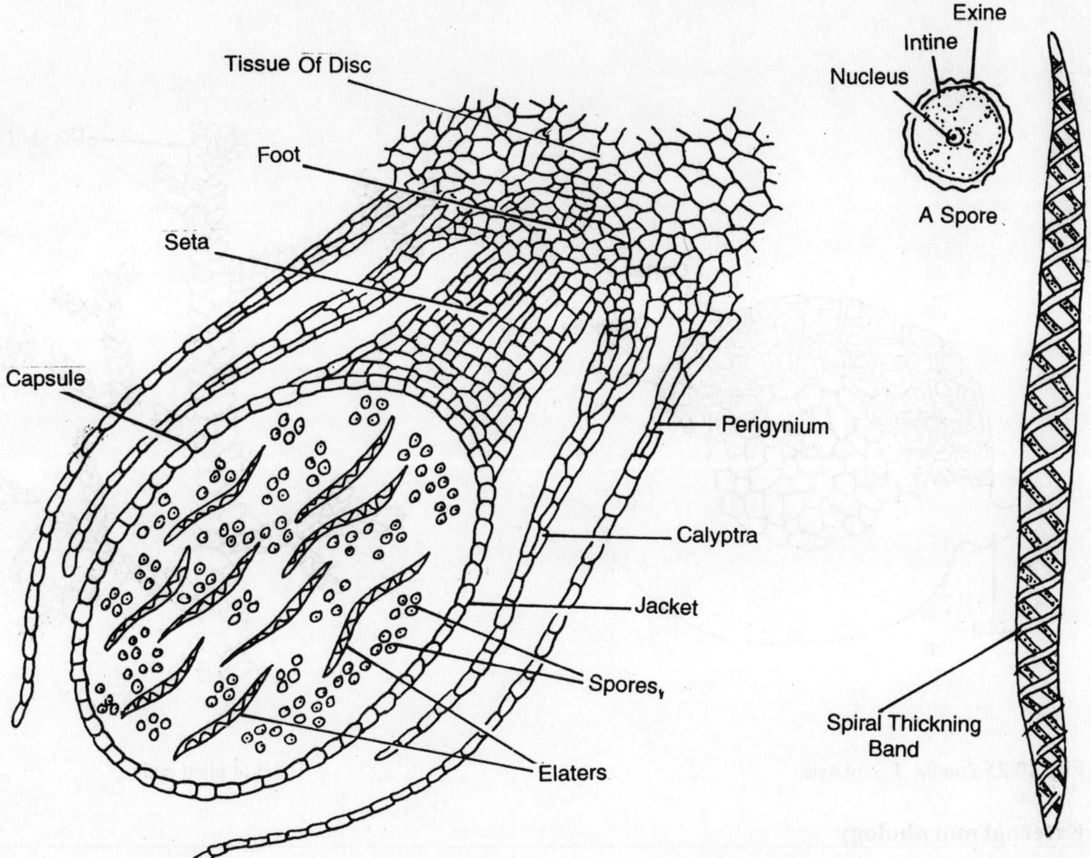

PORELLA

Common species : *P.platyphylla, P.rotundifolia etc.*

Occurrence : The genus *Porella* also known as *Madotheca* is a typical and widely distributed member of the order Jungermanniales. The genus has 180 species of which about 34 are recorded from India. *Porella* is usually found in moist shady places growing on logs, trunks of trees and water dripping rocks.

Study : 1. External morphology
2. Anatomy of axis
3. Reproduction
4. Structure of the sporophyte
5. Systematic position and identification

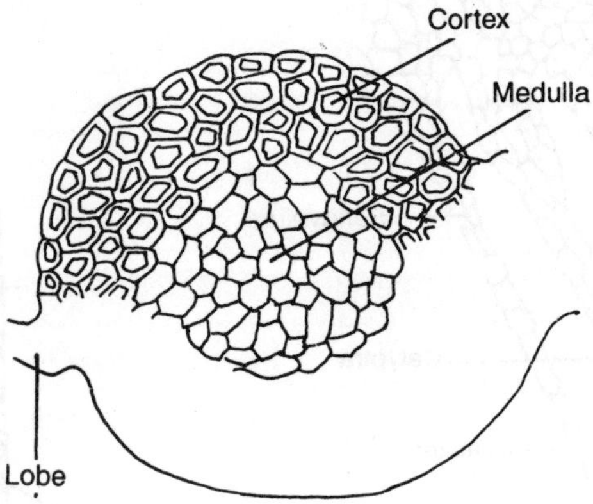

Fig. 20.25 Porella. T.S. of axis

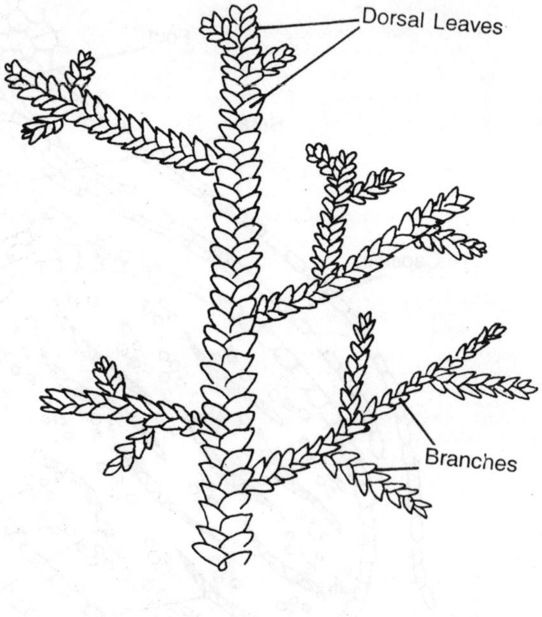

Habit of plant body

External morphology

1. The Gametophyte of *Porella* is greenish and leafy. When mature grows to a length of 15 cms.
2. It grows flat on the surface of substratum.
3. The plant is dorsiventrally differentiated.
4. The gametophyte consists of a central axis or stem and is covered by leaf like expansion.
5. The stem shows monopodial branching.
6. The stem and branches bear two types of leaves viz., dorsal leaves and ventral leaves.
7. The dorsal leaves also called paragastrica are found on the dorsal surface arranged in two rows.
8. The dorsal leaves completely cover the stem and on the dorsal surface their arrangement is said to be incubous ie., the anterior (upper) margin of each leaf overlaps the posterior (lower) margin of next leaf found above.
9. On the ventral surface the arrangement of dorsal leaves is said to be Succubous ie., the lower margin of each leaf overlaps the upper margin of leaf below.
10. Such an arrangment of leaves enables the easy flow of water.

11. Each dorsal leaf is bilobed, it has a large *Postical lobe* and a small *Antical lobe*.

12. Arranged on the ventral surface and covering them is found a row of leaves called ventral leaves or *Amphigastria*.

13. Totally on the axis there are three rows of leaves (two rows of dorsal leaves and one row of ventral leaf).

14. In the leaves of *Porella* as in other foliose liver worts, the midrib is absent.

15. Each leaf consists of a single layer of cells.

16. From the undersurface of gametophyte arise a few simple rhizoids which help mainly in fixation and rarely in absorption.

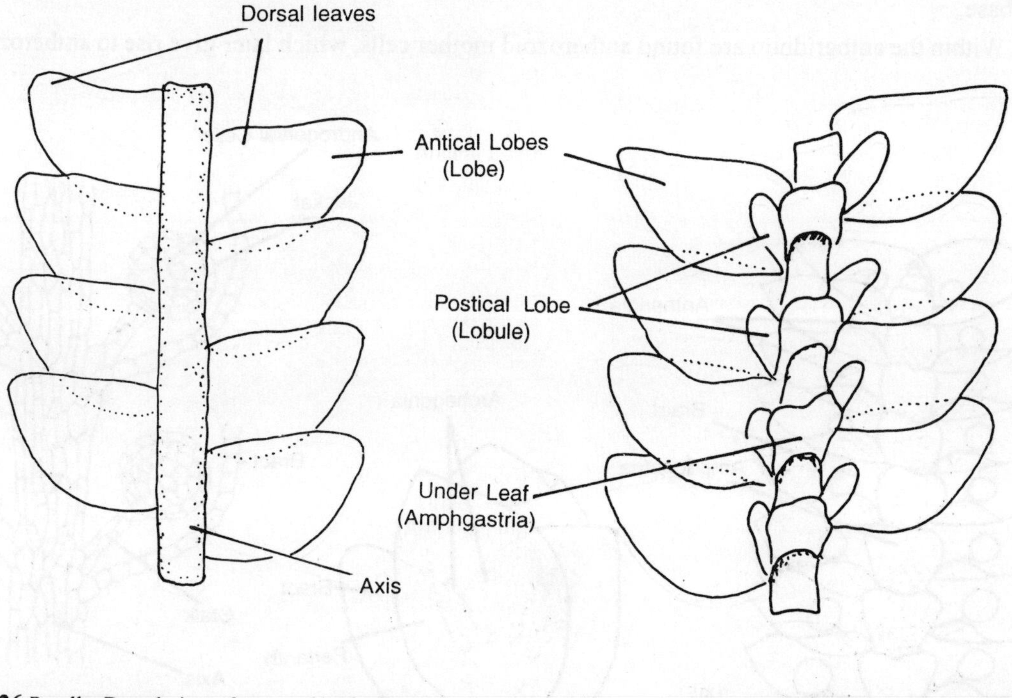

Fig. 20.26 Porella. Dorsal view of a part of leafy shoot   Ventral view of a part of leafy shoot

Anatomy of the Axis

1. The stem at younger regions has only a parenchymatous tissue.

2. The cells are thin walled and filled with chloroplasts.

3. The epidermal layer is not well differentiated.

4. In the older portions of the stem, there is differentiation into outer cortical region and inner medullary region.

5. The cortex is made up of smaller cells with slightly thickened walls.

6. The medulla consists of layer cells with thinner walls.

Vegetative propagation

1. Progressive death and decay of older parts up to a branching will result in the separation of a branch which will develop into a new individual.

Sexual reproduction

1. Sexual reproduction is of the oogamous type. Sex organs are antheridia (male) and archegonia (female).

2. *Porella* is dioecious, male plants are smaller than the female plants.

Male plant and antheridium

1. The plant produces a number of short lateral antheridial branches, which project out more or less perpendicular to the main axis.

2. Every antheridial branch is composed of a number of compactly arranged small leaves called perigonial leaves.

3. In the axil of each perigonial leaf arises an antheridium.

4. A well developed antheridium has a long stalk and the spherical antheridium proper.

5. The stalk is two celled in breadth.

6. The jacket of the antheridium is single layered at the apex, two layered at the sides and three layered at the base.

7. Within the antheridium are found antherozoid mother cells, which later give rise to antherozoids.

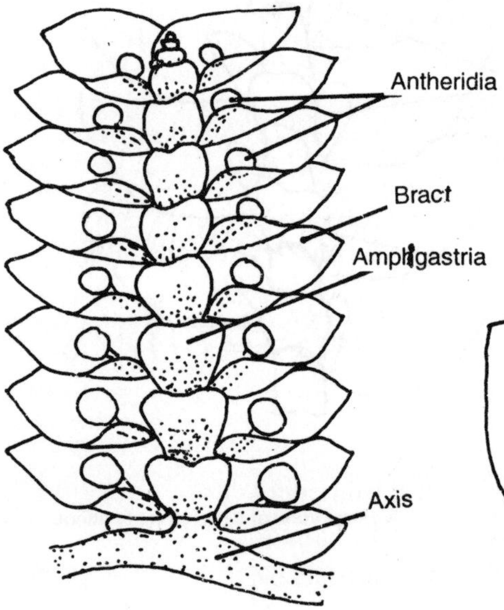

Fig. 20.27 Porella. Antheridial branch

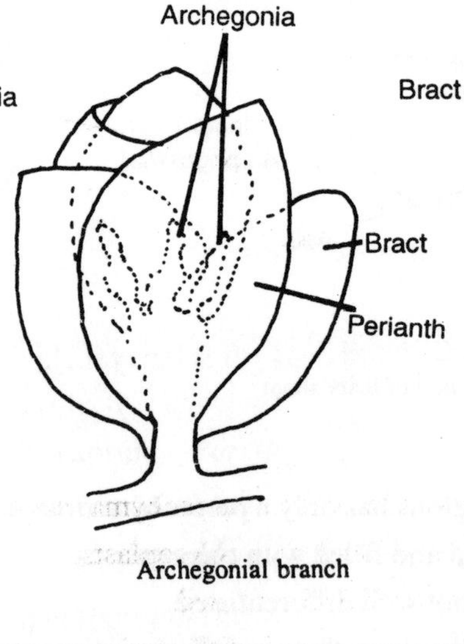

Archegonial branch

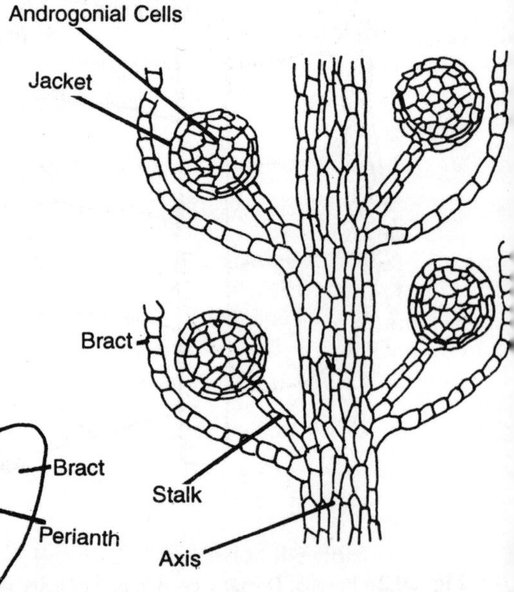

L.S. of antheridial branch

Female plant and archegonium

1. The female plant produces archegonia on special branches called archegonial branches.

2. Each archegonial branch is almost like a vegetative branch, but ends in a few larger leaves called the Perichaetial leaves or bracts.

3. The archegonia appear in terminal clusters.

4. A well developed archegonium is a flask shaped structure with a basal bulged venter an upper narrow neck.

5. The tip of the neck is covered by four cover cells and within the neck are present 6-8 neck canal cells.

6. The venter wall is 2-3 layered thick and it has an egg cell below and a venter cell above.
7. The neck is covered by five vertical rows of cells forming the wall (six rows are found in Marchantiales).
8. Fertilization results in the formation of zygote which develops into the sporophyte.

Structure of the sporophyte

1. The sporophyte of porella is divided into three parts namely foot, seta and capsule.
2. The foot is bulbous made up of parenchyma cells and is embedded in the gametophytic tissue.
3. The seta is stack like and carries the capsule slightly away from the gametophyte.
4. The capsule is globose and is surrounded by a jacket made up of many layers of cells.
5. Within the capsule are found a large number of spore mother cells, some of which become sterile and form the elaters.
6. The elaters are spindle shaped and have hygroscopic spiral thickenings. The elaters help in the dispersal of the spores.
7. The spore mother cells undergo reduction division to form tetrads of haploid spores.
8. The sporophyte has two protective envelopes viz., calyptra and perianth.
9. Calyptra is formed by the venter of the archegonium, while the perianth is formed by the fusing perichaetial leaves.
10. The perianth is bilipped.
11. The spores on liberation develop into new individuals of *Porella.*

Diagrams to be drawn

1. Habit sketches of male and female plants. (Dorsal and ventral views).
2. Cross section of the axis.
3. Section of an antheridial branch.
4. Section of an archegonial branch.
5. Long section of a mature sporophyte.

Systematic position and identification

A.
1. Plant body is a parenchymatous gametophyte.
2. It is leafy or thalloid.
3. Vasculature and roots absent.
4. Sex organs surrounded by a sterile jacket layer.
5. Life cycle exhibits a heteromorphic alternation of generations with a predominant gametophyte ... *Bryophyta.*

B.
1. Thalli are lobed and dorsiventrally flattened or leafy.
2. Chloroplasts do not have pyrenoids.
3. No columella in the sporophyte ... *Hepaticopsida.*

C.
1. Internal tissue differentiation absent in the gametophyte.
2. Scales are absent.
3. Rhizoids only one type (simple).
4. Archegonial neck has 5 rows of cells ...*Jungermanniales.*

D.
1.Gametophyte is leafhy
2. Three rows of leaves present on the axis.
3. Archegonia produced at the tips of archegonial branches ...*Jungermanniae.*

E.
1. Dorsal leaves have two lobes and arranged in incubous manner.
2. Simple rhizoids arise from the ventral surface. They are not many in number.

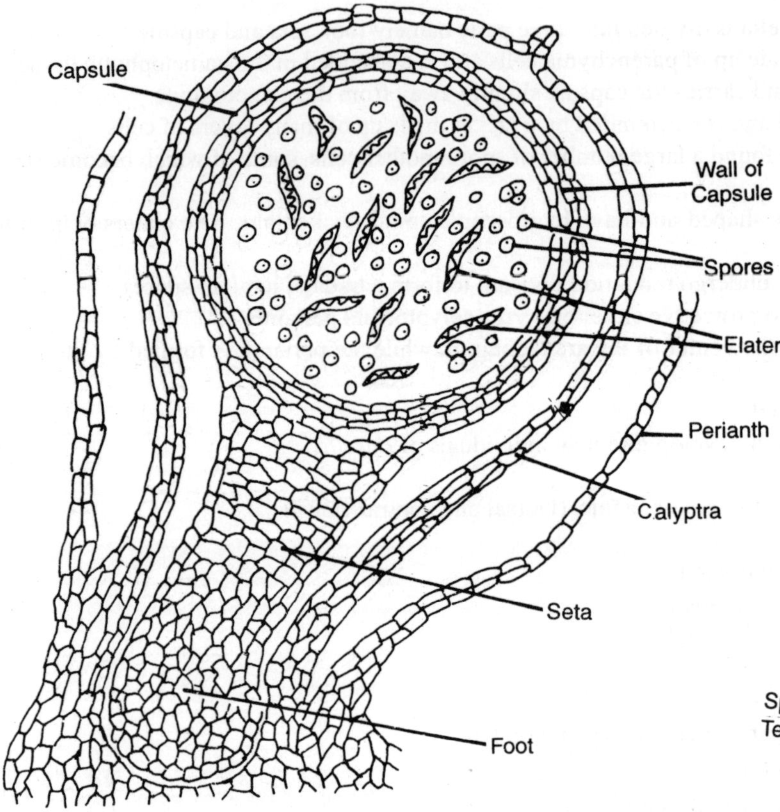

Fig. 20.28 Porella. L.S. of sporogonium

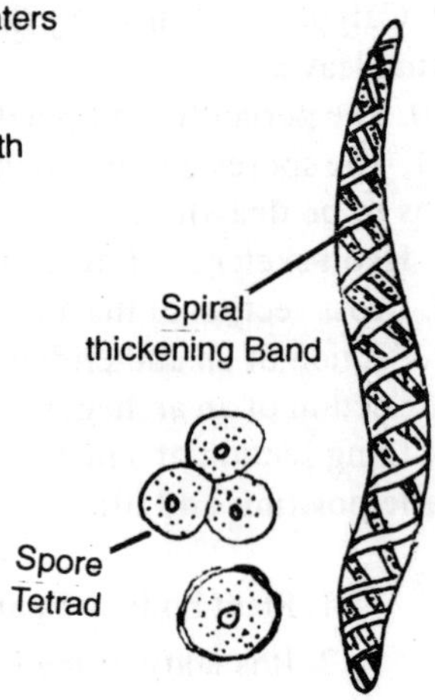

Spores and elater

F. 1. Antheridial jacket single layered at the apex but double layered at the sides.
2. Sporophyte has two protective envelopes.
3. No elaterophore in the capsule ...*Porella*.

PELLIA

Common species : *P.epiphylla, P.fabborniana* etc.

Occurrence : The gametophytes of *Pellia* grow on moist, cool, shady places, on trunks of trees or on wet rocks forming extensive mats.

Study : 1. External morphology of thallus
2. Internal structure
3. Reproduction
4. Structure of the sporophyte
5. Systematic position and identification

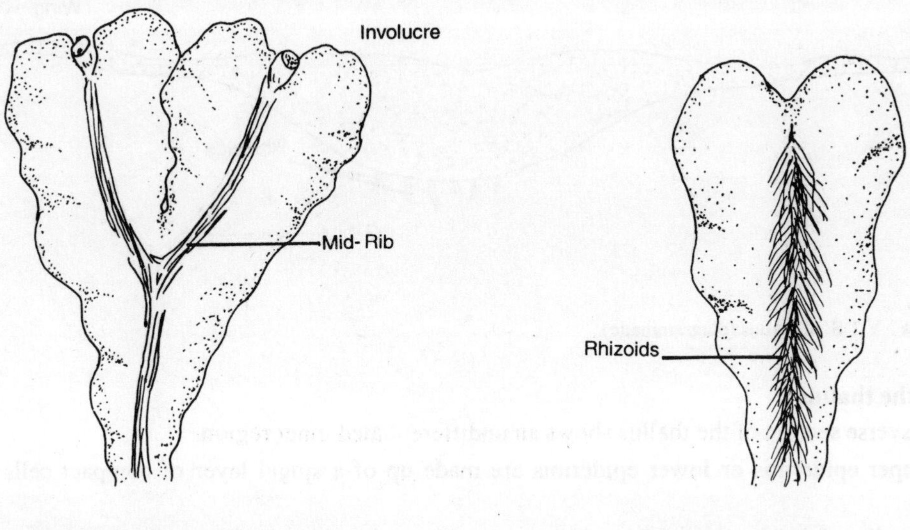

Fig. 20.29 Pellia . Dorsal surface of thallus Ventral surface of thallus

External morphology

1. The plant body is a dorsiventrally flattened dichotomously branched thallus.
2. The margins of the thallus are undulate and sinuous, often the lobes overlap one another.
3. There is a midrib in the centre, which if often purple in colour.
4. The dorsal surface of the thallus is much greener in colour than the ventral surface.
5. The thallus on its ventral surface bears only smooth rhizoids.
6. Pegged rhizoids and scales are absent
7. The growing points of the thallus are situated in the apical nothces present at the tips of lobes.
8. The presence or absence of moisture decides the morphoiogical appearance of the thallus.
9. Thalli growing near water are delicate, narrow and ribbon like, with a distinct mid-rib, while those growing on a dry soil are short thick and deep green without prominent mid-rib.

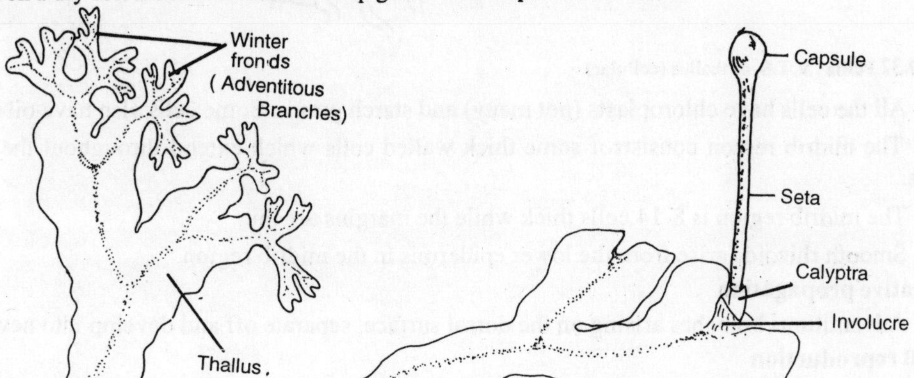

Fig. 20.30 Pellia . Thallus showing winter frongs Thallus with sporogonium

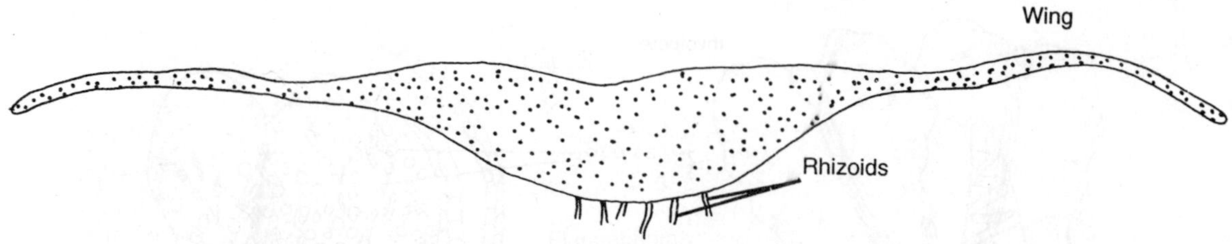

Fig. 20.31 Pellia . V.T.S. of thallus (diagrammatic)

Anatomy of the thallus

1. A transverse section of the thallus shows an undifferentiated inner region.

2. The upper epidermis or lower epidermis are made up of a singel layer of compact cells rich in chloroplasts.

3. All the cells in between the epidermal layers are alike and there is no differentiation into storage and photosynthetic zones.

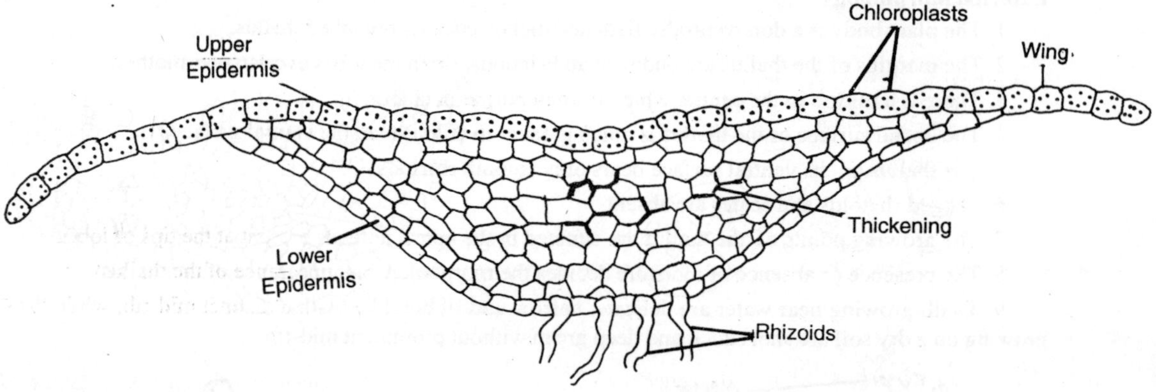

Fig. 20.32 Pellia . V.T.S. of thallus (cellular)

4. All the cells have chloroplasts (not many) and starch grains. Some cells also have oil contents.

5. The midrib region consistsof some thick walled cells which extend throughout the length of the thallus.

6. The midrib region is 8-14 cells thick while the margins are thin.

7. Smooth rhizoids arise from the lower epidermis in the midrib region.

Vegetative propagation

1. Adventitious branches arising on the dorsal surface, separate off and develop into new individuals.

Sexual reproduction

1. Sexual reproduction is oogamous. Sex organs are antheridia (male) and archegonia (female).

2. Species of *Pellia* may be monoecious as in *P.epiphylla* or dioecious as in *P.nessiana*.

3. Monoecious species are usually protandrous.

4. Sex organs are found embedded on the dorsal surface.

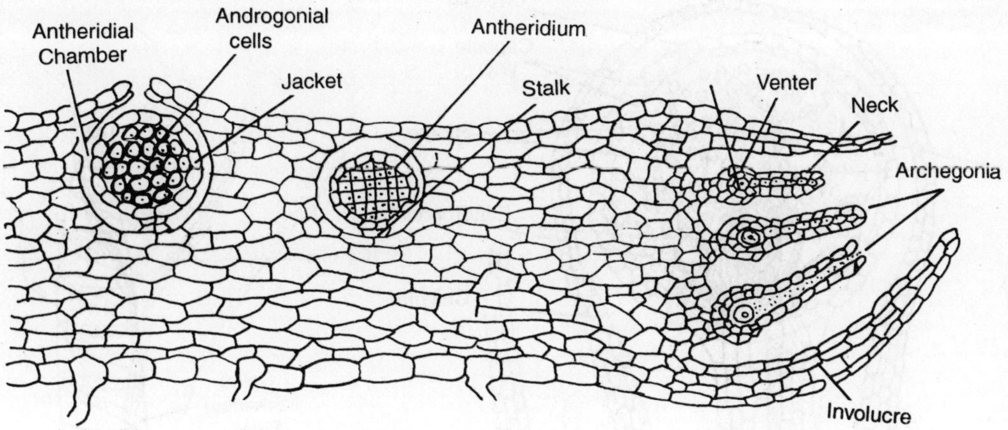

Fig. 20.33 Pellia . V.L.S. of thallus passing through sex organs

Antheridium

1. Male thalli in dioecious species are smaller than the female thalli.
2. Presence of small knob like outgrowths on the dorsal surface indicate, the presence of antheridia.
3. Antheridia are found embedded in small cavities.
4. Each cavity encloses a single antheridium and opens out by a small opening.
5. Each antheridium is a globose structure having a short multicellular stalk.
6. The wall of the antheridium is made up of a single layered jacket.

The antherozoid mother cells present inside the antheridium give rise to biflagellate coiled antherozoids.

Archegonium

1. The archegonia are found in groups of 6-12 on the dorsal surface, a little behind the apical region.
2. The archegonia are produced on a slightly elevated part called the receptacle.
3. All the archegonia are covered by means of a flap like envelope called the involucre.
4. The involucre arises from the thallus as an outgrowth.
5. Each archegonium is a flask shaped structure, and has a short stalk.
6. The archegonium is similar in structure to what is seen in *Porella*.
7. The wall of the neck is made up of five vertical rows of cells typical to Jungermanniales.

Structure of the sporophyte

1. The sporophyte is divided into foot, seta and capsule
2. The foot is well developed, somewhat bulbous and forms a collar like sheath surrounding the basal portion of the seta.
3. The seta has vertically elongated cells. It is short in the young sporophyte, but becomes long at maturity.
4. The capsule occupies the terminal part of seta and is globose in shape.
5. The wall of the capsule is made up of three layers of cells.

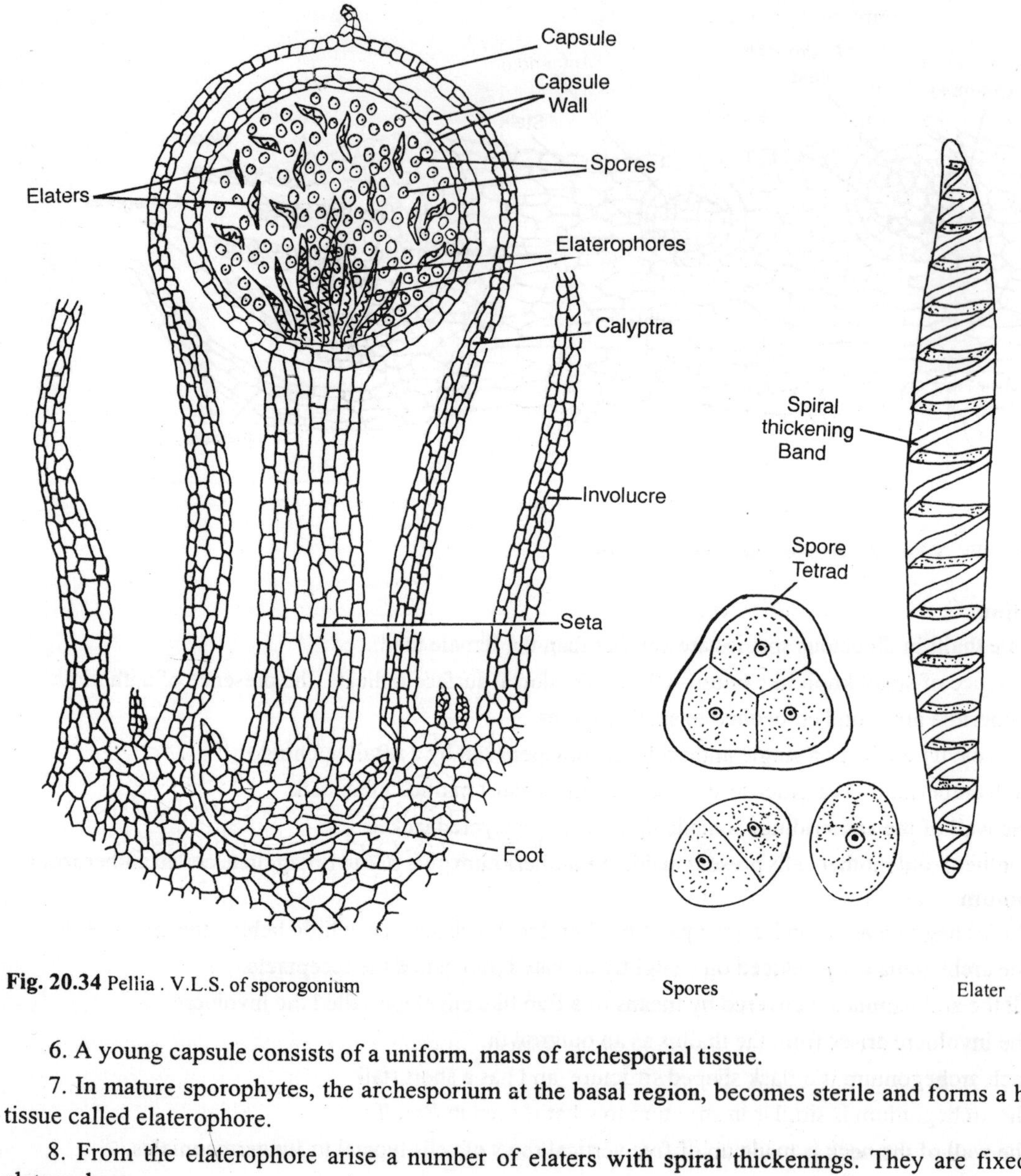

Fig. 20.34 Pellia . V.L.S. of sporogonium Spores Elater

6. A young capsule consists of a uniform, mass of archesporial tissue.

7. In mature sporophytes, the archesporium at the basal region, becomes sterile and forms a hump of tissue called elaterophore.

8. From the elaterophore arise a number of elaters with spiral thickenings. They are fixed to the elaterophore.

9. In the remaining part of capsule are found haploid spores as well as free elaters.

10. Free elaters are similar in structure to fixed elaters.

11. The spores are small, haploid and uninucleate.

12. The wall of the capsule consists of thin walled cells at four places. These mark the lines of dehiscence. The capsule opens out into four valves.

13. On liberation from the sporophyte, the spores develop into new gametophytic individuals.

14. The mature sporophyte has two envelopes viz calyptra and involucre. Calyptra is formed by the venter of the archegonium; while involucre is formed by the gametophytic tissue.

Diagrams to be drawn

1. Habit sketch - dorsal and ventral views.
2. Cross section of the thallus
3. Cross section of the thallus to show antheridia
4. Section of the thallus to show archegonia
5. Long section of the sporophyte
6. Elaterophore, elaters and spores (enlarged)

Systematic position and identification

A.
1. Plant body is a parenchymatous gametophyte
2. It is leafy or thalloid
3. Vasculature and roots absent
4. Sex organs surrounded by a sterile jacket layer
5. Life cycle exhibits, heteromorphic alterntion of generations with a predominant gametophyte ... *Bryophyta.*

B.
1. Thalli are lobed and dorsiventraly flattened or leafy.
2. Chloroplasts do not have pyrenoids
3. No columella in the sporophyte ... *Hepaticopsida*

C.
1. Internal tissue differentiation absent in the gametophyte.
2. Scales are absent
3. Rhizoids only of the simple type
4. Archegonial neck has 5 rows of cells in the wall... *Jungermanniales.*

D.
1. Apical cell does not get transformed into an archegonial initial
2. Plant body is thalloid ... *Metzerinae.*

E.
1. Antheridia develop singly in cavities.
2. Archegonia found in clusters
3. Elaterophore present ... *Pelliaceae*

F.
1. Thallus has sinores overlapping lobes
2. Archegonia enveloped by involucre
3. Knob like outgrowths present on the thallus to indicate the underlying antheridia
4. Elaterophore present at the base of the capsule
5. Capsule splits into 4 valves... *Pellia*

CHAPTER 21

ANTHOCEROTOPSIDA

The class Anthocerotopsida is a small but unique group among Bryophytes. It is characterised by the following features :

1. The gametophytic plant body is a dorsiventrally flattened thallus, with no internal tissue differentiation.
2. Rhizoids are present on the ventral surface; they are of the simple (smooth) type.
3. Pegged rhizoids and scales are absent.
4. The thallus consists of mucilagi[illegible]enious cavities inhabited by cyanophycean algae.
5. Each cell of the gametophyte has a single chloroplast with a pyrenoid
6. The sporophyte is capable of indefinite growth
7. The central region of the sporophyte is made up of columella derived from the endothecium.

The class antherocerotopsida includes a singel order Anthocerotales with a single family Anthocerotaceae. The family includes about 5-6 genera with nearly 300 species all of which are distributed widely but have a special preference to tropical and warm temperate regions. The six genera in the family are *Anthoceros, Phaeoceros, Aspiromitus, Notothylos, Dendroceros* and *Megaceros*. Of all the genera, Anthoceros is the largest with about 200 species.

ANTHOCEROS

Common species : *A.hallii, A.lavis, A.erectus, A.fusiformis, A.punctatus* etc.

Occurrence : The genus with its nearly 200 species is cosmopolitan in its distribution, but having a preference to temperate and tropical regions. About 25 species are found in India, growing on moist shady places or on the side bunds of rice fields.

Study : 1. External morphology
2. Internal structure
3. Vegetative propagation
4. Sexual reproduction
5. Structure of the sporophyte
6. Systematic position and identification

External morphology

1. The gametophytic plant body is a dorsiventrally flattened, lobed horizontally growing thallus.

2. The thallus is thicker at the centre than at the margins, but there is no distinct midrib.

3. The dorsal surface of the thallus may be smooth, warty or rough with ridges.

4. On the dorsal surface are found many stalk or horn like structures (hence the name Hornwort) representing the sporophytes.

5. Arising from the ventral surface are a number of smooth rhizoids.

6. Pegged rhizoids and scales are absent.

7. Small dark green patches are found on the ventral surface (this can be seen with a hand lens) representing the position of algal colonies inside.

8. The branching of the thallus is not strictly dichotomous.

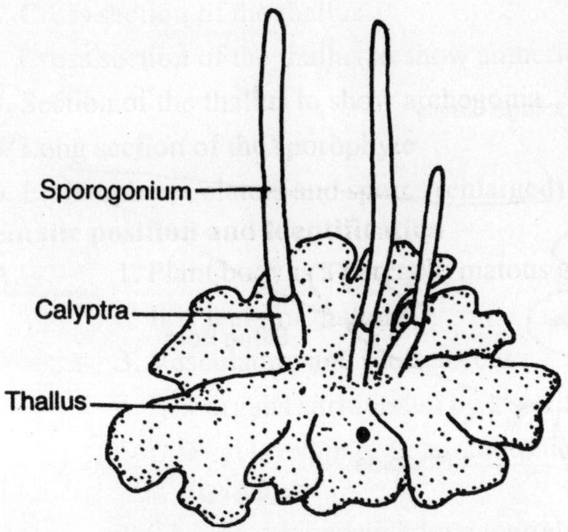

Fig. 21.1 Anthoceros. Thallus bearing sporophytes

Internal structure of the thallus

1. A vertical transverse section will show the following features.

2. The internal tissue does not show any differentiation into storage and assimilatory zones.

3. The entire inner region consists of isodiametric parenchyma cells.

4. However, on either surface (dorsal and ventral) the superficial layers are slightly different and can be designated as upper and lower epidermis.

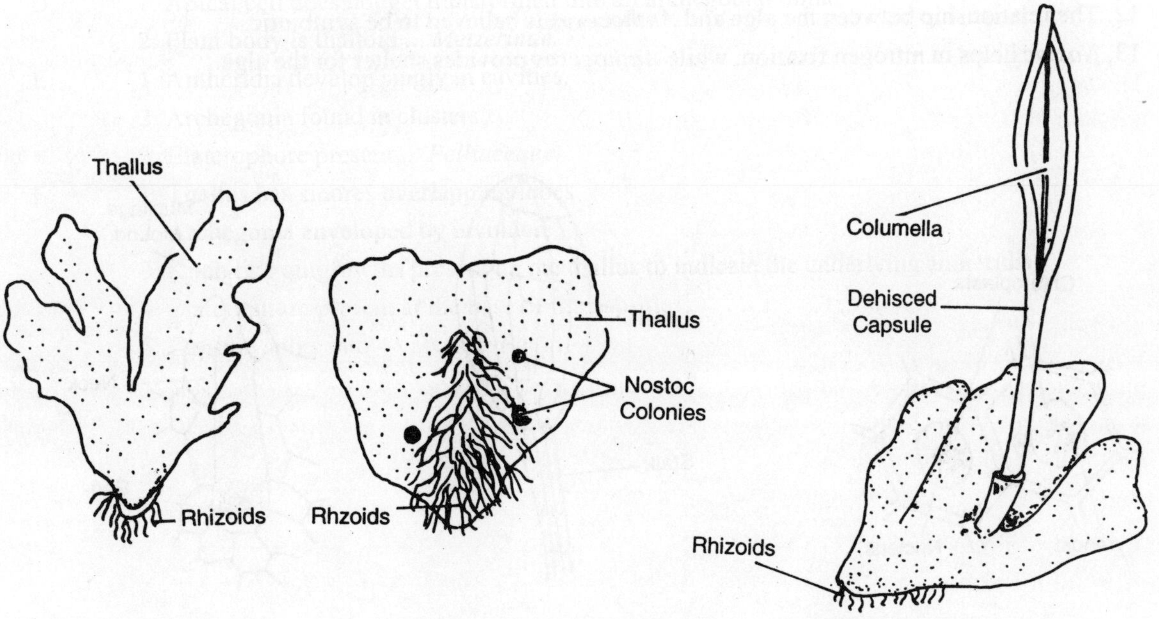

Fig. 21.2 Anthoceros. Ventarl surface of thallus showing Nostoc colonies Thallus bearing dehisced sporophyte

5. Epidermal cells are smaller in size and more regular in arrangement.
6. All the cells have a single lens shaped chloroplast with a pyrenoid.
7. Occasionally the number of chloroplasts may be more than one.
8. The cells are uninucleate.

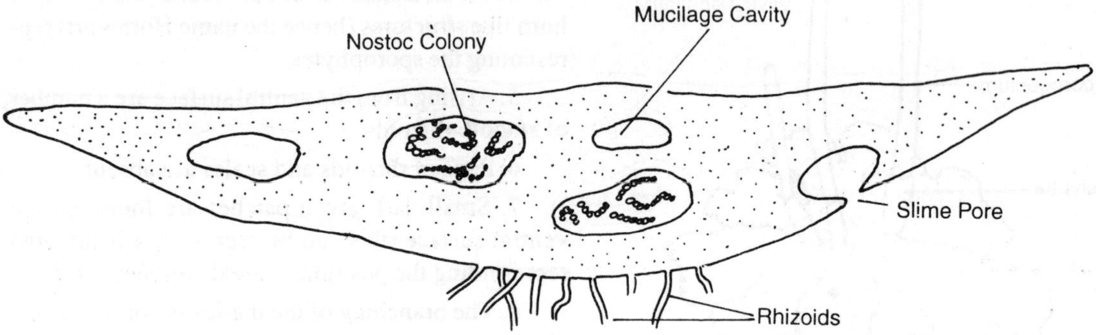

Fig. 21.3 Anthoceros. V.T.S. of thallus (diagrammatic)

9. There are no airchambers or cavities, but on the ventral surface are found certain small mucilage filled cavities.

10. These are intercellular cavities opening through small pores on the ventral surface.

11. These cavities are often inhabited by endophytic algae such as *Nostoc*, the *Nostoc* filaments entering the thallus through the pore.

12. The relationship between the alga and *Anthoceros* is believed to be symbiotic.

13. *Nostoc* helps in nitrogen fixation, while *Anthoceros* provides shelter for the alga.

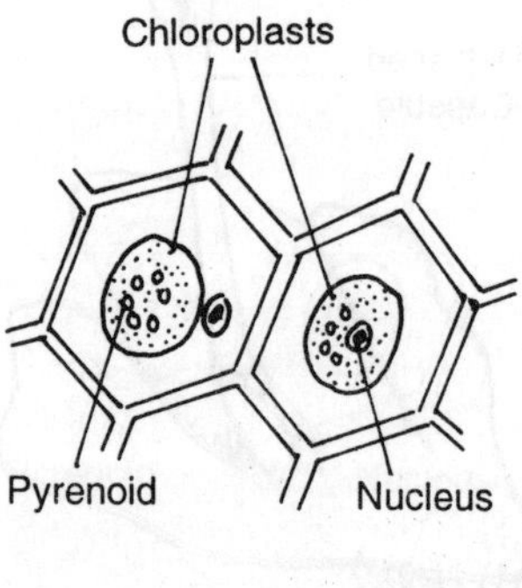

Fig. 21.4 Anthoceros. Cells enlarged showing internal structure

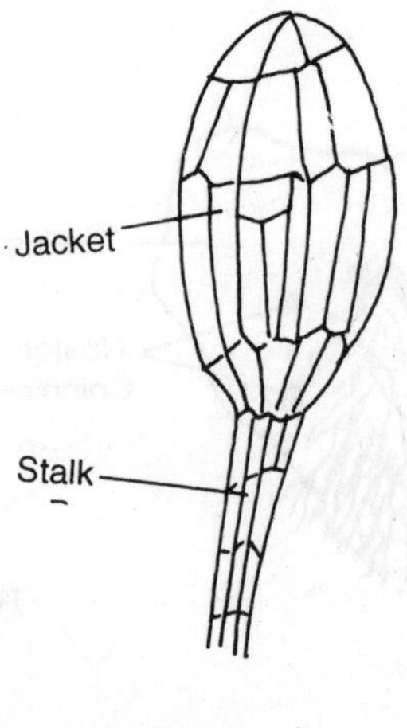

Single antheridium (external features)

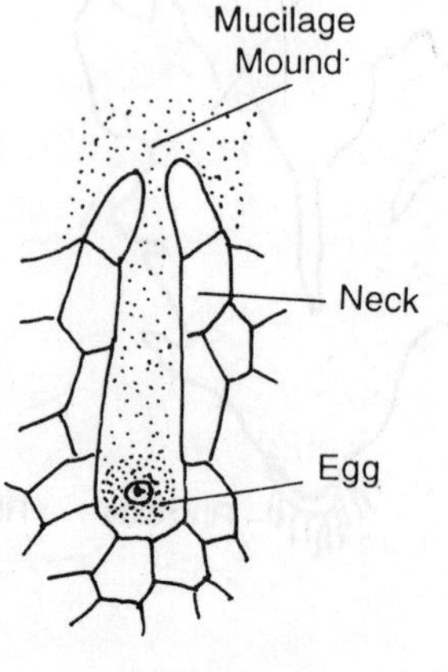

Mature archegonium

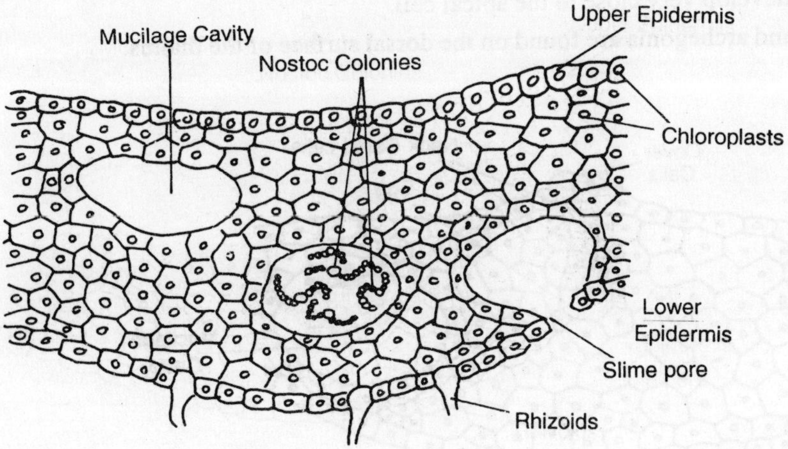

Fig. 21.5 Anthoceros. A portion of V.S. of thallus (cellular)

Vegetative propagation

1. By the progressive death and decay of older parts and consequent separation of dichotomies.

2. Small tubers are formed on the thallus which develop into new individuals.

3. During unfavourable conditions, the apex remains persistent, while the other parts die. The persistent apex develops into a new individual.

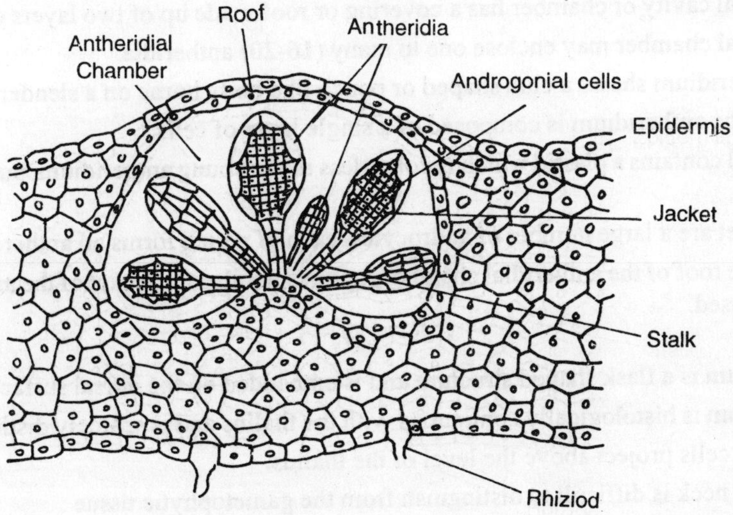

Fig. 21.6 Anthoceros. A portion of V.S. of thallus showing antheridia

Sexual reproduction

1. It is of the oogamous type. Sex organs are antheridia (male) and archegonia (female).

2. Species of *Anthoceros* may be monoecious (eg. *A.fusiformis, A.punctatus* etc) or dioecious (eg. *A.erectus, A.hallii* etc)

3. Sex organs develop very close to the apical cell.

4. Antheridia and archegonia are found on the dorsal surface of the thallus.

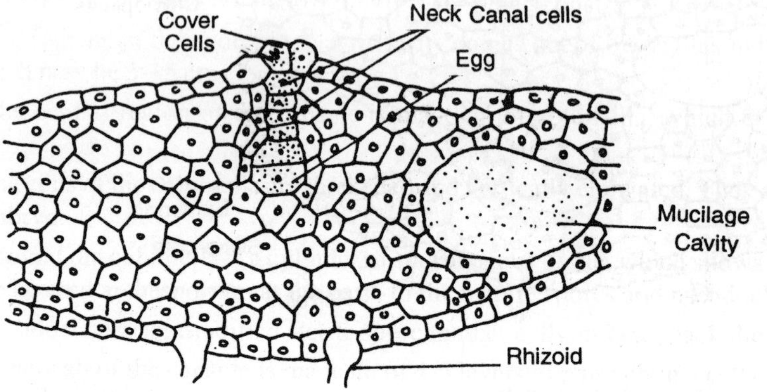

Fig. 21.7 Anthoceros. V.S. of thallus passing through archegonia (A part)

Antheridium

1. Antheridia are found embedded on the dorsal surface of the thallus, in small cavities called antheridial cavities.

2. Each antheridial cavity or chamber has a covering or roof made up of two layers of cells.

3. Each antheridial chamber may enclose one to many (16-20) antheridia.

4. A mature antheridium shows a club shaped or pouch like body borne on a slender or massive stalk.

5. The jacket of the antheridium is composed of a single layer of cells.

6. Each jacket cell contains a plastid which is colourless in the young antheridium, but becomes green at later stages.

7. Inside the jacket are a large number of androcytes, each of which forms an antherozoid.

8. At maturity, the roof of the antheridial chamber ruptures, so also the jacket of the antheridium and the antherozoids are released.

Archegonia

1. The archegonium is a flask shaped structure and is embedded on the dorsal surface of the thallus.

2. The archegonium is histologically continuous with the thallus and is scarcely distinguishable.

3. Only the cover cells project above the level of the thallus.

4. The wall of the neck is difficult to distinguish from the gametophytic tissue .

5. Within the neck are 4-6 neck canal cells.

6. The venter region has a venter canal cell and an egg cell.

7. There is not much difference is size between neck and venter. The neck is as broad as the venter.

8. Neck canal cells and venter canal cells degenerate to facilitate the entry of antherozoids.

9. The antherozoids fertilize the egg and it becomes a zygote.

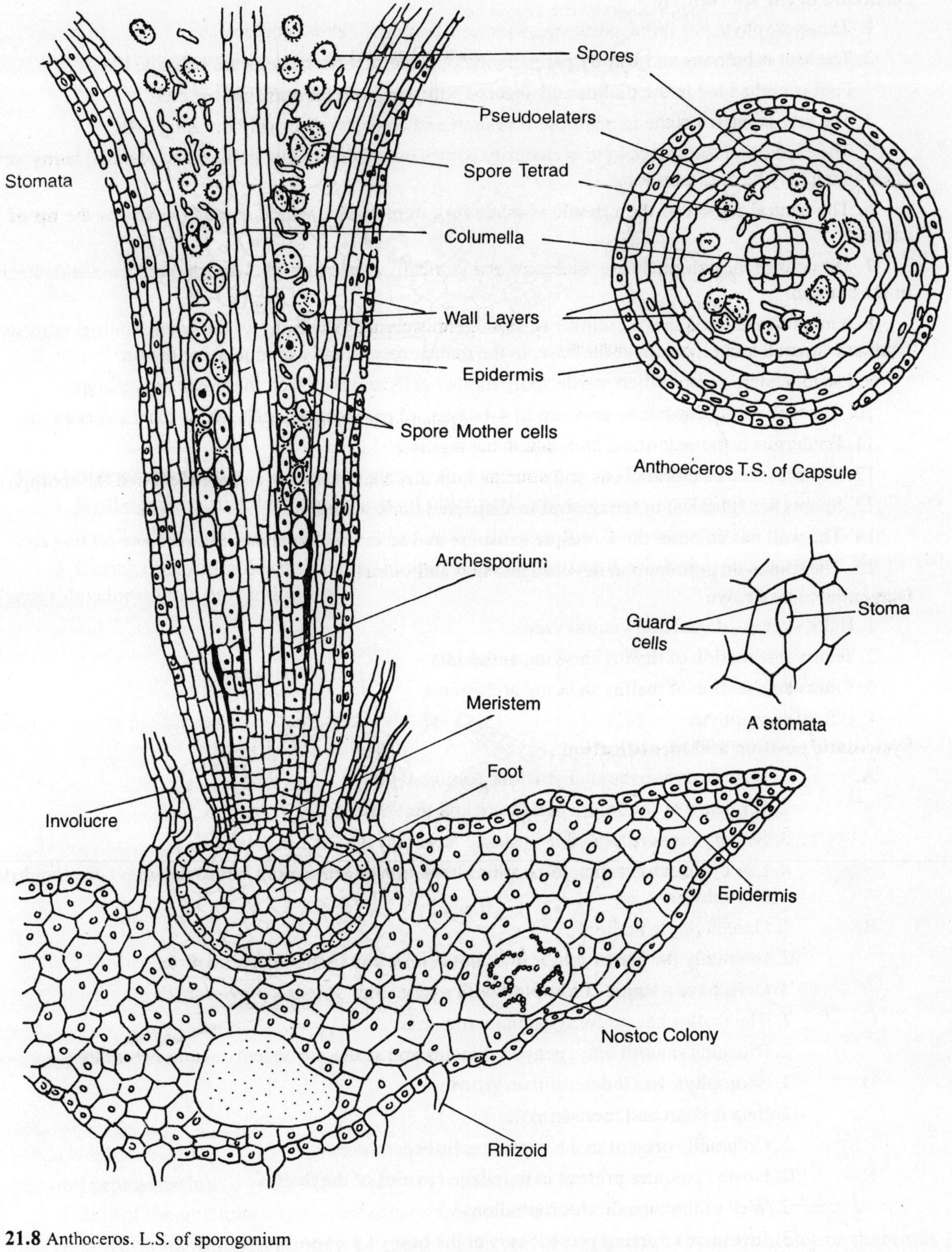

Fig. 21.8 Anthoceros. L.S. of sporogonium

Structure of the sporophyte

1. The sporophyte has three parts viz., foot, seta and capsule.
2. The foot is bulbous and wholly parenchymatous; the superficial cells are palisade like.
3. Foot is embedded in the thallus and absorbs nutrition from the surrounding cells.
4. The seta is intermediate in position. It is short and consists of actively dividing cells.
5. The capsule of the sporophyte at maturity comes out of the thallus in the form of small horny structures which may be 5-15 cms long.
6. The central region of the capsule is made up of columella, which extends nearly to the tip of the capsule.
7. The cells of the columella are thickened and vertically elongated. They are derived from endothecium of the embryo.
8. Around the columella is a cylinder of sporogenous tissue, which shows the whole sporogenous series from one layered archesporium at the base, to the mature spores and pseudoelats at the tip.
9. Pseudoelaters are modified sterile spore mother cells and they lack the spiral thickenings.
10. The wall of the capsule is made up of 4-6 layers of parenchyma cells which have chloroplasts.
11. Epidermis is the outermost layer and it has stomata.
12. The presence of chloroplasts and stomata indicates the nutritional independence of the sporophyte.
13. Spores are spherical or tetrahedral in shape and thick walled.
14. The wall has an outer thick opaque exospore and an inner endospore. Spores have oil droplets.
15. The spores on germination develop into new individuals.

Diagrams to be drawn

1. Habit sketch - dorsal and ventral views.
2. Transverse section of thallus showing antheridia.
3. Transverse section of thallus showing archegonia.
4. L.S. of sporophyte.

Systematic position and identification

A.
1. Plant body is a parenchymatous gametophyte.
2. It is leafy or thalloid; vasculature and root absent.
3. Sex organs surrounded by a sterile jacket layer.
4. Life cycle exhibits heteromorphic alternation of generations with a dominant gametophyte... Bryophytes.

B.
1. Gametophyte thalloid.
2. Internally the gametophyte does not exhibit any tissue differentiation.
3. Cells have a single chloroplast with a pyrenoid...*Anthocerotopsida.*

C.
1. The thallus has uniform isodiametric cells.
2. Rhizoids smooth only, pegged rhizoids and scales are absent ...*Authocerotales.*

D.
1. Sporophyte has indeterminate growth.
2. Seta is short and meristematic.
3. Columella present in the capsule...*Anthocerotaceae.*

E.
1. Nostoc colonies present in mucilage cavities of the thallus.
2. Wall of the capsule chlorophyllous.
3. Involucre covering present only at the base of the sporophyte ...*Anthoceros.*

CHAPTER 22

BRYOPSIDA

The class Bryopsida is the third largest the class of Bryophyta including 660 genera and 14,500 species. The characteristic features of Bryopsida are as follows:

1. The gametophyte is differentiated into two stages, viz., (a) prostrate, thalloid, branched protonema and (b) erect leafy gametophore.

2. The gametophore has a central axis on which are spirally arranged the leaves bearing sex organs.

3. Rhizoids are multicellular and have oblique cross walls.

4. Internally, the axis has a peripheral region and the central region; the latter performing conduction.

5. Sex organs are formed from superficial cells of the gametophore and show a definite apical growth in the early stage.

6. The sporophyte consists of foot, seta and capsule.

7. The capsule wall is multilayered with an epidermis having functional or non-functional stomata.

8. The archespenium is either derived from endothecium or from amphithecium usually enclosing a central columella.

9. Archsporium gives rise to only spores. Elaters are absent.

10. Spores on germination first develop into a filamentous structure called protonema. Later protonema gives rise to the adult gametophyte.

Classification

The class is divided into 5 subclasses. They are:

1. Sphagnidae – *Sphagnum*
2. Andreaidae
3. Bryidae – *Funaria*
4. Buxabaumidae
5. Polytrichidae – *Polytrichum.*

SPHAGNUM

Common species : *S. tenellum, S. molluscum, etc.*

Occurrence: *Sphagmum* is the monotypic genus of the family Sphagnaceae. It is commonly called the Peat moss or bog moss. The genus has 336 species of which 17 are recorded from India. Majority of the species grow in aquatic areas, where the water is soft and acidic. Some species are terrestrial growing on damp soil or dripping rocks.

Study:
1. External morphology
2. Anatomy of the axis.
3. Vegetative propagation
4. Sexual reproduction.
5. Structure of the sporophyte
6. Structure of the protonema.
7. Systematic position and identification

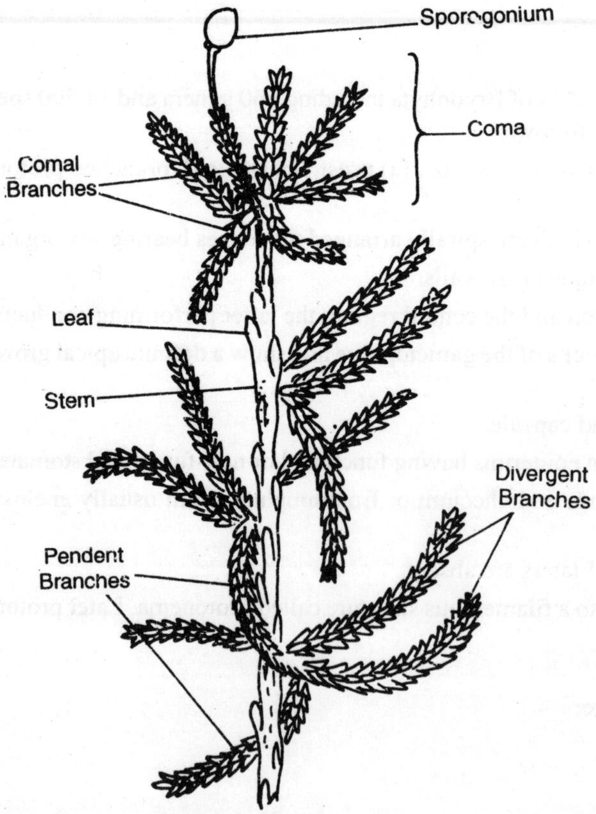

Fig. 22.1 Sphagnum. Plant body (Gametophore)

External Morphology:

1. The plant body of *Sphagnum* is the gametophyte often called the gametophore.

2. The colour of the gametophyte differs from species to species. It may be pale or dark green, orange, brown, rose, pink etc.

3. The gametophyte has a thick central axis called the stem. The stem shows lateral branching. The branches arise in tufts of 6-8.

4. The stem is anchored to the soil with the help of numerous rhizoids rising from the base.

5. Rhizoids are multicellular with oblique septa.

6. In aquatic forms, rhizoids usually disappear, especially in adult gametophores.

7. The main axis as well as the branches are densely covered with small sessile 'leaves'.

8. The 'leaves' are spirally arranged.

9. At the proximal end of the stem the short branches crowd themselves to form a head or coma.

10. The branches are of only one type in submerged species.

11. In floating species, the branches are of two types, viz., diverging branches or branches of limited growth and pendulous branches or branches of unlimited growth.

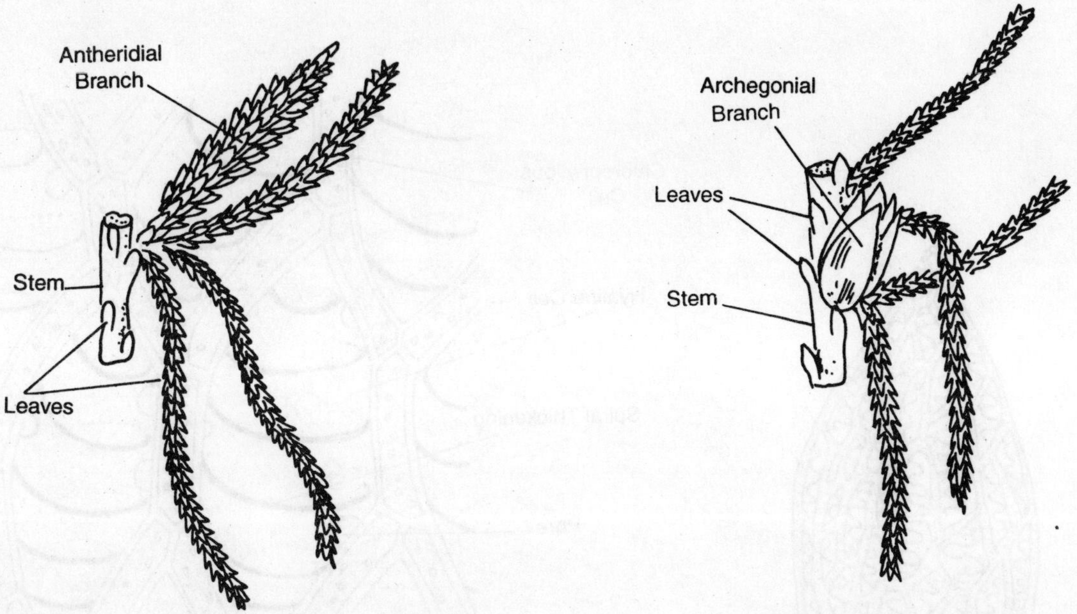

Fig. 22.2 Sphagnum. A portion of stem showing male branch   A portion of stem showing female branch

12. Diverging branches are short erect and grow up, while pendulous branches are long and drooping downwards. Pendulous branches are also known as Flagelliform branches.

13. The pendulous branches form a loose envelope around the stem, helping in the conduction of water.

14. This branch is exactly similar to the main stem in structure.

15. The leaves are spirally arranged, small, sessile and scaly. They have no midrib.

16. The leaves are one cell thick and have two types of cells, viz., chlorophyllous cells and hyaline cells.

17. The two types of cells are alternately arranged giving a net like appearance.

18. The chlorophyllous cells also called assimilatory cells are narrow and filled with chloroplasts. They help in photosynthesis. They are uninucleate.

19. The hyaline cells are rhomboidal in shape without any protoplast in them. They are dead cells.

20. These cells are colourless and filled with water.

21. The surfaces of the walls of hyaline cells have small circular or oral pores.

22. The inner surface of these cells are also traversed by annular and spiral thickenings.

23. These thickenings give mechanical support and prevent the cells from being folded.

Anatomy or the stem

1. In some submerged species there is absolutely no tissue differentiation.

2. In floating and terrestial species, some tissue differentiation is noticed.

3. A transverse section shows three regions, viz., cortex, hadrome and medulla.

4. There is a single layered epidermis as the outermost layer.
5. Next to the epidermis is the cortex (Hyalodermis). The cortical cells absorb and store water.
6. In some species, the epidermis at leaf insertions has flask shaped cells called *retort cells*.
7. The retort cells help in absorption and are usually inhabited by microorganisms.

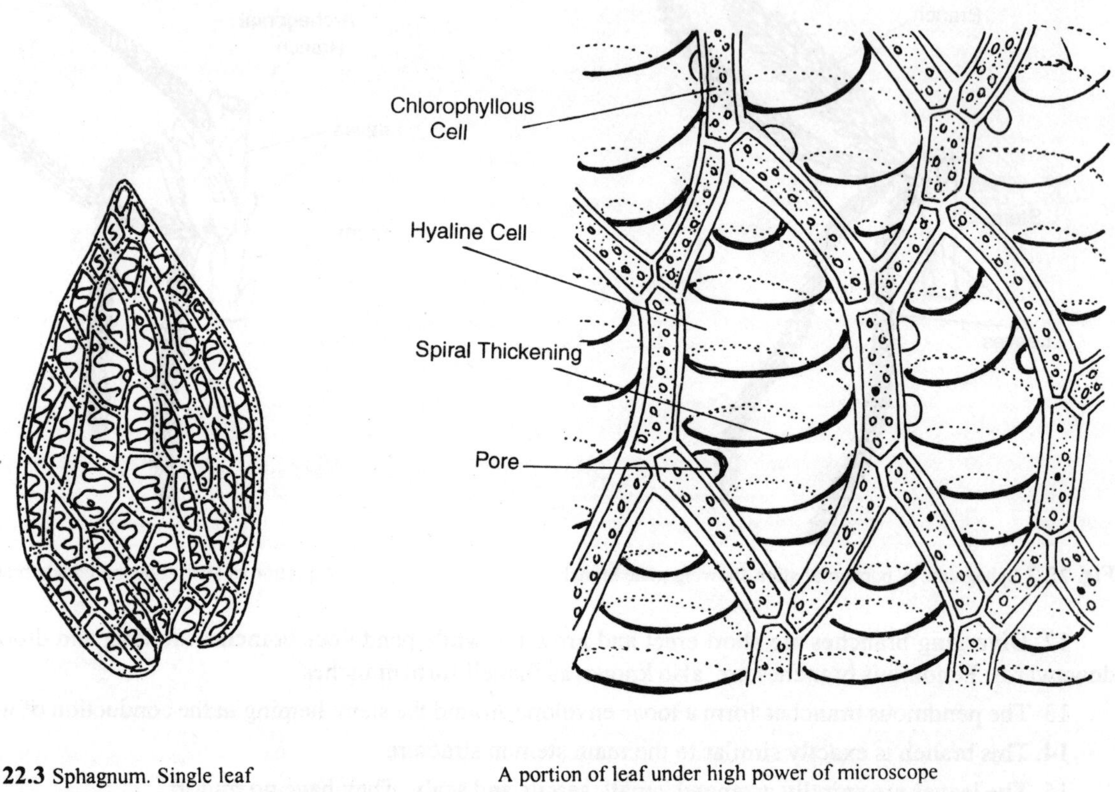

Fig. 22.3 Sphagnum. Single leaf A portion of leaf under high power of microscope

8. The region next to the cortex is the hardrome. The cells are thick walled and give mechanical support to the stem.
9. The central region consists of thin walled medulla.

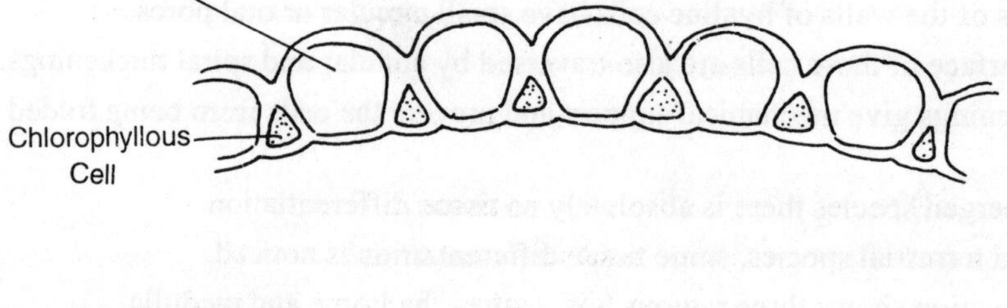

Fig. 22.4 Sphagnum. T.S. of leaf (A part)

Vegetative propagation

1. Takes place by fragmentation or by deliberate separation of lateral branches which grow into new individuals.

2. Sex organs are antehridia (male) and archegonia (female) developed on separate branches.

3. In monoecious species, antheridia develop earlier than archegonia.

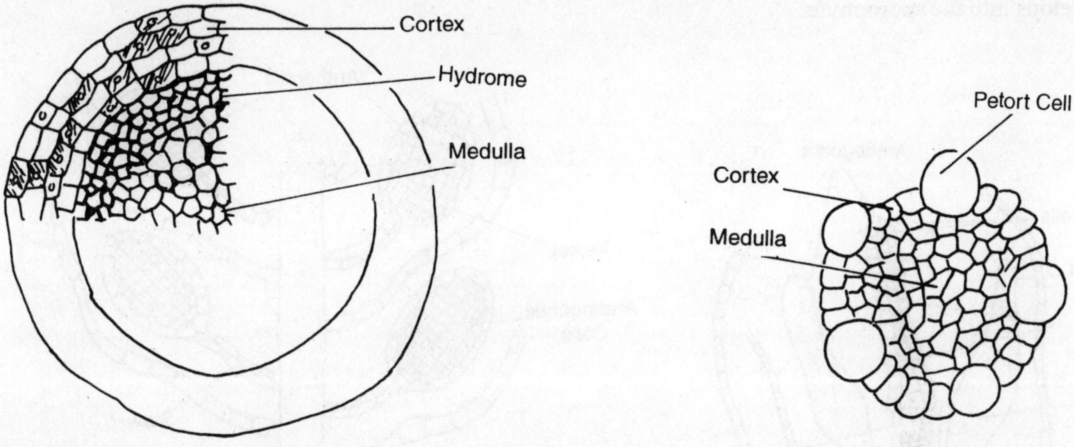

Fig. 22.5 Sphagnum. T.S. of old axis

T.S. of young axis

Antheridia

1. The antheridial branches are small and short.

2. They are covered by leaves which are red, purple, brown or yellow in colour.

3. A single antheridium arises from the axil of each leaf.

4. A mature antheridium has long multicellular stalk and an oval or globose antheridium proper.

5. The wall of the antheridium is made up of a single layered jacket.

6. The antheridium has a number of androcytes which metamorphose to form antherozoids. The antherozoids are coiled and biflagellate.

7. The antherozoids will be suspended in a mucilagenous mass.

Archegonia

1. The archegonia are formed on separate branches called the archegonial branches.

2. The archegonial branches are very small having few compactly arranged leaves.

3. The leaves are large and green in colour.

4. These leaves protect the archegonia and also the developing sporophyte.

5. The archegonia occur at the apex of the archegonial branch and range in number from two to five. Normally they are three in number.

6. Central archegonium is called the primary archegonium as it would have directly developed from the apical cell.

7. The others are called secondary archegonia as they are formed from the segments cut off by the apical cell.

8. A mature archegonium has a basal slightly bugled venter and a narrow twisted neck.

9. The venter and lower portion of the neck have a 2-3 cell thick wall.

10. Venter encloses a venter canal cell and an egg cell.

11. The neck encloses 8-9 neck canal cells and is covered by eight or more cover cells.

12. In an archegonium ready for fertilization, all the cells except the egg dissolve.

13. The antherozoids enter the archegonium with one of them fertilizing the egg to form the zygote which develops into the sporophyte.

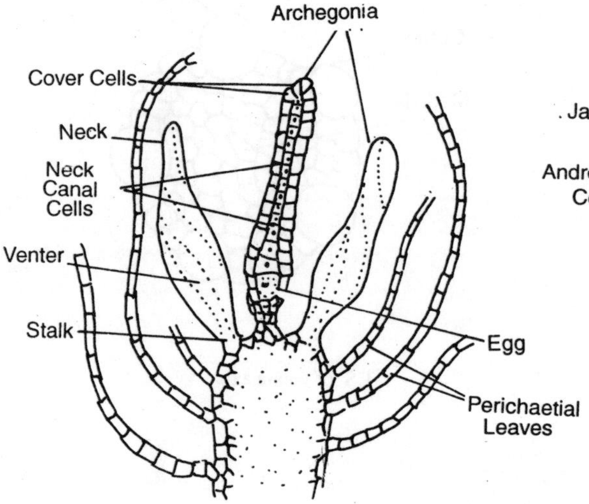

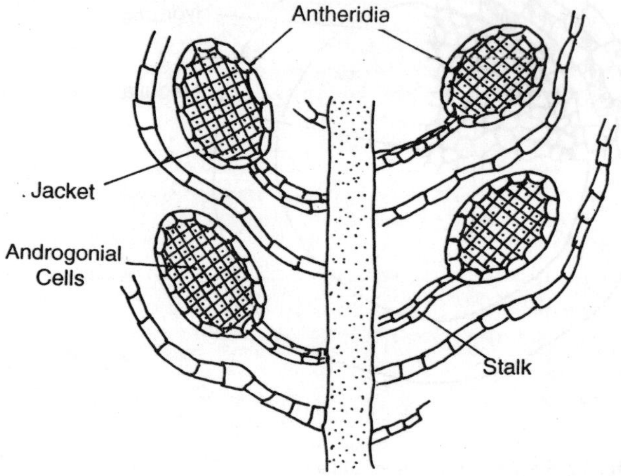

Fig. 22.6 Sphagnum. L.S. of female branch passing through archegonia   L.S. of a portion of male branch passing through antheridia

Structure of the sporophyte

1. The sporophyte is covered by two protective sheaths, viz., calyptra and perichaetium.

2. Calyptra is formed by the venter wall of the archegonium, while perichaetium is formed by the perichaetial leaves.

3. The sporophyte is divided into usual three parts fix., foot, seta and capsule.

4. The sporophyte is situated on a long leafless stalk (stalk of the archegonium) called pseudopodium.

5. The upper end of the pseudopodium enlarges to form a bulged structure called vaginula.

6. The foot is the basal most part of the sporophyte and is buried in the vaginula.

7. The foot is bulbous and made up of parenchyma cells.

8. The seta is highly reduced and is represented by a sinous or groove.

9. The terminal part of the sporophyte is the capsule.

10. The capsule is globose and has a central hump of dome shaped tissue called columella.

11. Overarching the columella is the horse shoe shaped spore sac.

12. The spore sac at first has spore mother cells. All of them are fertile and undergo reduction division to produce tetrads of haploid spores.

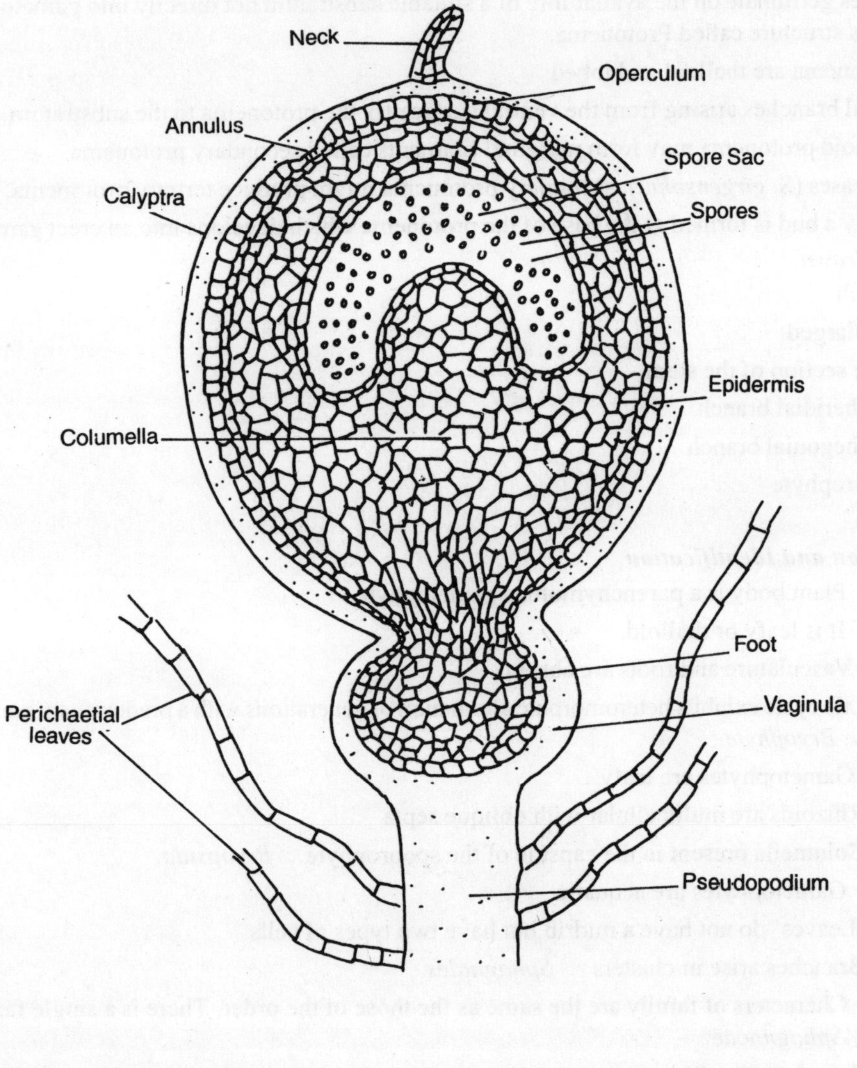

Fig. 22.7 Sphagnum. L.S. of sporogonium

13. A mature spore sac has only the spores and there are no elaters.

14. The wall of the capsule is made up of 4-5 layers of cells.

15. The outermost layer containing columnar cells is the epidermis.

16. At the apex of the capsule the wall has narrow, circular groove called the annulus.

17. The cells at the region of the annulus are thin walled.

18. The tip of the capsule above the annulus is called operculum.

19. In a mature sporophyte the annular cells break and the operculum comes out in the form of a hemispherical lid exposing the spores.

20. The spores are discharged because of the compressed air inside and the type of dispersal is called air gun mechanism.

21. The spores are spherical in shape and are surrounded by two wall layers namely, exine and intine.

22. The spores germinate on the availability of a suitable substratum not directly into gametophyte, but into a filamentous structure called Protonema.

23. The protonema are thalloid and lobed.

24. Rhizoidal branches arising from the ventral surface fix the protonema to the substratum.

25. The thalloid protonema may form marginal filaments called secondary protonema.

26. In some cases (S. *girgensohni*), secondary protonema might produce tertiary protonema.

27. Ultimately a bud is formed at the base of the protonema which develops into an erect gametophore.

Diagrams to be drawn

1. Habit sketch
2. A 'leaf' enlarged.
3. Transverse section of the stem.
4. L.S. of antheridial branch
5. L.S. of archegonial branch
6. L.S. of sporophyte
7. Protonema.

Systematic position and Identification

A. 1. Plant body is a parenchymotous gametophyte.
2. It is leafy or thalloid.
3. Vasculature and roots are absent.
1. Life cycle exhibits heteromorphic alternation of generations with a predominant gametophyte ... *Bryophyta.*

B. 1. Gametophytes are leafy.
2. Rhizoids are multicellular with oblique septa
3. Columella present in the capsule of the spoorophyte... *Bryopsida.*

C. 1. Gametophytes are acquatic
2. 'Leaves' do not have a midrib but have two types of cells.
3. Branches arise in clusters ... *Sphagnales*

D. 1. Characters of family are the same as the those of the order. There is a single family .. *Sphagnacae*
2. Branches are of two types
3. Sprophyte has a pseudopodium
4. Foot in enclosed in the vaginula
5. Seta is extremely reduced.
6. Leaves of the gametophyte have two types of cells...*Spha gnum.*

FUNARIA

Common species : *F. hygrometrica*

Occurrence: *Funaria* is commonly called the moss. It grows as bright green velvety patches on damp soil, wet walls etc. The genus includes 117 species of which 15 have been reported from India. The most widely distributed species is *F. hygrometrica*.

Study : 1. External morphology
2. Anatomy of stem
3. Sexual reproduction
4. Protonema
5. Systematic position and Identification

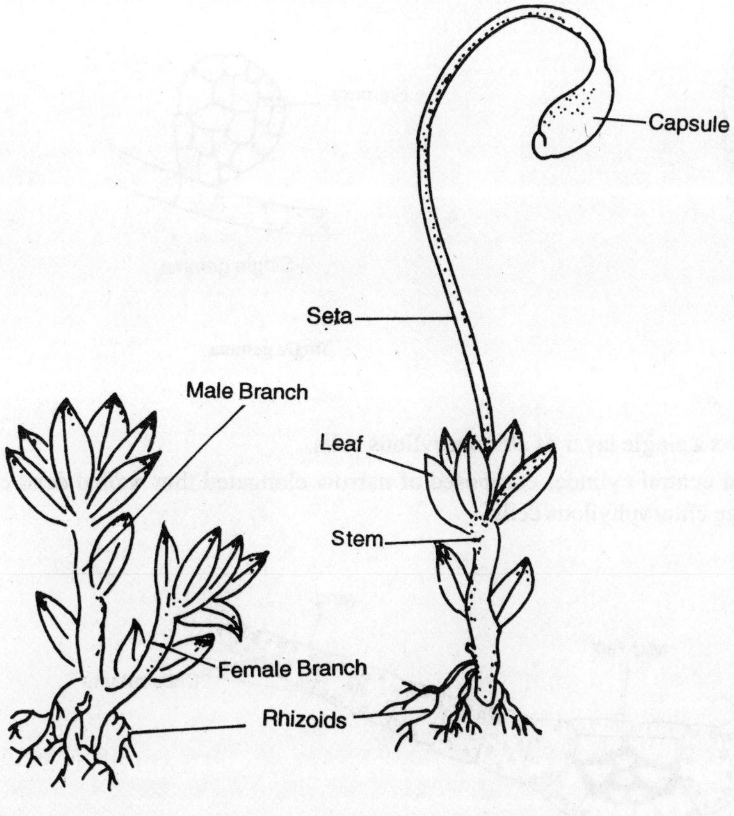

Fig. 22.8 Funaria. A plant showing branched axis   A plant bearing sporophyte

External Morphology:

1. The adult gametophytes arise from the prostrate thalloid protonema.

2. The gametophytes (gametophores) are erect, slender and reach a height of 2-4 cms.

3. The gametophyte has a central axis called stem or cauloid. The axis may or may not be branched.

4. The stem is anchored to the soil with the help of rhizoids. The rhizoids are multicellular with oblique cross walls.

5. The stem is surrounded by a number of spirally arranged 'leaves' or phylloids.

6. The leaves are sessile, simple, entire with distinct midrib.

7. The leaves are smaller at the lower region, larger and closely packed towards the upper portion.

8. The leaves at the apex are called perichaetial leaves.

9. The perichaetial leaves are very large and radially arranged in the form or a rosettle called the 'moss flower' within which are present sex organs.

Anatomy of stem

1. A cross section of the stem shows three distinct regions epidermis, cortex and central cylinder.

2. Epidermis is made up of a single layer of compactly arranged cells.

3. There are no stomata in the epidermis.

4. Epidermal cells at the base of the stem produce multicellular rhizoids.

5. Epidermal cells have many choloroplasts .

6. Cortex is made up of many layers of polygonal cells.

7. In a young stem, the outer cortical cells are also chlorophyllous.

8. In older stems the outer cortical cells become thick walled and give mechanical support.

9. Central cylinder is composed of elongated compactly arranged thin walled dead cells. These cells also offer mechanical support to the gametophore.

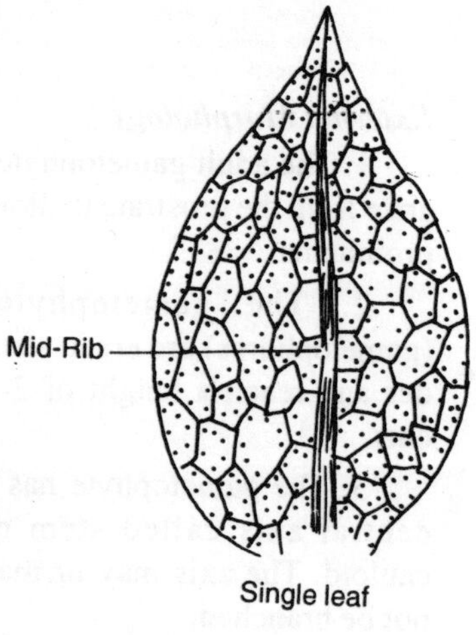

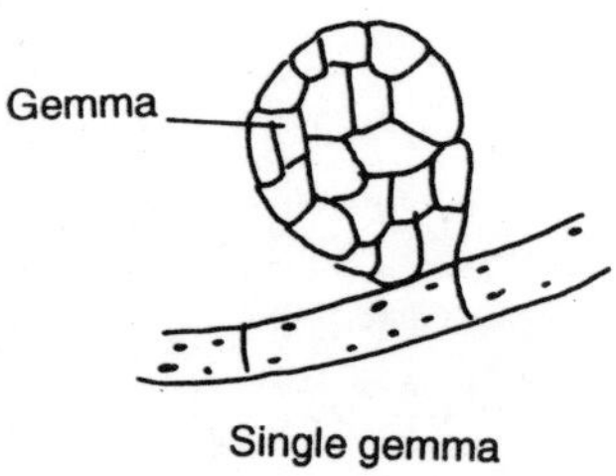

Fig. 22.9 Funaria. Single leaf

Single gemma

Anatomy of leaf

1. Anatomically, the leaf shows a single layer of chlorophyllous cells.

2. The midrib is made up of a central cylinder composed of narrow elongated thin walled dead cells. They are surrounded by a few large chlorophyllous cells.

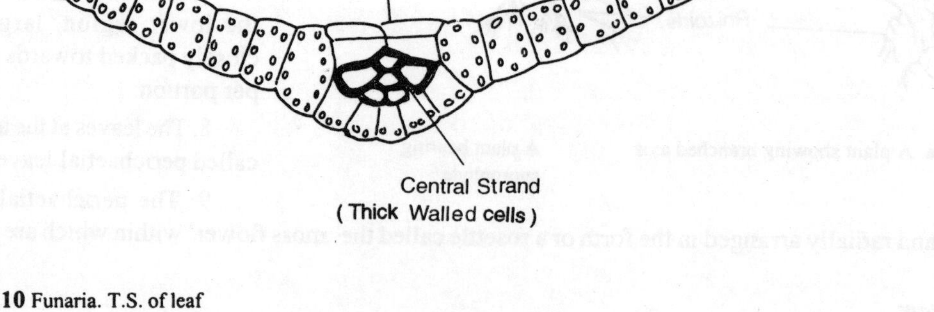

Fig. 22.10 Funaria. T.S. of leaf

Vegetative propagation

1. Gemmeae formed on the midrib of the leaf detach from the plant body and develop into a new individual.

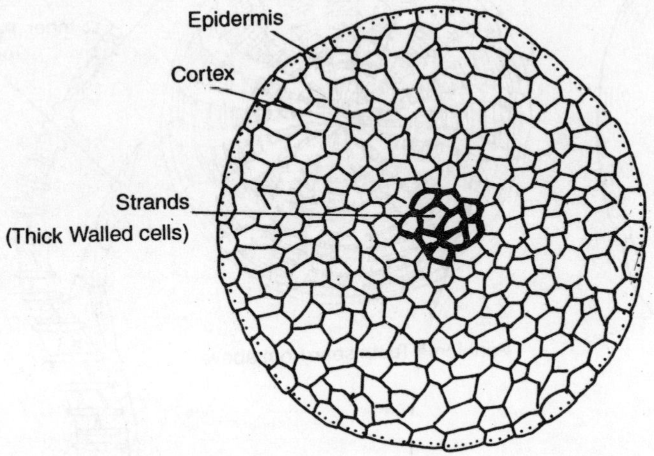

Fig. 22.11 Funaria. T.S. of axis

2. Bulbils formed on rhiozids develop into new gametophores.

Sexual reproduction

1. Sexual reproduction is oogamous. Sex organs are antheridia (male) and archegonia (female).

2. Some species are monoecious and some are diooecious.

3. In some species, antheridia and archegonia develop at the apices of different branches of a cluster of the same gametophore. This condition is called *Autoicous*.

4. If antheridia and archegonia are produced on different branch clusters of the same gametophore the condition is said to be paraoicous.

5. If the sex organs are irregularly scattered at the tip of a gametophore, the conditions said to be *synoicous*.

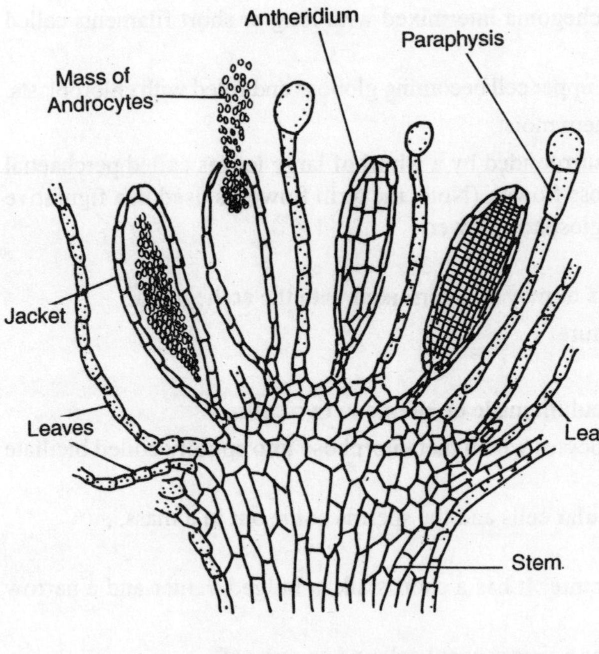

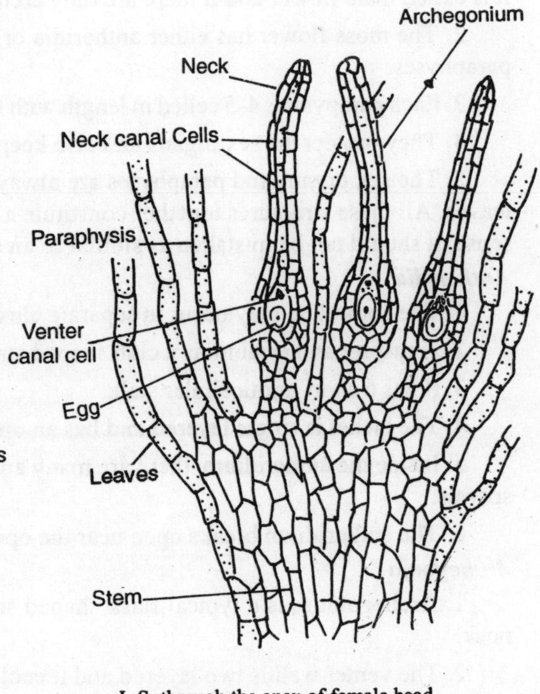

Fig. 22.12 Funaria. L.S. through the apex of male head L.S. through the apex of female head

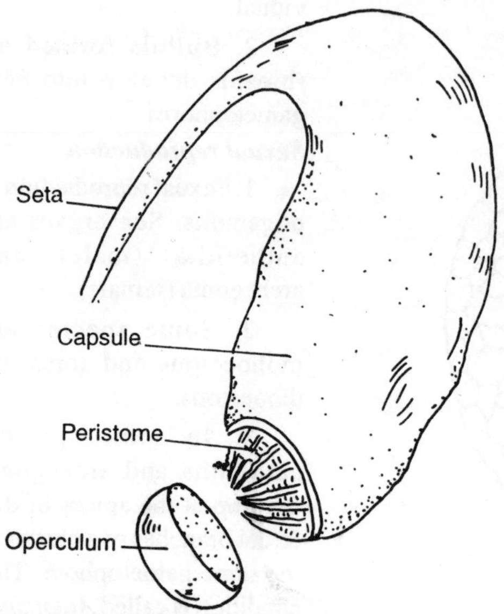

Fig. 22.13 Funaria. A mature capsule

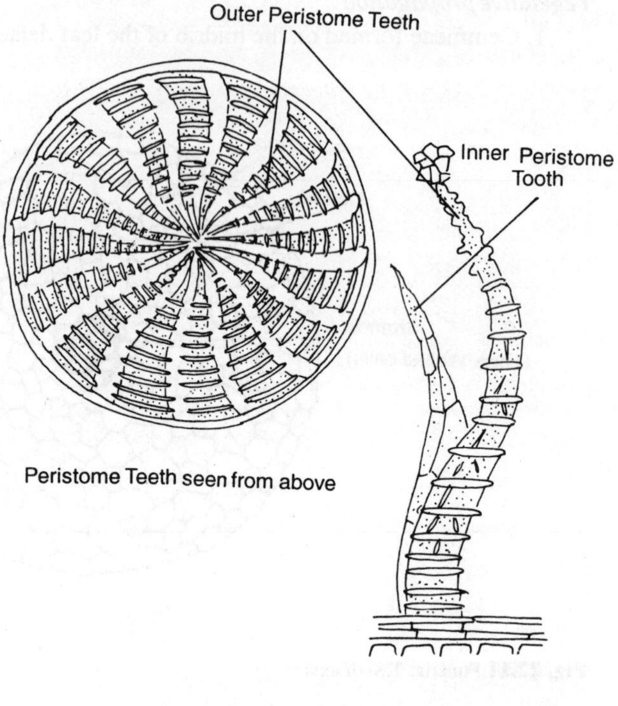

Two peristome teeth

Structure of a moss flower

1. A moss flower is nothing but a cluster of leaves surrounding the sex organs. If there are only antheridia, it is called male flower and if there are only archegonia it is said to be female flower.

2. The moss flower has either antheridia or archegonia intermixed with long or short filaments called paraphyses.

3. Each parapysis is 4-5 celled in length with the upper cell becoming globose and filled with chloroplasts.

4. They protect the sex organs and also keep them moist.

5. The sex organs and paraphyses are always surrounded by a whorl of large leaves called perchaetial leaves. All these structures together constitute a moss flower. (Note: the term flower is used in a figurative sense, it should not be mistaken as similar to an angiosperm flower).

Antheridia

1. The antheridia may occur in separate clusters or may be intermixed with the archegonia.

2. A mature antheridium is a club shaped structure.

3. It has a short multicellular stalk.

4. The jacket is single layered and has an operculum made up of one to two cells.

5. Inside the antheridium, there are many androcytes which metamorphose into spirally coiled biciliate sperms.

6. The antheridium breaks open near the opercular cells and the sperms come out in a mass.

Archegonia

1. Archegonium is a typical flask shaped structure. It has a short stalk, a bulged venter and a narrow neck.

2. The venter wall is two layered and it encloses a venter canal cell and an egg cell.

3. The neck is narrow and is made up of six rows of spirally twisted cells enclosing six or more neck canal cell.

4. The apex of the neck is covered by four large cover or lid cells.

5. The antherozoids enter the archegonium. The fertilized egg, i.e., zygote develops into the sporophyte.

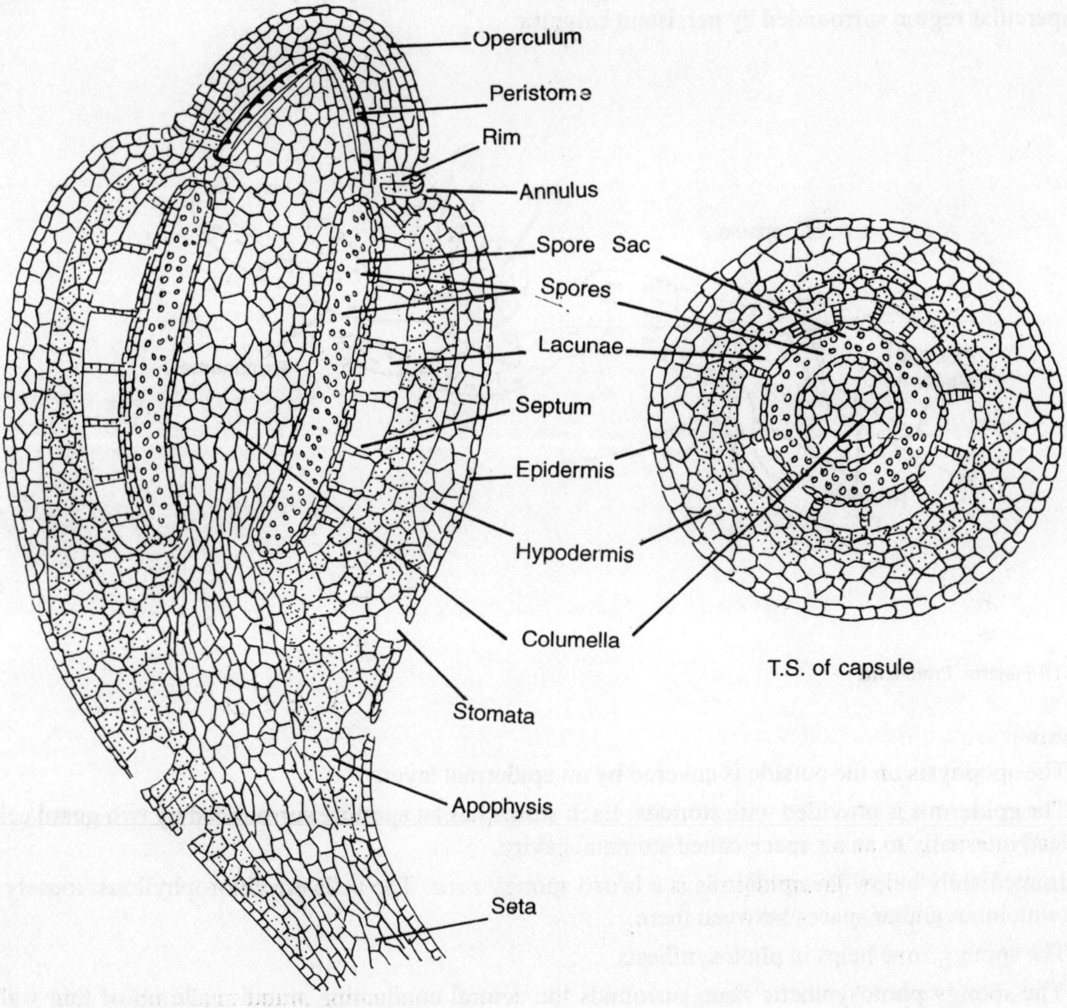

Fig. 22.14 Funaria. L.S. of capsule

Structure of sporophyte

1. A mature sporophyte is differentiated into 3 regions, viz., the basal foot, the middle seta and the terminal capsule.

2. The foot is very small and is embedded at the apex of the gametophore helping in absorption and anchorage.

3. Seta is very long and cylindrical and consists of central narrow elongated cells surrounded by a few layered cortex and single layered epidermis.

4. Seta helps in the conduction of water and mineral elements.

5. Seta also serves to keep the capsule in an elevated position.

6. The capsule is roughly beak shaped and bent towards one side.

7. In the beginning it is green in colour, later it turns to yellow and orange.

8. The capsule is differentiated into three regions, viz., the lower apophysis, the middle theca and the upper opercular region surrounded by persistent calyptra.

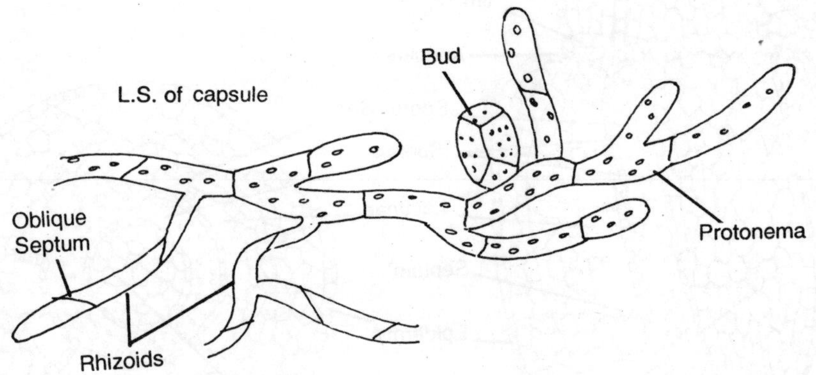

Fig. 22.15 Funaria. Protonema

Apophysis

1. The apophysis on the outside is covered by an epidermal layer.

2. The epidermis is provided with stomata. Each stoma has an aperture surrounded by two guard cells., which lead internally to an air space called stomatal cavity.

3. Immediately below the epidermis is a broad spongy zone. The cells are chlorophyllous, loosely arranged with intercellular spaces between them.

4. The spongy zone helps in photosynthesis.

5. The spongy photosynthetic zone surrounds the central conducting strand made up of thin walled narrow, elongated cells without chloroplasts.

6. The central strand is continuous. It is connected with the central strand of seta below and linked with the columella above by means of a bundle of filaments.

Theca

1. Theca consists of a central sterile structure called columella.

2. Columella is made up of thin walled parenchyma cells and projects above in the form of a cone in the opercular region.

3. Columella is surrounded by barrel shaped spore sac.

4. The inner wall of the spore sac is single layered while the outer wall is three layered.

5. The spore sac encloses numerous small haploid spores.

6. The outer wall of the spore sac gives rise to radiating filaments called trabeculae, each of which is made up of three to five chlorophyllous cells.

7. The trabeculae are separated from one another by large air chambers called lacunae.

8. The lacunae are surrounded by one to two layers of cholrenchyma cells.

9. The chlorenchyma region is surrounded by hypodermis composed of two or more layers of thin walled colourless parenchyma cells.

10. Encircling the hypodermis is the single layered epidermis. The outer walls of epidermal cells are thickened.

Opercular region

1. The opercular region is made up of outer operculum and the inner peristome surrounding the central parenchymatous tissue.

2. Peristome consists of two sets of peristomial teeth. The outer set called exostome and the inner set endostome.

3. The peristomial teeth are narrow, long and tapering. They are compactly arranged and convergent in the form of a cone.

4. The peristomial teeth are hygroscopic and help in the dispersal of the spores.

5. The operculum is made up of three layers of cells.

6. The operculum is attached to the theca region with the help of thin walled cells called annulus.

7. At maturity the annular cells tear open, with the operculum falling off. The peristomal teeth help in the dispersal of the spores.

Protonema

1. The spore released from the capsule on germination do not develop into the gametophore directly but develop first into a filamentous structure called protonema.

2. The spore on germination develops into primary protonema.

3. The primary protonema has two types of branches, viz., chloronema(chlorophyllous branches) and Rhizonema (rhizoidal branches).

4. Rhizonema penetrates the substratum while chloronema are aerial.

5. Rhizonema penetrates the substratum while chloronema are aerial.

6. Many lateral buds are formed on the protonema. These develop into adult gametophores.

Diagrams to be drawn

1. Habit sketch
2. T.S. of 'stem'
3. T. S. of 'leaf'
4. L.S. of female 'flower'
5. L. S. of male 'flower'
6. L. S. of sporophyte
7. Peristomial teeth (enlarged)
8. Protonema showing buds.

Systematic position and Identification

A. 1. Plant body is a parechymatous gametophyte

2. It is leafy or thalloid; vasculature and roots are absent.

3. Sex organs surrounded by a sterile jacket layer.

4. Life cycle exhibits heteromorphic alternation of generations with a predominant gametophyte … *Bryophyta*.

B. 1. Gametophytes are leafy

2. Rhizoids are multicellular with oblique septa

3. Columella present in the capsule of the sporophyte… *Bryopsida*

C. 1. Leaves have a prominent midrib.

2. Seta is slender and elongated.

3. Air cavities present in the capsule … *Bryidae*

D. 1. Plants terrestrial in habitat.

2. Leaves clustered at the apices of the gametophore

3. Two concentric rings of peristomial teeth present.

4. Calyptra is usually distended… *Funariales*

E. 1. Leaves many celled thick at the midrib, but one celled thick at sides.

2. Height of gametophytes rarely exceeds a few cms.

3. Calyptra has a long beak … *Funariaceae*

F. 1. Leaves arranged spirally on the axis.

2. Anatomically the axis is highly differentiated; it has epidermis cortex and a central conduct ing strand.

3. Leaves at the apex cluster to form capitate structures. …*Funaria*

POLYTRICHUM

Common species: *P. densifolumin, P. commune, P. xanthopilus, P. juniperinum,*

Occurrence: The genus with about 100 species is distributed universally, but majority of species show preference to cool temperate and tropical regions. Cool shady soil, marshes, peat bog, soil along pond banks etc. are the usual habitats where *Polytrichum* grows.

Study: 1. External morphology

2. Anatomy of stem

3. Anatomy of leaf.

4. Sexual reproduction

5. Structure of the sporophyte.

6. Protonema.

7. Systematic position and Identification.

External Morphology:

1. The plant body of *Ploytrichum* is a gametophyte.
2. It has an underground rhizome from which spring up leafy shoots.
3. In fair environmental conditions, the plant may each a height of 20 cms.
4. The stem is erect and cylindrical. It is branched.

5. In *P. commune*, a transitional region is found between the rhizome and the arial shoots.

6. The leaves are spirally arranged in three vertical rows.

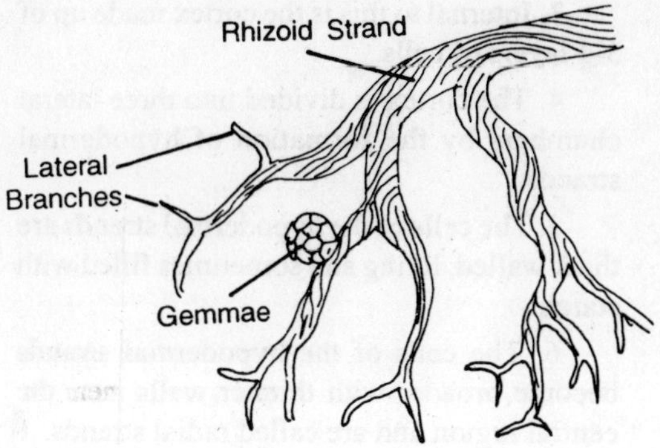

Fig. 22.16 Polytrichum. Rhizoidal strand with gemmae

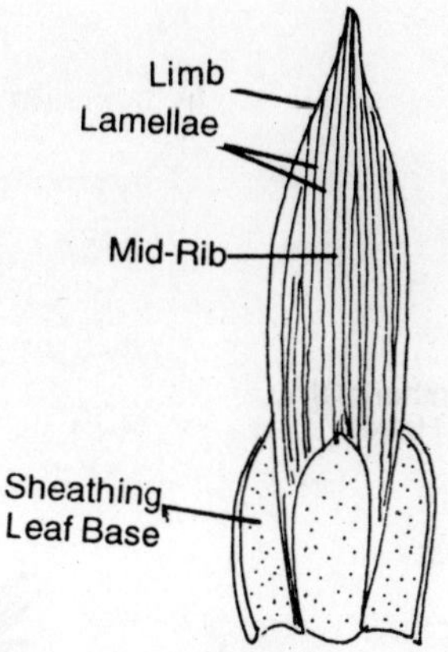

Single leaf

7. The leaves on the transitional zone and rhizome and brown or colourless, those on the aerial shoots are large and rigid.

8. The leaves are provided with sheathing membranous leaf base which narrows up into a limb.

9. The leaf has a broad midrib, which bears on it adaxial face many cell plates (lamellae). These are one celled thick and photosynthetic.

10. The rhizome is anchored to the soil with the help of rhizoids.

11. The rhizoids are multicellular with oblique cross walls.

12. The rhizoids entangle themselves into loosely knit manes, thus helping in the external capillary movement of water.

13. The water can rise through the wick like rhizodial manes.

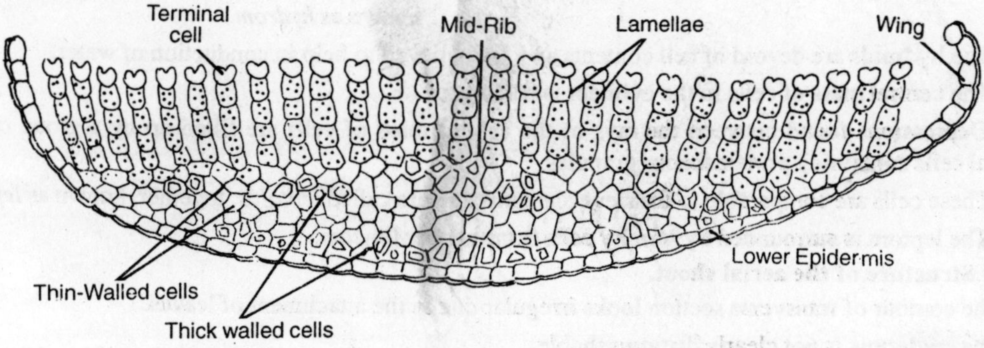

Fig. 22.17 Polytrichum. T.S. of leaf

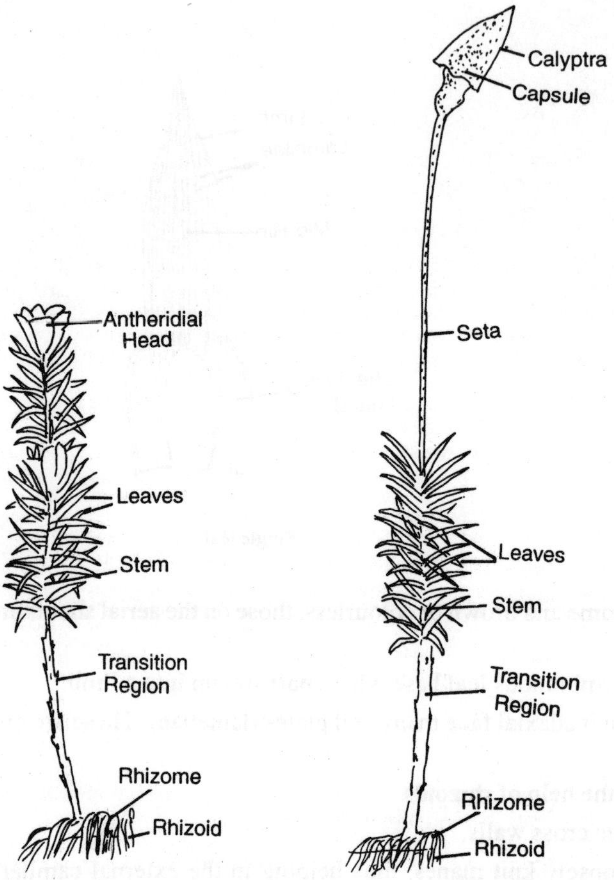

Fig. 22.18 Polytrichum. A male plant — A female plant with sporophyte

Anatomy of Rhizome.

1. A cross section of the rhizome is roughly circular in outline.

2. The superficial layer consists of thick or thin walled cells.

3. Internal to this is the cortex made up of 3-4 layers of cells.

4. The cortex is divided into three lateral chambers by the formation of hypodermal strands.

5. The cells of the hypodermal strands are thick walled, living and sometimes filled with starch.

6. The cells of the hypodermal strands become broader with thinner walls near the central region and are called radial strands.

7. The innermost layer of the cortex has suberised raidal and tangential walls and is comparable to endodermis.

8. The layer however is broken because of the presence of radial strands.

9. Next to the endodermis is found a 2-3 layered pericycle. This layer also is discontinuous.

10. Internal to the pericycle is the central cylinder.

11. This is mostly composed of thick walled but living cells often filled with starch.

12. These cells are individually called *steroids*, collectively constituting the *Stereom*.

13. Scattered among the cells of the stereom are present the *hydroids*, collectively known as *hydrom*.

14. The hydroids are devoid of cell contents and are believed to help in conduction of water.

15. The central mass of cells in the cylinder is triradiate.

16. Depressions found between the rays of the central mass of cells are filled up by a group of 6-8 polygonal cells containing proteinaceous granules.

17. These cells are comparable to the sieve elements and are called *leptoids*, together, known as *leptom*.

18. The leptom is surrounded by starchy cells forming the *Amytom*.

Internal Structure of the aerial shoot.

1. The contour of transverse section looks irregular due to the attachment of leaves.

2. The epidermis is not clearly distinguishable.

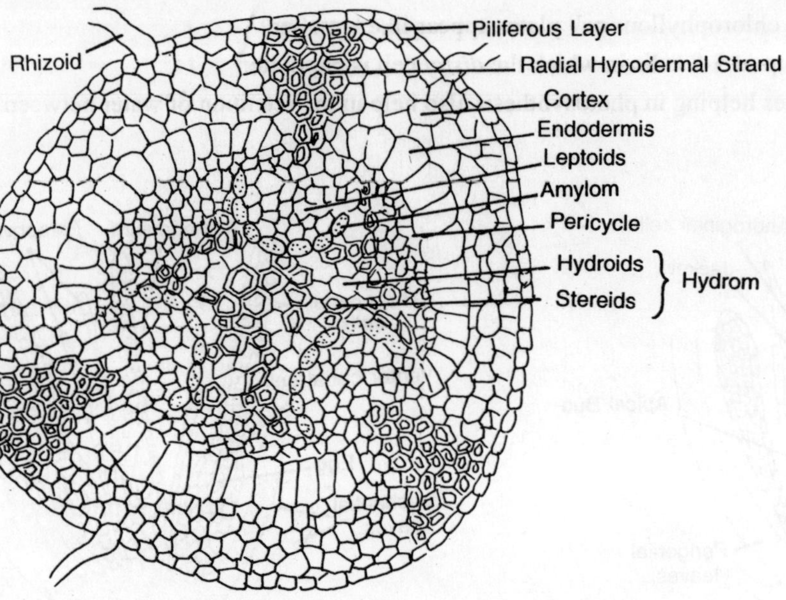

Fig. 22.19 Polytrichum. T.S. of rhizome

3. Internal to this is the cortex, divided into two zones, viz., outer prosenchymatous zone and inner parenchymatous zone.

4. The cortical cells are starch filled and many leaf traces are found in the cortex.

5. Internal to the cortex is an ill defined ruidmentary penicycle. It is discontinuous.

6. Next to the pericycle in the leptom mantle are present sieve tube like cells interrupted by starch filled cells.

7. The leptom mantle is lined internally by the hadrom sheath, where cells are suberised.

8. The hadrom sheath encloses the hadrom mantle conisiting of thin walled cells.

9. Centrally located is the hydrome cylinder made up of thick walled cells.

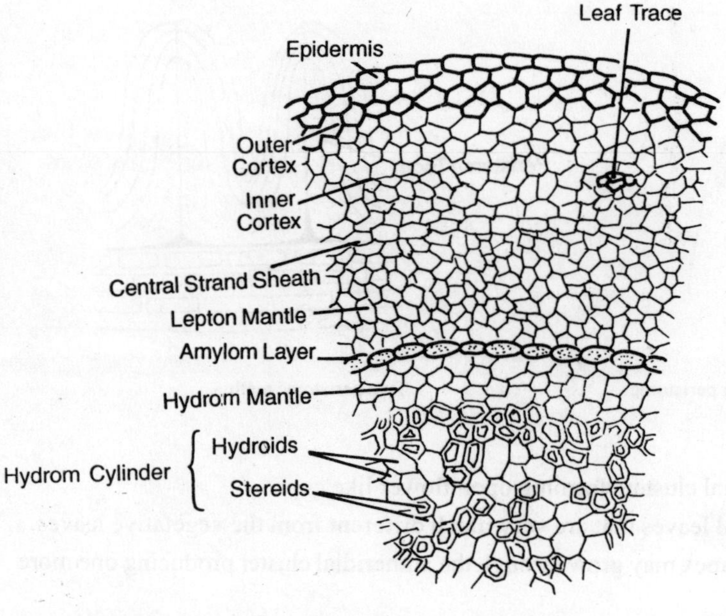

Fig. 22.20 Polytrichum. T.S. of aerial stem (a portion cellular)

10. This tissue helps in water conduction and is comparable to the xylem of vascular plants.

Internal structure of the leaf:

1. A vertical transverse section of the leaf shows a bifacial arrangement of cells.

2. The midrib is many celled thick and merges with the lamina at the margins.

3. The lower face of the leaf has a well marked single layered epidermis consisting of large thickwalled cells.

4. Closely attached to the lower epidermis is found a layer of erotic cells.

5. The tissue next to this is made up of parenchyma cells.

6. On the upper face, the leaf has many parallely arranged plates of chlorophyllous cells arising from the lower layer of parenchyma cells.

7. In a sectional view the chlorophyllous cell plates appear like lamellae.

8. Each lamella is made up of 5-8 cells in which the *distal* cell is papillate.

9. These cell plates besides helping in photosynthesis also help in the retention of water between them.

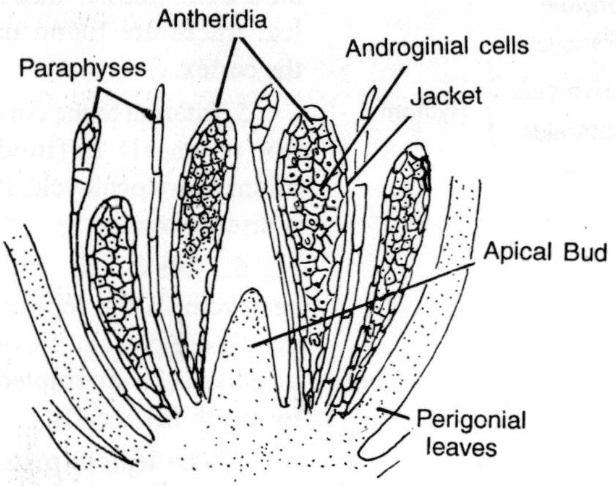

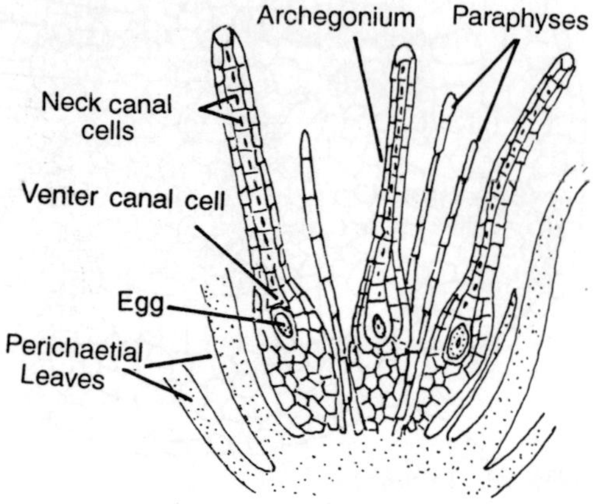

Fig. 22.21 Polytrichum. L.S. through the apex of male head

L.S. through the apex of female head

Sexual reproduction

1. Sexual reproduction is oogamous. Sex organs are antheridia (male) and archegonia (female).

2. *Polytrichum* is generally dioecious.

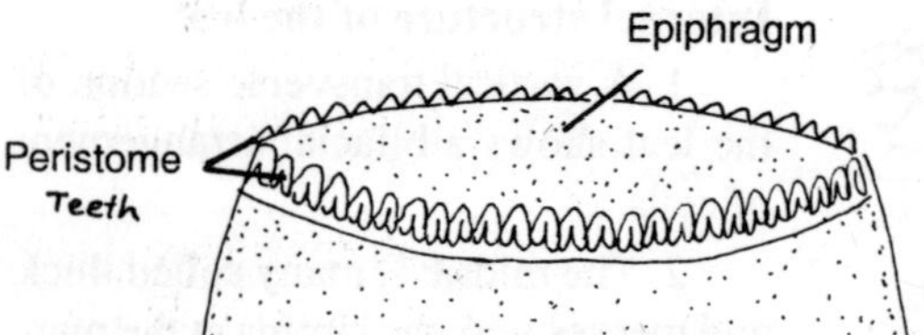

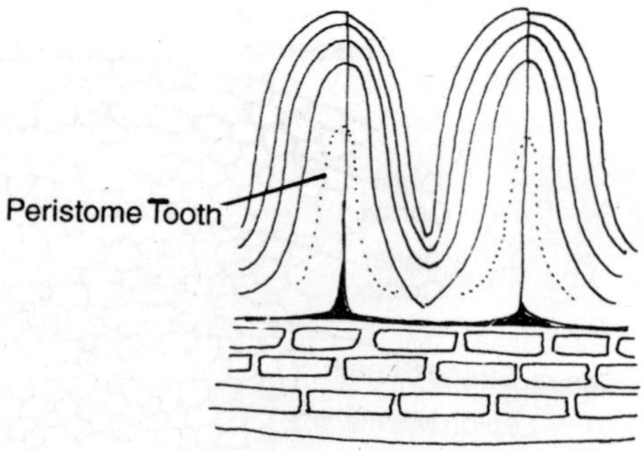

Fig. 22.22 Polytrichum. Dehisced capsule showing peristome

Two peristome teeth

Antheridia

1. These are generally borne in terminal clusters forming open flower like cups.

2. The cup is surrounded by perigonial leaves that are very much different from the vegetative leaves.

3. In some cases, the vegetative shoot apex may grow through the antheridial cluster producing one more group of antheridia.

4. Within the cup are found a number of antheridia intermingled with paraphyses.

5. Each antheridium is a clubshaped structure having a short stalk.

6. The jacket of the antheridium is made up of a single layer of cells.

7. The antheridium has a number of androcytes which metamorphose to give rise to coiled biciliate antherozoids.

Archegonia

1. Archegonia are also borne on terminal clusters.

2. They are surrounded by pink coloured perichaetial leaves.

3. The leaves are arranged in a rosette looking like a flower.

4. Usually three archegonia are found in each cluster as in *P. commune*.

5. With the development of the archegonia, the growth of the shoot stops.

6. Scattered among the archegonia are found a number of hair like structures called paraphyses.

7. Each archegonium has a stalk, a bulbous venter and a spirally twisted neck.

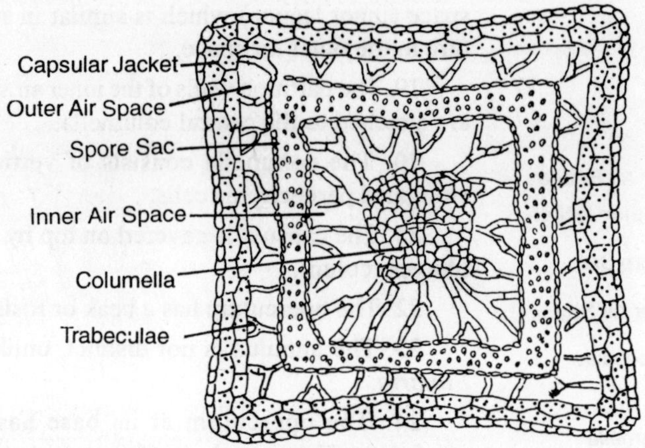

Fig. 22.23 Polytrichum. T.S. of capsule passing through spore sac

8. Rest of the structural details of archegonia are same as seen in *Funaria*.

9. Fertilization is as seen *Funaria*.

Sporophyte

1. The sporophyte has three parts, viz., basal foot, middle seta and the terminal capsule.

2. The foot is parenchymatous and is embedded in the archegonial cluster.

3. The seta is long and cylindrical. Anatomically it shows the following features.

4. The outermost layer of cells is thick walled.

5. Internal to this is a band of sclerotic cells which gradually merge into the inner thin walled loosely arranged chlorophyllous cells.

6. In the centre of the seta runs a conducting strand.

7. The seta terminates at the base of the capsule into a swollen structure called the apophysis.

8. The apophysis is separated from the rest of the capsule by constriction.

9. Anatomically the apophysis has an epidermis interrupted with stomata.

10. The epidermis encloses a mass of green chorenchymatous cells.

11. In long section, the capsule shows an outer wall composed of several layers of cells.

12. The outermost layer of the wall is the epidermis.

13. Internal to the wall is an air space (outer lacuna) separated into air chambers by means of parenchymatous partitions.

14. The cells of these partitions are chlorophyllous and are connected internally to the outer wall of the spore sac.

15. Next to the outer air chamber is found the spore sac which is covered on either side by means of a two layered wall.

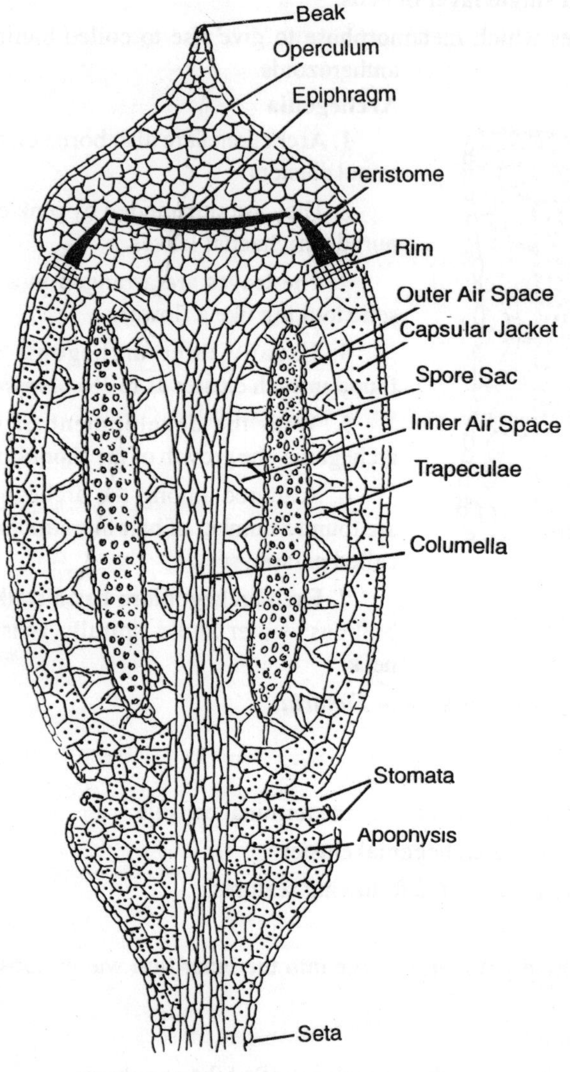

Fig. 22.24 Polytrichum. L.S. of capsule

16. The wall encloses the archesporium of 4-6 layers of cells which ungergo reduction division to form haploid spores.

17. There are no elaters.

18. Internal to the spore sac is the inner air space (inner lacuna) which is similar in structure to the outer air space.

19. The partition walls of the inner air space are attached to the central columella.

20. The columella consists of vertically elongated parenchyma cells.

21. The capsule is covered on top by a lid like operculum.

22. The operculum has a beak or rostrum.

23. The annulus is not distinct, unlike in *Funaria*.

24. The operculum at its base has the epiphragm. This is made up of a transverse band of thin membranous cells without intercellular spaces.

25. This stretches like a drum head (tympanum) covering the opening of the capsule.

26. Outside the epiphragm is the peristome.

27. The peristome has 36-64 teeth which are connected to the capsule below and to the epiphragm above.

28. The spores are minute in size, haploid and uninucleate.

29. On the availability of the suitable substratum, the spores germinate to develop into protonema.

Protonema

1. The spore germinates to form a branched, septate filamentous protonema.

2. Some of the branches grow vertically above (chloronema) while some others grow down and help in anchorage.

3. Subsequently buds appear near the base of chloronema.

4. These buds develop into leafy gametophytes of *Polytrichum*.

Diagrams to be drawn

1. Habit sketch
2. Anatomy of Rhizome
3. Anatomy of aerial shoot.
4. Anatomy of leaf.
5. L.S. of male flower.

6. L. S. of female flower.
7. L. S. Of sporophyte.
8. Protonema.

Systematic Position and Identification:

A.
1. Plant body is a parenchymatous gametophyte.
2. It is leafy or thalloid.
3. Vasculature and roots are absent
4. Sex organs surrounded by sterile jacket layer.
5. Life cycle exhibits a heteromorphic alternation of generations with a predominant gametophyte ... Bryophyta.

B.
1. Gametophytes are leafy.
2. Rhizoids are multicellular with oblique septa.
3. Columella present in the capsule of the sporophyte... *Bryopsida.*

C.
1. Leaves with distinct midrib.
2. Anatomically leaves show a high degree of tissue differentiation....*Polytrichidae.*

D.
1. Gametophytes perennial and quite tall.
2. Leaves have chlorophyllous cell plates on the dorsal surface.
3. Peristomial teeth 36-64 in number ....*Polytrichales.*

E.
1. Single family. Same characters as those of the order... *Polytrichaceae.*

F.
1. Capsule is angular in shape.
2. Rhizoids form a tangled mane.
3. Two rows of air spaces found in the capsule.
4. Operculum has a beak ... *Polytrichum.*